Tissue Engineering

METHODS IN MOLECULAR MEDICINE™

John M. Walker, SERIES EDITOR

140. **Tissue Engineering,** edited by *Hansjörg Hauser and Martin Fussenegger, 2007*
139. **Allergy Methods and Protocols,** edited by *Meinir G. Jones and Penny Lympany, 2007*
138. **Neutrophil Methods and Protocols,** edited by *Mark T. Quinn, Frank R. DeLeo, and Gary M. Bokoch, 2007*
137. **Microtubule Protocols,** edited by Jun Zhou, *2007*
136. **Arthritis Research:** *Methods and Protocols, Vol. 2, edited by Andrew P. Cope, 2007*
135. **Arthritis Research:** *Methods and Protocols, Vol. 1, edited by Andrew P. Cope, 2007*
134. **Bone Marrow and Stem Cell Transplantation,** edited by *Meral Beksac, 2007*
133. **Cancer Radiotherapy,** edited by *Robert A. Huddart and Vedang Murthy, 2007*
132. **Single Cell Diagnostics:** *Methods and Protocols, edited by Alan Thornhill, 2007*
131. **Adenovirus Methods and Protocols, Second Edition, Vol. 2:** *Ad Proteins, RNA, Lifecycle, Host Interactions, and Phylogenetics,* edited by *William S. M. Wold and Ann E. Tollefson, 2007*
130. **Adenovirus Methods and Protocols, Second Edition, Vol. 1:** *Adenoviruses, Ad Vectors, Quantitation, and Animal Models,* edited by *William S. M. Wold and Ann E. Tollefson, 2007*
129. **Cardiovascular Disease:** *Methods and Protocols, Volume 2: Molecular Medicine,* edited by *Qing K.Wang, 2006*
128. **Cardiovascular Disease:** *Methods and Protocols, Volume 1: Genetics,* edited by *Qing K. Wang, 2006*
127. **DNA Vaccines:** *Methods and Protocols, Second Edition,* **edited by** *Mark W. Saltzman, Hong Shen, and Janet L. Brandsma, 2006*
126. **Congenital Heart Disease:** *Molecular Diagnostics,* **edited by** *Mary Kearns-Jonker, 2006*
125. **Myeloid Leukemia:** *Methods and Protocols,* edited by *Harry Iland, Mark Hertzberg, and Paula Marlton, 2006*
124. **Magnetic Resonance Imaging:** *Methods and Biologic Applications,* edited by *Pottumarthi V. Prasad, 2006*
123. **Marijuana and Cannabinoid Research:** *Methods and Protocols,* edited by *Emmanuel S. Onaivi, 2006*
122. **Placenta Research Methods and Protocols:** *Volume 2,* edited by *Michael J. Soares and Joan S. Hunt, 2006*
121. **Placenta Research Methods and Protocols:** *Volume 1,* edited by *Michael J. Soares and Joan S. Hunt, 2006*
120. **Breast Cancer Research Protocols,** edited by *Susan A. Brooks and Adrian Harris, 2006*
119. **Human Papillomaviruses:** *Methods and Protocols,* edited by *Clare Davy and John Doorbar, 2005*
118. **Antifungal Agents:** *Methods and Protocols,* edited by *Erika J. Ernst and P. David Rogers, 2005*
117. **Fibrosis Research:** *Methods and Protocols,* edited by *John Varga, David A. Brenner, and Sem H. Phan, 2005*
116. **Inteferon Methods and Protocols,** edited by *Daniel J. J. Carr, 2005*
115. **Lymphoma:** *Methods and Protocols,* edited by *Timothy Illidge and Peter W. M. Johnson, 2005*
114. **Microarrays in Clinical Diagnostics,** edited by *Thomas O. Joos and Paolo Fortina, 2005*
113. **Multiple Myeloma:** *Methods and Protocols,* edited by *Ross D. Brown and P. Joy Ho, 2005*
112. **Molecular Cardiology:** *Methods and Protocols,* edited by *Zhongjie Sun, 2005*
111. **Chemosensitivity:** *Volume 2, In Vivo Mod-els, Imaging, and Molecular Regulators,* edited by *Rosalyn D. Blumethal, 2005*
110. **Chemosensitivity:** *Volume 1, In Vitro Assays,* edited by *Rosalyn D. Blumethal, 2005*
109. **Adoptive Immunotherapy:** *Methods and Protocols,* edited by *Burkhard Ludewig and Matthias W. Hoffman, 2005*
108. **Hypertension:** *Methods and Protocols,* edited by *Jérôme P. Fennell and Andrew H. Baker, 2005*
107. **Human Cell Culture Protocols,** *Second Edition,* edited by *Joanna Picot, 2005*
106. **Antisense Therapeutics,** *Second Edition,* edited by *M. Ian Phillips, 2005*
105. **Developmental Hematopoiesis:** *Methods and Protocols,* edited by *Margaret H. Baron, 2005*
104. **Stroke Genomics:** *Methods and Reviews,* edited by *Simon J. Read and David Virley, 2004*
103. **Pancreatic Cancer:** *Methods and Protocols,* edited by *Gloria H. Su, 2004*
102. **Autoimmunity:** *Methods and Protocols,* edited by *Andras Perl, 2004*
101. **Cartilage and Osteoarthritis:** *Volume 2, Structure and In Vivo Analysis,* edited by *Frédéric De Ceuninck, Massimo Sabatini, and Philippe Pastoureau, 2004*
100. **Cartilage and Osteoarthritis:** *Volume 1, Cellular and Molecular Tools,* edited by *Massimo Sabatini, Philippe Pastoureau, and Frédéric De Ceuninck, 2004*
99. **Pain Research:** *Methods and Protocols,*edited by *David Z. Luo, 2004*
98. **Tumor Necrosis Factor:** *Methods and Protocols,* edited by *Angelo Corti and Pietro Ghezzi, 2004*
97. **Molecular Diagnosis of Cancer:** *Methodsand Protocols, Second Edition,* edited by *Joseph E. Roulston and John M. S. Bartlett, 2004*
96. **Hepatitis B and D Protocols:** *Volume 2, Immunology, Model Systems, and Clinical*

METHODS IN MOLECULAR MEDICINE™

Tissue Engineering

Edited by

Hansjörg Hauser

Martin Fussenegger

HUMANA PRESS ✱ TOTOWA, NEW JERSEY

Softcover reprint of the hardcover 2nd edition 2007
999 Riverview Drive, Suite 208
Totowa, New Jersey 07512

www.humanapress.com

This publication is printed on acid-free paper. ♾
ANSI Z39.48-1984 (American Standards Institute) Permanence of Paper for Printed Library Materials.

Production Editor: Christina Thomas
Cover design by Karen Schulz

Cover illustration: Chapter 3, Human Embryonic Stem Cells For Tissue Engineering
Figure 1.D, 13 HESC line
Author, Daniel Kitsberg

10 9 8 7 6 5 4 3 2 1
eISBN 978-1-59745-443-8

Editorial Overview

Classical tissue engineering is based on seeding cells into biodegradable polymer scaffolds or gels, culturing and expanding them in bioreactors, and finally implanting the resulting tissue into the recipient organism, where the maturation of the new organ takes place. Capitalizing on this basic concept, tissue engineering has rapidly evolved in the past decade into an integrating discipline in which every organ forms a science of tissue engineering: Each of these sciences are interfacing with different scientific communities, including biotechnology, biopharmaceutical manufacturing, chemical engineering, cell biology, developmental biology, gene therapy, medical sciences, and organic chemistry. With so many "tissue engineers" at work on this globe, the day of successful implantation of a fully functional artificial organ seems to be near. Yet, much knowledge on molecular crosstalk among cell communities is still missing as are technologies for precise multiscale control of vascularization, innervation, differentiation, and shape of multicellular organoid structures. Managing and covering all specialized methods implemented in current tissue engineering activities is a mission impossible. However, precise understanding of diverse technologies and methods used to drive tissue engineering into a clinical reality remains a key success factor. Like tissue engineering itself, this book is intended to gather experts of various disciplines to share recent advances in tissue engineering-related methodologies. Our goal is to provide a comprehensive volume that integrates a wide, but not all-inclusive, spectrum of methods required to implement current and future progress in tissue engineering. The knowledge collected in this volume defines the impressive progress made in many aspects of tissue engineering and also reminds us of how much remains to be overcome in this important field.

Hansjörg Hauser
Martin Fussenegger

Contents

Contributors

CATHARINA A. M. ADELÖW • *Laboratory for Regenerative Medicine and Pharmacobiology, Institute of Bioengineering, Swiss Federal Institute of Technology (EPFL), Lausanne, Switzerland*

PATRICK AU • *Department of Radiation Oncology, Edwin L. Steele Laboratory, Massachusetts General Hospital, Harvard Medical School, Boston, MA, and Harvard-MIT Division of Health Sciences and Technology, Massachusetts Institute of Technology, Cambridge, MA*

H. BANNASCH • *Department of Plastic and Hand Surgery, University of Freiburg Medical Center, Freiburg i. Br., Germany*

ANDREA BARBERO • *Institute for Surgical Research and Hospital Management, University Hospital Basel, Basel, Switzerland*

GIUSEPPE BASTA • *Department of Internal Medicine, Section of Internal Medicine and Endocrine and Metabolic Sciences, University of Perugia, Perugia, Italy*

PAOLO BIANCO • *Department of Experimental Medicine and Pathology, La Sapienza University, Rome, Italy, San Raffaele Biomedical Science Park, Rome, Italy, and Department of Health and Human Services, Craniofacial and Skeletal Diseases Branch, National Institute of Dental and Craniofacial Research, National Institutes of Health, Bethesda, MD*

OLIVER BRÜSTLE • *Institute of Reconstructive Neurobiology, University of Bonn – Life & Brain Center and Hertie Foundation, Bonn, Germany*

RICCARDO CALAFIORE • *Department of Internal Medicine, Section of Internal Medicine and Endocrine and Metabolic Sciences, University of Perugia, Perugia, Italy*

CHRISTOPHER CANNIZZARO • *Harvard-MIT Division for Health Sciences and Technology, Massachusetts Institute of Technology, Cambridge, MA*

SERGHEI CEBOTARI • *Department of Thoracic and Cardiovascular Surgery, Hanover Medical School, Hanover, Germany*

PIERRE CHARBORD • *Equipe Microenvironnement de l'Hématopoïèse et Cellules Souches, Faculté de Médecine INSERM ESPRI-EA3855, Tours, France*

TEUN P. DE BOER • *Department of Medical Physiology, University Medical Center Utrecht, Utrecht, the Netherlands*

BRUNO DELORME • *Equipe Microenvironnement de l'Hématopoïèse et Cellules Souches, Faculté de Médecine INSERM ESPRI-EA3855, Tours, France*

FRANK EDENHOFER • *Institute of Reconstructive Neurobiology and Stem Cell Engineering Group, University of Bonn – Life & Brain Center and Hertie Foundation, Bonn, Germany*

ANDREW K. EKAPUTRA • *Division of Bioengineering, National University of Singapore, Singapore*

NICOLA ELVASSORE • *Department of Chemical Engineering, University of Padua, Padua, Italy*

Elisa Figallo • *Harvard-MIT Division for Health Sciences and Technology, Massachusetts Institute of Technology, Cambridge, MA, and Department of Chemical Engineering, University of Padua, Padua, Italy*

M. Föhn • *Department of Plastic and Hand Surgery, University of Freiburg Medical Center, Freiburg i. Br., Germany*

Peter Frey • *Professor for Pediatric Urology, Centre Hospitalier Universitaire Vaudois (CHUV), Switzerland, and Institute of Bioengineering, Swiss Federal Institute of Technology (EPFL), Lausanne, Switzerland*

Dai Fukumura • *Department of Radiation Oncology, Edwin L. Steele Laboratory, Massachusetts General Hospital, Harvard Medical School, Boston, MA*

Martin Fussenegger • *Institute for Chemical and Bio-Engineering, ETH Zurich, Zurich, Switzerland*

Sharon Gerecht • *Harvard-MIT Division for Health Sciences and Technology, Massachusetts Institute of Technology, Cambridge, MA*

Hansjörg Hauser • *Department of Gene Regulation and Differentiation, National Research Center for Biotechnology GBF, Braunschweig, Germany*

Axel Haverich • *Department of Thoracic and Cardiovascular Surgery, Hanover Medical School, Hanover, Germany*

Andres Hilfiker • *Department of Thoracic and Cardiovascular Surgery, Hanover Medical School, Hanover, Germany*

Simon P. Hoerstrup • *Clinic of Cardiovascular Surgery, Zurich University Hospital, Zurich, Switzerland, and Department of Surgical Research and Clinic for Cardiovascular Surgery, Division of Cardiovascular Regenerative Medicine, University Hospital and University of Zurich, Zurich, Switzerland*

Dietmar W. Hutmacher • *Division of Bioengineering, National University of Singapore, Singapore*

Rakesh K. Jain • *Department of Radiation Oncology, Edwin L. Steele Laboratory, Massachusetts General Hospital, Harvard Medical School, Boston, MA*

Jens M. Kelm • *Institute for Chemical and Bio-Engineering, ETH Zurich, Zurich, Switzerland, Clinic of Cardiovascular Surgery, Zurich University Hospital, Zurich, Switzerland, and Department of Surgical Research and Clinic for Cardiovascular Surgery, Division of Cardiovascular Regenerative Medicine, University Hospital and University of Zurich, Zurich, Switzerland*

Daniel Kitsberg • *Stem Cell Technologies (SCT) Ltd, Jerusalem, Israel*

Theo Kofidis • *Division of Thoracic and Cardiovascular Surgery, Hanover Medical School, Hanover, Germany*

Sergei A. Kuznetsov • *Department of Health and Human Services, Craniofacial and Skeletal Diseases Branch, National Institute of Dental and Craniofacial Research, National Institutes of Health, Bethesda, MD*

Christopher X. F. Lam • *Division of Bioengineering, National University of Singapore, Singapore*

ARTUR LICHTENBERG • *Department of Thoracic and Cardiovascular Surgery, Hanover Medical School, Hanover, Germany*

IVAN MARTIN • *Institute for Surgical Research and Hospital Management, University Hospital Basel, Basel, Switzerland*

TOBIAS MAY • *Department of Gene Regulation and Differentiation, National Research Center for Biotechnology GBF, Braunschweig, Germany*

ANITA MOL • *Department of Surgical Research and Clinic for Cardiovascular Surgery, Division of Cardiovascular Regenerative Medicine, University Hospital and University of Zurich, Zurich, Switzerland*

WOLFGANG MORITZ • *Clinic of Visceral Surgery, Zurich University Hospital, Zurich, Switzerland*

KNUT MÜLLER-STAHL • *Division of Thoracic and Cardiovascular Surgery, Hanover Medical School, Hanover, Germany*

CHRISTINE MUMMERY • *Hubrecht Laboratory, Utrecht, the Netherlands*

LARS K. NIELSEN • *Department of Chemical Engineering, The University of Queensland, Brisbane, Australia*

LARS NOLDEN • *Institute of Reconstructive Neurobiology, University of Bonn – Life & Brain Center and Hertie Foundation, Bonn, Germany*

HYOUNGSHIN PARK • *Harvard-MIT Division for Health Sciences and Technology, Massachusetts Institute of Technology, Cambridge, MA*

ROBERT PASSIER • *Hubrecht Laboratory, Utrecht, the Netherlands*

MICHAEL PEITZ • *Institute of Reconstructive Neurobiology and Stem Cell Engineering Group, University of Bonn – Life & Brain Center and Hertie Foundation, Bonn, Germany*

MILICA RADISIC • *Institute of Biomaterials and Biomedical Engineering and Department of Chemical Engineering and Applied Chemistry, University of Toronto, Toronto, ON, Canada*

MARA RIMINUCCI • *Department of Experimental Medicine, University of L'Aquila, L'Aquila, Italy, and San Raffaele Biomedical Science Park, Rome, Italy*

PAMELA GEHRON ROBEY • *Department of Health and Human Services, Craniofacial and Skeletal Diseases Branch, National Institute of Dental and Craniofacial Research, National Institutes of Health, Bethesda, MD*

DOERTHE SCHMIDT • *Clinic of Cardiovascular Surgery, Zurich University Hospital, Zurich, Switzerland, and Department of Surgical Research and Clinic for Cardiovascular Surgery, Division of Cardiovascular Regenerative Medicine, University Hospital and University of Zurich, Zurich, Switzerland*

DETLEF SCHUMANN • *Division of Bioengineering, National University of Singapore, Singapore*

JOSH TAM • *Department of Radiation Oncology, Edwin L. Steele Laboratory, Massachusetts General Hospital, Harvard Medical School, Boston, MA, and Harvard-MIT Division of Health Sciences and Technology, Massachusetts Institute of Technology, Cambridge, MA*

NINA TANDON • *Department of Biomedical Engineering, Columbia University, New York, NY*

NICHOLAS E. TIMMINS • *Department of Surgery and Research, University Hospital Basel, Basel, Switzerland*

IGOR TUDORACHE • *Department of Thoracic and Cardiovascular Surgery, Hanover Medical School, Hanover, Germany*

ANJA VAN DE STOLPE • *Hubrecht Laboratory, Utrecht, the Netherlands*

STIENEKE VAN DEN BRINK • *Hubrecht Laboratory, Utrecht, the Netherlands*

MARCEL A. G. VAN DER HEYDEN • *Department of Medical Physiology, University Medical Center Utrecht, Utrecht, the Netherlands*

MARGA VAN ROOIJEN • *Hubrecht Laboratory, Utrecht, the Netherlands*

GORDANA VUNJAK-NOVAKOVIC • *Department of Biomedical Engineering, Columbia University, New York, NY*

DORIEN WARD • *Hubrecht Laboratory, Utrecht, the Netherlands*

DAGMAR WIRTH • *Department of Gene Regulation and Differentiation, National Research Center for Biotechnology GBF, Braunschweig, Germany*

1

In Vitro Expansion of Tissue Cells by Conditional Proliferation

Tobias May, Hansjörg Hauser, and Dagmar Wirth

Summary

Cell therapies rely on the implantation of well-characterized functional cells with defined properties. Often, the cells of interest do not proliferate in vitro and thus cannot be expanded to the amount needed for characterization, purification, manipulation, or cloning. Here, we describe a method that allows the reversible expansion of cells by the introduction of a proliferator gene controlled by a regulatable expression module. The module is transferred by DNA transfer or by lentiviral transduction. The addition of a clinically accepted regulator [Doxycycline (Dox)] induces proliferator expression and expansion of the cells ad infinitum. Removal of the regulator eliminates the effect of the proliferator and leaves the cells in a non-proliferating status. The method has been applied to different mouse and human tissues. This chapter describes the method for the well-examined example of mouse embryonic fibroblast (MEF) expansion.

Key Words: Cell expansion; Tet-system; Autoregulated expression; Lentiviral transduction; Conditional immortalization.

1. Introduction

Immortalized cell lines represent a useful tool to study biological processes but are also indispensable for biotechnological applications. Among the advantages are their infinitive expansion capacity and the reproducible properties. Cell lines exist from various species and tissues. Many properties of these immortalized cells reflect those of the primary cells they have been derived from. However, the fact that immortalized cells have an infinitive life span and thus significantly changed proliferation properties makes them different from primary cells.

From: *Methods in Molecular Medicine, 2nd ed.: Tissue Engineering*
Edited by: H. Hauser and M. Fussenegger © Humana Press Inc., Totowa, NJ

Infinitive growth is a consequence of the immortalization process that is achieved by selection for random mutations or by stable expression of certain oncogenes.

Because of the in vitro cultivation, many cells rapidly lose specific differentiation properties. This is thought to be because of the alteration of environmental factors (soluble and cell bound) as well as the loss of a 3D architecture. Many cell lines show increased genetic instability, because of either the expression of the immortalizing oncogene or the random mutations due to long-term cultivation and selection for rapid proliferation.

Several cell types can be expanded by growth in specific media and conditions that allow them to keep their differentiation potential. However, this is not possible to the extent needed for experimental approaches or therapies for most cell types. One alternative might be to express genes that lead to the required expansion. The expression of these proliferator genes would have to be obliterated after the expansion period without leaving disturbing remainders. This approach has two critical components: one concerns the proliferator gene. It should specifically induce proliferation in a timely restricted manner and should not have influence on other cellular properties. In particular, it should not induce permanent (epi)genetic changes. Ideal candidates are genes that naturally control the growth of a given cell population. Such candidates have been successfully used for expansion. One example concerns the expansion of hematopoietic stem cells, which was accomplished by the constitutive expression of the intracellular domain of Notch ***(1)***. As such proliferator genes are usually highly specific for a given cell type, a more general approach for experimental evaluation is to use broadly active oncogenes such as the SV40 virus-derived T antigen (TAg).

The second critical component concerns the way in which conditional expression of the proliferator gene is achieved. Several approaches have been undertaken to restrict the expression of the gene to the period of cell expansion. One of the first strategies for a controlled expansion used a temperature-sensitive mutant of the TAg ***(2, 3)***. This approach has several drawbacks—among them are (i) high clonal variability concerning the proliferation control ***(4)***, (ii) it is restricted to TAg, and (iii) the regulatory switch being suited only for in vitro use. A strategy that circumvents these problems employs recombinase-mediated excision by the Cre/loxP or the Flp/FRT system to eliminate the immortalizing gene ***(5–9)***. The disadvantage of these systems is the need for expression of the recombinase, the completeness of its action, and the risk of recombinase gene integration. An alternative approach is to make use of transcriptionally regulated expression of the immortalizing gene(s) by the Tet system ***(10–12)***.

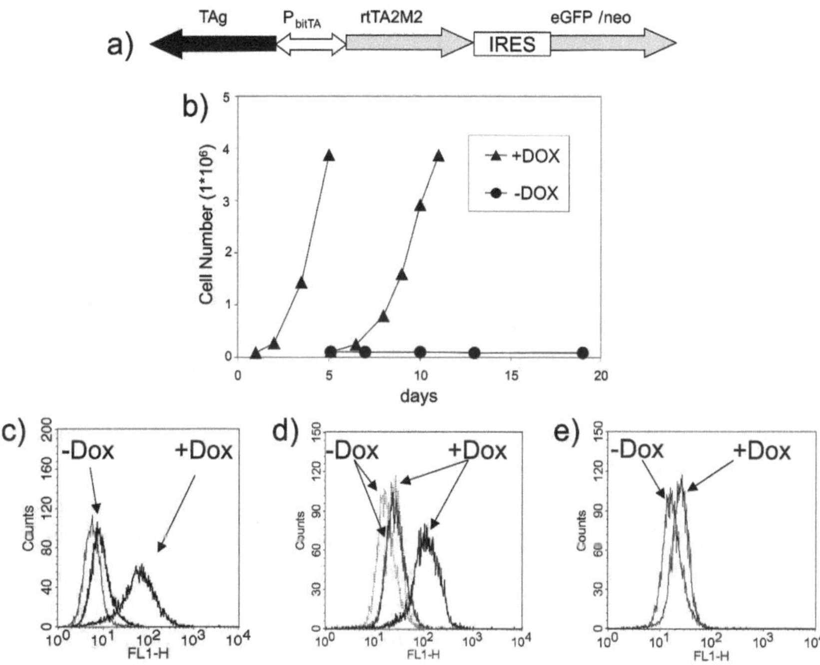

Fig. 1. Regulation potential of the autoregulatory immortalization vector. (a) Schematic presentation of the plasmid vector pRITA. The bidirectional Tet-dependent promoter P_{bitTA} drives the expression of two mRNAs. One encodes the TAg, and the other encodes the reverse transactivator $rtTA2^{S}M2$ ***(16)*** and a fusion protein of eGFP and neomycin phosphotransferase (eGFP/neo). (b) Cells are usually cultivated with Dox (triangles) that results in an exponential growth behavior. For monitoring Dox-dependent proliferation, the cells were split and cultivated in parallel with (triangles) or without (circles) Dox, respectively. Three independent cultures were cultivated and counted in duplicate. Cell numbers are scaled to 1×10^6. (c) For GFP analysis, the cells were washed and trypsinized followed by flow cytometry analysis. The arrows indicate the cultivation conditions. Balb/c 3T3 cells served as a control (light gray). (d) For determination of TAg levels, the cells were permeabilized after trypsinization. Subsequently, indirect immunostaining of intracellular TAg was performed, and cells were analyzed using flow cytometry. As controls, Balb/c 3T3 cells (not shown) and the secondary antibody alone with both activated and repressed cells (dotted lines) were analyzed. (e) The controls are shown in a separate histogram. (Reproduced from **ref. 11** with permission from Oxford University Press.)

The method described here concerns a recently developed transcriptional control system. The strategy is based on a transcriptionally regulated cassette that allows the autoregulated doxycycline (Dox)-dependent expression of a proliferator gene, the reverse transactivator (rtTA2), and a fusion protein of eGFP and neomycin phosphotransferase (eGFP/neo) (*see* **Fig. 1**) ***(11)***. In the presence of Dox, the inducible promoter is active, and all genes are expressed. This cassette allows the selection of immortalized cell clones in the presence of G418. Although cells can be immortalized by physical transduction (DNA/$Ca_3(PO_4)_2$ coprecipitation, lipofection-based protocols), the efficiency of these approaches is frequently not sufficient, mostly because of the cellular refractoriness toward physical transduction protocols. We therefore describe the integration of the autoregulated expression cassette into lentiviral particles. This results in highly efficient transduction of the fully regulatable cassette in a broad spectrum of mammalian cells and tissues.

To use this system for a broad range of cell types, we have used TAg as a proliferator gene. TAg has been used successfully for the establishment of cell lines of different origin (e.g., epithelial, endothelial) and of different species including mouse and rat (human cells can also be immortalized with TAg but less efficient) ***(13, 14)***. The vector system was designed in a way that permits easy replacement of TAg by tissue- and species-specific proliferators at wish.

Cells that were conditionally immortalized using the autoregulated cassettes depicted in **Fig. 1** above show a strict regulation of growth, although a basal expression of TAg can be monitored. Interestingly, these cells can be repeatedly switched on and off, with all Dox-induced changes being fully reversible as monitored by gene-expression profiles ***(11, 15)***. Thus, the transcriptional control provides a reliable system for fully reverting the proliferating phenotype to a stationary mode, and this represents a new tool for the establishment of cell lines with improved properties.

2. Materials

2.1. Cell Culture

1. A standard cell-culture equipment is required. For the lentiviral transduction, the respective safety requirements have to be considered. Cells are handled according to standard protocols. In this chapter, all cell-culture media and solutions, which are required for the methods described in the Materials, are listed Section in alphabetical order.
2. $CaCl_2$: 2.5 M solution in H_2O, sterile filtered, stored at −20° C.
3. Crystal violet solution: 5 g crystal violet, 8.5 g NaCl, 143 ml formaldehyde, 500 ml ethanol adjusted to 1000 ml with H_2O, stored at room temperature.

4. Dulbecco's modified Eagle's medium (DMEM): Sigma no. D-7777, high glucose with sodium pyruvate dissolved in aqua dest, sterile filtered, aliquoted and stored at 4°C.
5. DMEM-F: DMEM supplemented with 2 mM glutamine, 100 U penicillin, 100 μg/ml streptomycin, 1 mM non-essential amino acids, 0.1 mM ß-mercaptoethanol, and 10% fetal calf serum (FCS), stored at 4°C.
6. DMEM-V: DMEM supplemented with 2 mM glutamine, 100 U penicillin, 100 μg/ml streptomycin, and 10% FCS, stored at 4°C.
7. Dox: 2 mg/ml stock solution in 70% ethanol, sterile filtered. The stocks are wrapped with aluminium foil and stored at −20°C.
8. Fixation solution (SA-ß-Gal): 2% formaldehyde, 0.1% glutaraldehyde in phosphate-buffered saline (PBS) pH 7, freshly prepared.
9. Freezing solution: FCS with 5% dimethylsulfoxide (DMSO), stored at 4°C.
10. G418: 100 mg/ml stock solution in H_2O, sterile filtered, stored at −20°C.
11. Glutamine stock solution: 0.2 M glutamine in H_2O, aliquoted, sterile filtered, stored at −20°C.
12. HEBS: 280 mM NaCl, 50 mM 4-(2-hydroxyethyl)-piperazine-1-ethane sulfonic acid (HEPES), 1.5 mM Na_2HPO_4, pH 7.1; aliquoted and stored at −20°C.
13. HEPES stock solution: 1 M pH 7.12, sterile filtered, stored at 4°C.
14. Lysis buffer (intracellular staining): 0.5% Triton-X100, 0.5 mM ethylenediaminetetraacetic acid (EDTA), 1% bovine serum albumin (BSA) dissolved in PBS pH 7.
15. PBS: 140 mM NaCl, 27 mM KCl, 7.2 mM Na_2HPO_4, 14.7 mM KH_2PO_4, pH 6.8–7; autoclaved and stored at 4°C.
16. PBS*: PBS supplemented with 2% FCS, filtered with 0.45-μm filter, stored at 4°C.
17. Penicillin stock solution: 10,000 U/ml penicillin in H_2O, aliquoted, sterile filtered, stored at −20°C.
18. Polybrene stock solution: 4 mg/ml sterile filtered, stored at −20°C.
19. Propidium iodide stock solution: 5 mg/ml in PBS, sterile filtered, stored at 4°C.
20. Staining solution (SA-ß-Gal): 5 mM potassium hexacyanoferrate (II) ($K_4[Fe(CN)6]$), 5 mM potassium hexacyanoferrate (III) ($K_3[Fe(CN)_6]$), 2 mM $MgCl_2$, 150 mM NaCl, 40 mM NaH_2PO_4, 1 mg/ml X-gal (5-bromo-4-chloro-3-indolyl-β-D-galactopyranoside) (dissolved in dimethylformamid), and 40 mM citric acid pH 6.0.
21. Streptomycin stock solution: 10 mg/ml streptomycin in H_2O, aliquoted, sterile filtered, stored at −20°C.
22. TEP: 6 mM EDTA in PBS, 0.1–0.2% trypsin, sterile filtered, stored at 4°C.
23. Virus production medium: DMEM-V supplemented with 20 mM HEPES pH 7.12, stored at 4°C.

2.2. Preparation and Maintenance of Mouse Embryonic Fibroblasts

1. A pregnant mouse (days 13–14 post fertilization).
2. 70% ethanol for disinfection.

3. PBS.
4. Scissors, tweezers (autoclaved), scalpel.
5. Aluminium foil-covered styrofoam.
6. Ice.
7. Petri dish.
8. 24-well plates, 25-cm^2 flasks.
9. Sterile pipets.
10. 15- and 50-ml capped tubes.
11. Rocker.
12. Table-top centrifuge.
13. 37°C water bath.
14. Cryo vials.
15. −70°C freezer, liquid nitrogen.
16. PBS.
17. DMEM-F.
18. TEP.

2.3. Establishment of Conditionally Immortalized Fibroblasts

2.3.1. Establishment of Conditionally Immortalized Fibroblasts by Transfection

1. DMEM-F supplemented with (2 μg/ml) Dox.
2. DMEM-F supplemented with (2 μg/ml) Dox and 400 μg/ml G418.
3. TEP.
4. $CaCl_2$ solution.
5. HEBS.
6. 10 μg pRITA DNA.
7. 24-well plates, 6-well plates, 55-cm^2 plates.
8. Sterile Eppendorf tubes.
9. Vortex.
10. Mouse embryonic fibroblasts (MEFs).

2.3.2. Production of Lentiviral Particles

1. Lentiviral helper plasmids as purchased from Invitrogen (ViraPower™ Lentiviral Expression System): 14 μg pLP1 (gag/pol), 4.6 μg pLP2 (rev), and 7.8 μg pLP/VSVg (VSVg).
2. 20 μg immortalizing plasmid (LentiRITA).
3. Virus production medium.
4. 50-ml capped plastic tubes.
5. Sterile Eppendorf tubes.
6. Vortex.
7. $CaCl_2$ solution.

8. HEBS.
9. 293T cells (ATCC CRL-11268).
10. 140-cm^2 plates.

2.3.3. Establishment of Conditionally Immortalized Fibroblasts by Infection

1. DMEM-F supplemented with (2 μg/ml) Dox.
2. PBS.
3. 30-ml plastic syringe.
4. 0.45-μm syringe filter.
5. Polybrene.
6. MEFs.
7. Viral supernatant.
8. 6-well plates.

2.4. Selection of Conditionally Immortalized MEFs

1. DMEM-F supplemented with (2 μg/ml) Dox.
2. DMEM-F supplemented with (2 μg/ml) Dox and 400 μg/ml G418.
3. TEP.
4. PBS.
5. Autoclaved yellow tips.
6. Light microscope.
7. 96-well plates, 24-well plates, 12-well plates, 6-well plates, 55-cm^2 plates.

2.5. Characterization of Conditionally Immortalized MEFs

2.5.1. Clonogenicity Assay

1. 55-cm^2 plates.
2. DMEM-F supplemented with (2 μg/ml) and without Dox.
3. PBS.
4. Crystal violet solution.

2.5.2. Analysis of eGFP Expression via Flow Cytometry

1. DMEM-F supplemented with (2 μg/ml) and without Dox.
2. 6-well plates.
3. PBS.
4. TEP.
5. PBS*.
6. Propidium iodide.
7. 15-ml capped tubes.
8. Flow cytometry device.

2.5.3. Intracellular Staining of TAg

1. DMEM-F supplemented with (2 μg/ml) and without Dox.
2. 55-cm^2 plates, 96-well plates round bottom.
3. PBS.
4. TEP.
5. PBS*.
6. Lysis buffer (intracellular staining).
7. Antibody-recognizing TAg: clone PAb 416 (Merck Biosciences, cat. no. DP02).
8. Anti-mouse immunoglobulin G (IgG) antibody R-phycoerythrin (PE) labeled (Jackson Immuno Research).

2.5.4. SA-β-Gal Staining

1. DMEM-F supplemented with (2 μg/ml) and without Dox.
2. 6-well plates.
3. PBS.
4. Fixation solution.
5. Staining solution.

3. Methods

3.1. The Immortalization Cassette pRITA

1. For immortalization, an autoregulated, Dox-dependent expression vector, pRITA, is used, which allows conditional immortalization upon a single plasmid transduction (*see* **Fig. 1**) *(11)*. In this vector, a bidirectional tet-dependent promoter drives the expression of two mRNAs (*see* **Fig. 1a**). One encodes TAg. The other encodes the reverse transactivator rtTA2^{S}M2 *(16)* and the reporter selection fusion protein eGFP/neo linked via the EMCV IRES element. In the absence of Dox, a low basal expression of both TAg and eGFP is observed. Addition of Dox leads to activation of the positive feedback loop and thereby to the expression of TAg and eGFP (*see* **Fig. 1c and d**). For creation of the lentiviral immortalization construct, the expression cassette of pRITA was cloned via standard cloning procedures into a self-inactivating lentivirus (TREAutoR3) *(17)*.

3.2. Preparation, Maintenance, and Storage of MEFs

3.2.1. Preparation

1. A pregnant mouse (days 13–14 post fertilization) is killed by cervical dislocation and put on an aluminum foil-covered styrofoam plate.
2. The belly is rinsed with 70% ethanol, and the fur over the belly is lifted. A triangular cut is made, and the fur is pulled downward. Incisions in the skin are made from the center in several directions.

3. The uterus with the embryos is removed and soaked in PBS in Petri dish on ice. For every embryo, a separate Petri dish is used. Scissors and tweezers are rinsed with sterile aqua dest.
4. The embryos are separated under the binocular using two tweezers; all residual skins are removed, as well as the head and the red, blood-containing organs.
5. The embryos are transferred to a 24-well plate with 1 ml PBS to rinse the embryo. Then, the embryo is transferred to a Petri dish containing 5 ml TEP. It is cut into small pieces using a scalpel.
6. The pieces are transferred to a cell-culture flask containing 5 ml TEP and incubated for 30 min at 37°C on a rocker.
7. The suspension is transferred into a 50-ml capped centrifuge tube, and 40 ml DMEM-F is added. The cells are pelleted upon centrifugation (5 min at 200 × g rpm in a table-top centrifuge). The supernatant is carefully removed, and the cells are washed by resuspending the cell pellet in 50 ml DMEM-F using an Eppendorf pipette with a 1-ml blue tip. Washing is repeated, and cells are suspended in 10 ml DMEM-F.
8. 1 ml suspended cells is seeded into 4 ml DMEM-F in a 25-cm^2 tissue culture flask each and incubated at 37°C, 5% CO_2.
9. Upon cultivation, cells growing out of the tissue fragments as a monolayer can be observed with the microscope.

3.2.2. Maintenance

1. For passaging, MEFs (90% confluent) are washed with 5 ml PBS and 500 μl TEP is added.
2. Cells are incubated for 1–5 min at 37°C until the cells detach.
3. 10 ml DMEM-F is added; the cells are suspended and seeded into five small tissue culture flasks.
4. The culture can be split twice per week and maintained typically for five passages before the cell growth rate decreases and the cell morphology changes to a larger, flattened cell type.

3.2.3. Cryopreservation

1. Nearly confluent cells are trypsinized as described in **section 3.2.1**.
2. Five milliliters of DMEM-F is added to inactivate trypsin.
3. The cell suspension is transferred to 15-ml tube and spun for 5 min at 200 × g in table-top centrifuge.
4. The supernatant is discarded; the cell pellet is resuspended in freezing solution.
5. The suspension is distributed into cryo vials, 1 ml each, and put on ice for 30 min.
6. Cells are transferred to a –70°C freezer for 2–3 days and transferred to liquid nitrogen thereafter.
7. For thawing, a vial is warmed quickly in a 37°C water bath and the cell suspension is immediately added to fresh medium in a 15-ml capped tube.

8. The cells are spun for 5 min at 200 × g in a table-top centrifuge, resuspended in fresh medium and seeded in tissue culture flask and further maintained as described in **Section 3.2.2**.

3.3. Establishment of Conditionally Immortalized Fibroblasts by Transfection

1. 1×10^5 MEFs are seeded per 6 well in 2.5 ml DMEM-F containing Dox (2 μg/ml).
2. On the next day, the medium is renewed 4 h prior to transfection.
3. For transfection, 10 μg pRITA DNA is suspended in 100 μl 250 mM $CaCl_2$ in a sterile Eppendorf tube.
4. The DNA solution is added dropwise to 100 μl HEBS under continuous vortexing and then kept for 10 min at room temperature for precipitation.
5. The suspension is added to the culture medium of the plated cells.
6. As a control, a DNA-free precipitate should be added to cells.
7. After 4–20 h, the medium is replaced.
8. After 48 h, the cells are selected as described in **sections 3.5.1** and **3.5.2**.
9. *See* **note 2**.

3.4. Conditional Immortalization Upon Lentiviral Transduction

3.4.1. Production of Lentiviral Particles

1. 6×10^6 293T cells are plated per 140-cm^2 plate.
2. The day after DNA transfection of 293T cells is performed as described in **section 3.3**. For virus production, the lentiviral helper functions have to be cotransfected along with the transfer plasmid that harbors the immortalization expression cassette. The helper functions encode gag/pol, rev, and the VSVg surface protein.
3. The following day, the media are aspirated and 15-ml virus production media is added.
4. After 24 h, the supernatant containing the lentiviral particles is collected and filtrated with a 0.45-μm filter to get rid of cell debris. The lentiviral particles are stored at −70°C until use.
5. For a second production, again 15-ml virus production media is added to the producer cells. The supernatant containing the lentiviral particles is collected after 24 h as described in **section 3.6.4**.

3.4.2. Establishment of Conditionally Immortalized Cells by Lentiviral Transfer

1. 1×10^5 MEFs are plated per 6 well in 2.5 ml DMEM-F containing Dox (2 μg/ml).
2. On the following day, undiluted viral supernatant is supplemented with 8 μg/ml polybrene and added (0.1 ml/cm^2) to MEFs and incubated for 8 at 37°C, 5% CO_2.

3. The infected cells are washed with PBS, and DMEM-F supplemented with Dox (2 μg/ml) is added.
4. Two days after infection, the cells are trypsinized and selected as described in **sections 3.5.1** and **3.5.2**.
5. *See* **note 3**.

3.5. Selection of Conditionally Immortalized Cells

3.5.1. Selection with G418

1. The transfected or infected cells as obtained from **section 3.3 or 3.4.2** are plated in DMEM-F containing 0.4 mg/ml G418.
2. The selection medium is renewed every 3–4 days until the mock-transfected control is dead (usually 14–21 days).
3. The resulting G418-resistant colonies can be pooled or picked if desired. For picking, a light microscope is placed in a laminar flow and the colonies are collected with a P200 pipette with a yellow tip.
4. The resistant cell clones are transferred to a 96 well plate and the selection pressure is removed.
5. Cells are expanded for further use (*See* also **note 1**).

3.5.2. Selection for Growth Advantage

1. In case the MEFs to be immortalized are already G418 resistant (neo^R is often used for the generation of knock-out mice), transduced cells can be selected solely because of growth advantage that is conferred by TAg.
2. For this purpose, the MEFs are transduced as described in **sections 3.3**. and **3.4.2**.
3. When the transfected or infected cells reach confluence, they are trypsinized and transferred to a 55-cm^2 plate.
4. Cells are cultivated until they are confluent and then split 1/20. This procedure is repeated two to three times.
5. Then, the mock-transfected control cells stop their proliferation and enter senescence, whereas the immortalized cells form colonies.
6. The resulting colonies are isolated (*see* **section 3.5.1**.) to get rid of contaminating untransfected cells that could undergo spontaneous immortalization. Then, the cells are expanded for further use.

3.6. Characterization of Conditionally Immortalized Cells

3.6.1. Clonogenicity Assay

1. 1000–10,000 immortalized cells are plated per 55-cm^2 plate with (2 μg/ml) and without Dox.
2. The media are renewed every 3–4 days.
3. After 2 weeks, the medium is removed, the cells are washed with PBS, and the cells are stained with 4-ml crystal violet solution for 10 min. The activated cells

(with Dox) form huge colonies that can be macroscopically counted. The repressed cells (without Dox) show only minimal proliferation with small clusters of a few cells that display an enlarged and flattened morphology.

3.6.2. Flow Cytometry Analysis of eGFP Expression

1. The immortalized cells are cultivated with (2 μg/ml) and without Dox for at least 3 days.
2. The cells are washed twice with PBS and detached with TEP.
3. Cells are suspended in PBS*, transferred to a 15-ml capped plastic tube, and spun down at 200 × g for 5 min, using a table-top centrifuge.
4. The cell pellet is suspended with PBS* containing 50 μg/ml propidium iodide to stain the dead cells.
5. The cells are analyzed flow cytometry. A side scatter by (SSC/FSC forward Scatter) dot blot is applied to exclude cell debris (FSC < 200). The FL3 channel is used to analyze the propidium iodide staining which excludes dead cells from the analysis. The FL1 channel is used for the detection of GFP.
6. *See* **notes 4**, **5** and **6**.

3.6.3. Flow Cytometry Analysis of TAg Expression

1. The immortalized cells are cultivated with (2 μg/ml) and without Dox for at least 3 days. The cells should be subconfluent (50–70%) at the time of the staining.
2. Cells are washed with PBS, trypsinized, and suspended in PBS*.
3. The cells are counted, and at least 1×10^5 cells per sample are transferred into a 96-well plate with a round bottom. The following controls should be prepared: unstained cells and cells stained only with secondary PE-labeled antibody and cells that were stained with isotype control antibody along with the secondary antibody.
4. PBS* (200 μl) is added to the samples, and cells are spun down at 200 × g (5 min, table-top centrifuge).
5. For lysis, PBS is removed and 200 μl lysis buffer is added. Then the cells are incubated on ice for 15 min.
6. The antibody recognizing the SV 40 TAg is diluted in lysis buffer (10 μg/ml). 25 μl diluted solution is added, and the cells are incubated at room temperature for 30 min.
7. 25 μl PE-labeled anti-mouse secondary antibody is added. The cells are incubated in the dark (cover the 96-well plate with aluminium foil) for 30 min at room temperature.
8. The samples are centrifuged at 200 × g, and the cells are washed once with PBS*.
9. The cells are analyzed using flow cytometry as described in **section 3.5.3**.

3.6.4. SA-β-Gal Staining *(18)*

1. The immortalized cells are cultivated with and without Dox for at least 4 days.
2. The medium is removed, and the cells are washed with PBS.

3. The cells are covered with fixation buffer for 5 min at room temperature.
4. The fixation buffer is removed, and the cells are washed twice with PBS for 5 min.
5. The cells are stained overnight at 37°C using $0.15\,ml/cm^2$ staining solution.
6. *See* **note 7** and **8.**

Notes

1. Once a conditionally immortalized MEF cell line is established, its proliferation is absolutely dependent on the presence of Dox. For induction of the expression cassette, $2\,\mu g/ml$ Dox is sufficient. As Dox is thermolabile and light sensitive, the media have to be replenished every 3–4 days.
2. As an alternative to $Ca_3(PO_4)_2$ transfection, pRITA can also be transduced via lipofection-based protocols, which might be more efficient for other cell types. For MEFs, lipofection and $Ca_3(PO_4)_2$ transfection give comparable results.
3. For the transduction of oncogenes via lentiviral transduction, safety precautions have to be made.
4. The immortalization cassette pRITA encodes the selection/reporter gene eGFP/neo in addition to TAg. Comparable with TAg, the expression of eGFP can be regulated through the adjustment of the Dox concentration. This can be monitored using flow cytometry (*see* **section 3.6.2.**). In principle, the regulated expression can also be monitored via fluorescence microscopy. However, the fluorescence of the fusion protein is usually not strong enough to clearly visualize it. In addition, the autofluorescence of both states (proliferating and growth arrested) differs, thereby complicating the analysis.
5. The advantage of the fusion protein eGFP/neo is that transgene expression can be followed on single cell level (eGFP) and, apart from this, can be used for the selection of successfully transfected cells. However, the expression level that is effective differs for both functions: the amount of eGFP/neo molecules needed to confer resistance toward G418 is much lower than the amount required to give a fluorescence signal. This phenomenon occasionally leads to immortalized cells that are $G418^R$ but do not show eGFP expression. This effect has no influence on the proliferation control of the pRITA expression cassette.
6. The withdrawal of the inducer Dox immediately results in a strict and permanent growth arrest of the conditionally immortalized cells. At the same time, the cells undergo morphological changes. They increase size and become flattened and show a higher granularity. These morphological changes can be monitored by flow cytometry. The immortalized (activated) cells show a low granularity, which is detected in the SSC. Upon Dox withdrawal, the cells increase their size over time. In addition, the cells become very heterogenous with respect to their granularity.
7. The generally accepted marker for identification of senescence is SA-ß-Gal staining ***(18)***. We observed the first faintly stained cells 4 days after Dox withdrawal. After 7–9 days, the growth arrested immortalized cells exhibit a strong blue staining. The

strength of the staining is dependent on the cell density. In the very confluent state, even the activated cells may show a low percentage of SA-ß-Gal-positive cells.

8. The pH of the staining solution is critical for the specificity of the SA-ß-Gal staining. The pH has to be adjusted to 6. This is below the pH optimum for the bacterial ß-galactosidase (pH 7), which is usually used as a reporter gene. In addition, a lysosomal ß-galactosidase exists, which is present in eukaryotic cells and active at a pH 4.

References

1. Varnum-Finney, B., Xu, L., Brashem-Stein, C., Nourigat, C., Flowers, D., Bakkour, S., Pear, W. S., and Bernstein, I. D. (2000) Pluripotent, cytokine-dependent, hematopoietic stem cells are immortalized by constitutive Notch1 signaling. *Nat. Med.* **6**, 1278–1281.
2. Jat, P. S. and Sharp, P. A. (1989) Cell lines established by a temperature-sensitive simian virus 40 large-T-antigen gene are growth restricted at the nonpermissive temperature. *Mol. Cell Biol.* **9**, 1672–1681.
3. Jat, P. S., Noble, M. D., Ataliotis, P., Tanaka, Y., Yannoutsos, N., Larsen, L., and Kioussis, D. (1991) Direct derivation of conditionally immortal cell lines from an H-2Kb-tsA58 transgenic mouse. *Proc. Natl. Acad. Sci. U. S. A.* **88**, 5096–5100.
4. May, T., Wirth, D., Hauser, H., and Mueller, P. P. (2005) Transcriptionally regulated immortalization overcomes side effects of temperature-sensitive SV40 large T antigen. *Biochem. Biophys. Res. Commun.* **327**, 734–741.
5. Westerman, K. A. and Leboulch, P. (1996) Reversible immortalization of mammalian cells mediated by retroviral transfer and site-specific recombination. *Proc. Natl. Acad. Sci. U. S. A.* **93**, 8971–8976.
6. Rybkin, I. I., Markham, D. W., Yan, Z., Bassel-Duby, R., Williams, R. S., and Olson, E. N. (2003) Conditional expression of SV40 T-antigen in mouse cardiomyocytes facilitates an inducible switch from proliferation to differentiation. *J. Biol. Chem.* **278**, 15927–15934.
7. Berghella, L., De Angelis, L., Coletta, M., Berarducci, B., Sonnino, C., Salvatori, G., Anthonissen, C., Cooper, R., Butler-Browne, G. S., Mouly, V., Ferrari, G., Mavilio, F., and Cossu, G. (1999) Reversible immortalization of human myogenic cells by site-specific excision of a retrovirally transferred oncogene. *Hum. Gene Ther.* **10**, 1607–1617.
8. Cai, J., Ito, M., Westerman, K. A., Kobayashi, N., Leboulch, P., and Fox, I. J. (2000) Construction of a non-tumorigenic rat hepatocyte cell line for transplantation: reversal of hepatocyte immortalization by site-specific excision of the SV40 T antigen. *J. Hepatol.* **33**, 701–708.
9. Narushima, M., Kobayashi, N., Okitsu, T., Tanaka, Y., Li, S. A., Chen, Y., Miki, A., Tanaka, K., Nakaji, S., Takei, K., Gutierrez, A. S., Rivas-Carrillo, J. D., Navarro-Alvarez, N., Jun, H. S., Westerman, K. A., Noguchi, H., Lakey, J. R. T.,

Leboulch, P., Tanaka, N., and Yoon, J. W. (2005) A human ß-cell line for transplantation therapy to control type 1 diabetes. *Nat. Biotechnol.* **23**, 1274–1282.

10. Efrat, S., Fusco-DeMane, D., Lemberg, H., al Emran, O., and Wang, X. (1995) Conditional transformation of a pancreatic beta-cell line derived from transgenic mice expressing a tetracycline-regulated oncogene. *Proc. Natl. Acad. Sci. U. S. A.* **92**, 3576–3580.
11. May, T., Hauser, H., and Wirth, D. (2004) Transcriptional control of SV40 T-antigen expression allows a complete reversion of immortalization. *Nucleic Acids Res.* **32**, 5529–5538.
12. Marinkovic, D., Marinkovic, T., Kokai, E., Barth, T., Moller, P., and Wirth, T. (2004) Identification of novel Myc target genes with a potential role in lymphomagenesis. *Nucleic Acids Res.* **32**, 5368–5378.
13. Noble, M., Groves, A. K., Ataliotis, P., Ikram, Z., and Jat. P. S. (1995) The H-2KbtsA58 transgenic mouse: a new tool for the rapid generation of novel cell lines. *Transgenic Res.* **4**, 215–225.
14. Obinata, M. (2001) Possible applications of conditionally immortalized tissue cell lines with differentiation functions. *Biochem. Biophys. Res. Commun.* **286**, 667–672.
15. May, T., Mueller, P. P., Weich, H., Froese, N., Deutsch, U., Wirth, D., Kroger, A., and Hauser, H. (2005) Establishment of murine cell lines by constitutive and conditional immortalization. *J. Biotechnol.* **120**, 99–110.
16. Urlinger, S., Baron, U., Thellmann, M., Hasan, M. T., Bujard, H., and Hillen, W. (2000) Exploring the sequence space for tetracycline-dependent transcriptional activators: novel mutations yield expanded range and sensitivity. *Proc. Natl. Acad. Sci. U. S. A.* **97**, 7963–7968.
17. Markusic, D., Oude-Elferink, R., Das, A. T., Berkhout, B., and Seppen, J. (2005) Comparison of single regulated lentiviral vectors with rtTA expression driven by an autoregulatory loop or a constitutive promoter. *Nucleic Acids Res.* **33**, e63.
18. Dimri, G. P., Lee, X., Basile, G., Acosta, M., Scott, G., Roskelley, C., Medrano, E. E., Linskens, M., Rubelj, I., and Pereira-Smith, O. (1995) A biomarker that identifies senescent human cells in culture and in aging skin in vivo. *Proc. Natl. Acad. Sci. U. S. A.* **92**, 9363–9367.

2

Stem Cell Engineering Using Transducible Cre Recombinase

Lars Nolden, Frank Edenhofer, Michael Peitz, and Oliver Brüstle

Summary

Embryonic stem (ES) cells have become a major focus of scientific interest both as a potential donor source for regenerative medicine and as a model system for tissue development and pathobiology. Tight and efficient methods for genetic engineering are required to exploit ES cells as disease models and to generate specific somatic phenotypes by lineage selection or instruction. In 1990s, the application of site-specific recombinases (SSRs) such as Cre has revolutionized mammalian genetics by providing a reliable and efficient means to delete, insert, invert, or exchange chromosomal DNA in a conditional manner. Despite these significant advances, the available technology still suffers from limitations, including unwanted side effects elicited by the random integration of Cre expression vectors and leak activity of inducible or presumptive cell type-specific Cre expression systems. These challenges can be met by combining the Cre/loxP recombination system with direct intracellular delivery of Cre by protein transduction, thus enabling rapid and highly efficient conditional mutagenesis in ES cells and ES cell-derived somatic progeny. Modified recombinant variants of Cre protein induce recombination in virtually 100% of human ES (hES) and mouse ES (mES) cells. Here, we present methods for generating purified transducible Cre protein from *Escherichia coli* and its transduction into ES cells and their neural progeny.

Key Words: Protein transduction; Site-specific recombinase; Cre; Stem cell therapy; Affinity chromatography; Fusion protein.

1. Introduction

Gaining precise control over gene activity in mammalian cells has become increasingly important for the dissection of molecular mechanisms regulating cellular function. Recent developments in molecular biology provide researchers

From: *Methods in Molecular Medicine, 2nd ed.: Tissue Engineering*
Edited by: H. Hauser and M. Fussenegger © Humana Press Inc., Totowa, NJ

with a large variety of genetic tools enabling conditional control over gene function. In particular, the Cre/loxP recombination system has proven to be a powerful tool for conditional mutagenesis. The site-specific recombinase (SSR) Cre can be used to conditionally induce loss- or gain-of-function of genes in mammalian cells by recombination of previously integrated recombination recognition sites, designated as loxP *(1)*. The recombination reaction results in deletion, inversion, insertion, or translocation of loxP-modified sequences depending on the relative orientation of the loxP sites *(2)*. Gene ablation, for example, can be achieved by flanking the promoter or essential exon(s) of the gene-of-interest with loxP sites and subsequent induction of Cre activity. Alternatively, gain-of-function can be achieved by Cre-mediated deletion of loxP-flanked transcriptional stop sequences placed between the promoter and the gene-of-interest. Numerous studies involve Cre-mediated recombination as a means to conditionally mutate loxP-modified alleles in mammalian cells in vitro and in vivo *(2,3)*. Despite its many advantages, the Cre/loxP recombination system has basic limitations such as inefficient delivery, toxicity, and position-specific side effects resulting from randomly integrated Cre expression vectors. So far, Cre recombinase activity was induced in cultured cells mainly by transfection *(4)*, viral transduction *(5–7)*, or ligand-dependent activation *(8–10)*. Recently, protein transduction developed to a new paradigm for the manipulation of cells *(11–13)*. This technology is based on the observation that short peptides, referred to as protein transduction domains (PTDs), are able to confer cell permeability when linked to cargo moieties. Protein transduction has been successfully applied to various proteins such as reporter proteins [β-galactosidase *(14)* and green fluorescent protein *(15)*], transcription factors, and therapeutically active compounds *(16–18)*. The molecular mechanism of cellular uptake is largely unclear, although recent findings point to a combination of several cellular processes such as caveolin- or clathrin-mediated endocytosis and macropinocytosis *(19)*.

We and other groups recently reported that cell-permeable versions of Cre recombinase can efficiently induce recombination in mammalian cells by direct protein delivery *(20–24)*. HTNCre *(21)* is a recombinant fusion protein consisting of a basic protein translocation peptide derived from HIV TAT (TAT), a nuclear localization sequence (NLS), the Cre protein, and an amino-terminal histidine tag for efficient purification from *Escherichia coli* (*see* **Fig. 1**). Compared with classical DNA-based alternatives such as transfection or viral transduction, conditional mutagenesis employing cell-permeable Cre recombinase has several distinct advantages. First, Cre protein transduction into cultured mammalian cells is highly efficient and widely applicable, for example, HTNCre induces recombination of loxP-modified alleles in more than

Fig. 1. Schematic representation of the cell-permeable His-TAT-NLS-Cre (HTNCre) fusion protein ***(21)***. H6, 6× histidine-tag; TAT, protein transduction peptide derived from HIV TAT; NLS, nuclear localization sequence. The amino acid sequence of the amino terminus is depicted.

90% of undifferentiated embryonic stem (ES) cells ***(21)*** and 80% of ES cell-derived neural precursors ***(25)***. Even primary cells such as B and T cells ***(21)***, embryonic fibroblasts, and post-mitotic neurons (F. E., unpublished results) can be efficiently recombined by application of HTNCre. Second, Cre protein treatment seems not to interfere with cellular function, in particular with germ line competency of ES cells (F. E., unpublished results). Third, direct delivery of active Cre protein overcomes the problem of leakiness, which is frequently associated with inducible Cre systems such as hormone-inducible activation ***(26)*** and tetracycline-controlled transcription ***(27)***. Fourth, Cre protein transduction avoids the risk of insertional mutagenesis caused by undesired random integration of transgenic DNA. Finally, application of Cre protein transduction is a robust, reliable, and relatively uncomplicated method because it does not involve tedious selection procedures or laborious viral preparations.

The application of this technology requires expertise in the expression and purification of histidine-tagged proteins from bacteria. This chapter provides comprehensive protocols for the induction of HTNCre fusion proteins in *E. coli*, their subsequent purification employing Ni(II)-affinity chromatography and corresponding biochemical analysis of protein fractions. Finally, we provide ready-to-use protocols for the successful transduction of purified HTNCre protein into mouse ES (mES) and human ES (hES) cells as well as ES cell-derived neural precursors.

2. Materials

2.1. Expression and Purification of Recombinant Cell-Permeable Cre Protein

2.1.1. Protein Expression

1. LB medium: 10 g tryptone from caseine, tryptic digest (Roth), 5 g yeast extract (Roth), 5 g NaCl. Suspend in 1 l double-distilled water and autoclave at 121°C for 20 min. Store at 4°C in the dark.

2. TB medium for protein expression: 12 g tryptone from caseine, tryptic digest (Roth), 24 g yeast extract (Roth), 4 ml glycerol (100%), 2.31 g KH_2PO_4, 12.54 g K_2HPO_4. Suspend in 1 l double-distilled water and autoclave at 121°C for 20 min. Store at 4°C in the dark. Immediately before use, the medium is supplemented with 1% (w/v) glucose and antibiotics.
3. Glucose 50% (w/v): dissolve glucose in double-distilled water. Sterilize by filtration and store at room temperature.
4. Isopropyl-(β)-D-thiogalactopyranoside (IPTG) 1 M: dissolve IPTG (Roth) in double-distilled water and sterilize by filtration. Store in 1 ml aliquots at −20°C. Always use freshly thawed IPTG solution for induction; do not refreeze solution.
5. Culture flasks: 5 l baffled Erlenmeyer flasks.

2.1.2. Protein Purification

1. 50% Ni-NTA agarose (Qiagen).
2. Econo-Pac columns (Bio-Rad).
3. Benzonase (Novagen).
4. Lysozyme solution: dissolve lysozyme (Sigma) in ×1 buffer A to a concentration of 80 mg/ml. Always prepare freshly before use.
5. ×10 buffer A: 500 mM NaH_2PO_4, 10 mM Tris–HCl, pH 8.0. Sterilize by filtration and store at room temperature.
6. TT buffer: 2 M disodium tartrate, ×1 buffer A, 15 mM imidazole. Sterilize by filtration and store at 4°C.
7. Washing buffer: 500 mM NaCl, ×1 buffer A, 15 mM imidazole. Store at 4°C.
8. Elution buffer: 500 mM NaCl, ×1 buffer A, 250 mM imidazole. Store at 4°C.

2.1.3. Preparation of HTNCre Glycerol Stock

1. ZelluTrans Roth® dialysis tubing (Roth): ready-to-use dialysis tubing is prepared according to the manufacturer's instructions and stored in 50% ethanol at 4°C.
2. 4-(2-Hydroxyethyl)-piperazine-1-ethane sulfonic acid (HEPES)-buffered sodium chloride (SH): 600 mM sodium chloride, 20 mM HEPES, pH 7.4.
3. Buffered glycerol (GSH): 50% glycerol, 500 mM sodium chloride, 20 mM HEPES, pH 7.4.
4. Bradford reagent (Sigma) with bovine serum albumin (BSA) standard.
5. Low protein binding filter (e.g., Millex® GV, 0.22 μm, Millipore).

2.2. Sodium Dodecyl Sulfate–Polyacrylamide Gel Electrophoresis and Western Blot Employing Anti-Penta His Antibody

1. ×4 separating buffer: 1.5 M Tris–HCl pH 8.8, 0.4% sodium dodecyl sulfate (SDS). Store at room temperature.
2. ×4 stacking buffer: 0.5 M Tris–HCl pH 6.8, 0.4% SDS. Store at room temperature.
3. 30% acrylamide/bis-acrylamide solution (37.5:1; Roth), *N*,*N*,*N*,*N*-tetramethyl-ethylenediamine (TEMED; Bio-Rad).

4. 10% ammonium peroxodisulfate (10% APS): dissolve APS in double-distilled water and immediately freeze in 200 μl aliquots at −20°C. Always use freshly thawed solution.
5. Water-saturated isobutanol: shake equal volumes of water and isobutanol and allow the phases to separate. Use the top layer. Store at room temperature.
6. ×10 running buffer: 30.3 g Tris–HCl, 144.2 g glycine, 10 g SDS. Store at room temperature.
7. ×1 running buffer: mix 100 ml ×10 running buffer and 900 ml water.
8. Molecular weight marker: ×6 His ladder (Qiagen) and prestained marker (NEB).
9. ×2 SDS sample buffer: 90 mM Tris–HCl pH 6.8, 20% glycerol, 0.02% bromophenol blue, 100 mM dithiothreitol (DTT). Store at room temperature.
10. Nitrocellulose membrane (Bio-Rad).
11. Anti-His horseradish peroxidase (HRP) conjugate (Qiagen).
12. Western transfer buffer: 25 mM Tris base, 150 mM glycine, 10% methanol. Store at room temperature.
13. Tris-buffered saline (TBS) buffer: 10 mM Tris–HCl pH 7.5, 150 mM NaCl. Store at room temperature.
14. TBS–Tween buffer: 20 mM Tris–HCl pH 7.5, 150 mM NaCl, 0.05% (v/v) Tween 20. Store at room temperature.
15. TBS–Tween/Triton buffer: 20 mM Tris–HCl, 150 mM NaCl, 0.05% (v/v) Tween 20, 0.2% (v/v) Triton-X100. Store at room temperature.
16. ×10 blocking reagent buffer (Qiagen).
17. Blocking buffer: add 0.1 g blocking reagent (Qiagen) to 20 ml ×1 blocking reagent buffer, heat to 70°C, and stir until dissolved. Add 200 μl 10% Tween 20. The solution is sufficient for processing of one 8 cm × 10 cm minigel. Always prepare solution freshly.
18. SuperSignal® West Pico Chemiluminescent Substrate (Pierce).

2.3. Cre Protein Transduction into ES Cells

2.3.1. Cre Protein Transduction into mES Cells

1. Trypsin/ethylenediaminetetraacetic acid (EDTA) solution (×10; Gibco). For ready-to-use solution, dilute the ×10 solution 1:10 with phosphate-buffered saline (PBS).
2. ES medium: Dulbecco's modified Eagle's medium (DMEM) high glucose (Gibco) containing 15% (v/v) fetal bovine serum, 1% non-essential amino acids, 2 mM L-glutamine, 1 mM sodium pyruvate, 0.1 mM β-mercaptoethanol (all from Gibco), and 1000 U/ml leukemia inhibitory factor (LIF).
3. Low protein binding filter (e.g., Millex® GV, 0.22 μm, Millipore).

2.3.2. Cre Protein Transduction into hES Cells

1. KO/SR medium: knockout DMEM containing 20% (v/v) knockout serum replacement, 1% (v/v) non-essential amino acids, 1 mM L-glutamine, 0.1 mM

β-mercaptoethanol (all from Gibco), and 4 ng/ml basic fibroblast growth factor (bFGF, Invitrogen). Store at 4°C up to 2 weeks.

2. Accutase II solution (PAA Laboratories).

2.4. Cre Protein Transduction into hES Cell-Derived Neural Precursors

1. EB medium: 80% (v/v) knockout DMEM, 20% (v/v) knockout serum replacement, 1% (v/v) non-essential amino acids, 1 mM L-glutamine (all from Gibco).
2. ITS medium: 500 ml DMEM : F12 medium (Gibco), 500 μl insulin stock solution, 2.5 ml transferrin stock solution, 30 μl 500 μM sodium selenite. Sterilize by filtration and store at 4°C up to 4 weeks.
3. Insulin stock solution: dissolve 1 g insulin in 40 ml 10 mM NaOH and sterilize by filtration. Store in aliquots at −20°C.
4. Sodium selenite 500 μM: dissolve 0.0865 g sodium selenite in 10 ml double-distilled water and sterilize by filtration. Store in aliquots at −20°C.
5. Transferrin stock solution (10 mg/ml): dissolve 1 g transferrin (Sigma) in 100 ml double-distilled water. Sterilize by filtration and store in aliquots at −20°C.
6. N2 medium: DMEM : F12 medium (Gibco) supplemented with ×1 N2 supplement (Gibco). Store at 4°C for up to 4 weeks.
7. ×100 polyornithine (PO) solution: dissolve 100 mg PO (Sigma) in 67 ml double-distilled water and store in aliquots of 10 ml at −20°C until use. For preparation of ×1 PO solution, dilute the ×100 stock solution 1:100 in double-distilled water and sterilize by filtration. Store at 4°C up to 2 weeks.
8. Laminin from Engelbreth-Holm-Swarm murine sarcoma 1 mg/ml (Sigma).
9. ×10 trypsin/EDTA solution (Gibco): prior use, the ×10 solution is diluted 1:10 with PBS.

3. Methods

3.1. Expression and Purification of Recombinant Cell-Permeable Cre Protein

3.1.1. HTNCre Protein Expression

For expression of recombinant cell-permeable Cre protein (HTNCre), the *E. coli* expression strain TUNER (DE3)pLacI (Novagen) is transformed with the Cre expression plasmid pTriEx-HTNC *(21)*. We routinely express HTNCre in a 6-l scale, but the culture volumes can certainly be scaled up or down.

1. Inoculate 200 ml LB medium supplemented with 1% (w/v) glucose, 50 μg/ml carbenicillin, and 34 μg/ml chloramphenicol with the transformed *E. coli* expression strain and incubate with shaking overnight at 37°C.
2. Inoculate TB medium supplemented with 1% (w/v) glucose, 100 μg/ml ampicillin, and 34 μg/ml chloramphenicol with the overnight culture in a ratio of 1:50 and

incubate with vigorous shaking at 37°C until an optical density of 1.2–1.5 (measured at a wavelength of 600 nm) is reached.
3. Induce expression by adding 1 M IPTG solution to a final concentration of 0.5 mM and continue incubation for 1 h.
4. Pellet cells by centrifugation (4°C, 4000 ×*g*, 10 min) and freeze the bacterial pellets at −20°C. The cell pellets can be stored at −20°C for several weeks, but it is advisable to use them as soon as possible.

3.1.2. HTNCre Purification

For subsequent analysis of the purification process (*see* **section 3.2.**), take a sample of 25–50 μl of the clear supernatant (step 6), the flow through (step 8), the two washing fractions (step 9), and the elution fraction (*see* **section 3.1.3.**, step 3). Mix the samples with equal volumes of ×2 SDS sample buffer and store at −20°C.

1. The pellets are thawed and resuspended in 10 ml ×1 buffer A per liter expression culture by vigorous stirring (*see* **note 1**).
2. Add lysozyme stock solution (80 mg/ml) to a final concentration of 2 mg/ml and incubate at room temperature with continuous stirring for 20 min.
3. Add 25 U/ml benzonase to the suspension and incubate for 15 min.
4. Sonicate the suspension on ice for 2 min (power 50%, cycles 5), add 1 ml ice-cold TT buffer per ml suspension, mix well, and incubate on ice for 5 min.
5. Centrifuge the lysate for 30 min at 4°C and 35,000 ×*g*.
6. Decant the supernatant, add 2 ml 50% Ni-NTA agarose per liter of expression culture, mix well, and pour suspension into 50-ml tubes (*see* **note 2**).
7. Incubate the suspension with continuous rotation at 4°C for 1 h. All following steps should be carried out at 4°C.
8. Pour the suspension into Econo-Pac® columns (Bio-Rad) and allow the liquid to flow through completely. The agarose beads form a layer on the filter matrix of the columns (*see* **note 3**).
9. Wash the agarose beads twice with 5-column bed volumes washing buffer (*see* **note 4**) and proceed with elution of the protein (*see* **section 3.1.3.**).

3.1.3. Preparation of a Ready-To-Use HTNCre Glycerol Stock

Large-scale dialysis against PBS or other media can be avoided by diluting HTNCre from a highly concentrated stock solution into the desired medium followed by sterile filtration. The protein stock can be stored at −20°C for several months without loss of Cre activity. All dialyzing steps should be carried out with buffer/sample ratios of at least 50.

1. Purify HTNCre as described in **section 3.1.2.** and elute the protein in 3-column bed volumes elution buffer (*see* **note 5**).

2. Immediately after elution, mix gently and add elution buffer until most of the precipitated protein disappears. Do not dilute too much as this will also reduce the final concentration of the glycerol stock.
3. Pellet remaining precipitated protein in eluted fraction by centrifugation and sterilize supernatant by filtration (0.22-μm filter).
4. Determine protein concentration using Bradford reagent; it should be between 60 and 120 μM (approximately 2.5–5 mg/ml).
5. Wash dialysis tubing thoroughly with double-distilled water and then incubate it in HEPES-buffered sodium chloride for 5 min.
6. Dialyze cleared fraction first for 2–4 h against HEPES-buffered sodium chloride (SH) and dialyze again overnight to completely remove imidazole.
7. Remove eventually precipitated protein as described in step 3.
8. Dialyze first for 6–8 h and a second time overnight against buffered glycerol.
9. Determine protein concentration of the glycerol stock. The glycerol stock should have a concentration of 200–450 μM.
10. The HTNCre glycerol stock solution can be stored at −20°C for at least 6 months without significant loss of activity. The HTNCre expression and purification protocol are usually very robust and reliable; however, we recommend to assess the transduction and recombination efficiency of every new batch (*see* **note 6**).

3.2. Sodium Dodecyl Sulfate–Polyacrylamide Gel Electrophoresis and Western Blot Employing Anti-Penta His Antibody

To analyze the purification process, the collected protein samples are loaded on a SDS–polyacrylamide gel electrophoresis (PAGE) gel and separated electrophoretically. The separated proteins can then be either stained in the gel by Coomassie dye or blotted on a nitrocellulose membrane with subsequent antibody detection.

1. Prepare two 1-mm thick 10% separating gels by mixing 6.3 ml water, 5.0 ml 30% acrylamide/bisacrylamide solution, 3.8 ml ×4 separating buffer, 60 μl 10% APS solution, and 16 μl TEMED. Pour the gels, leaving some space for the stacking gel and overlay with water-saturated isobutanol. Allow the gel to polymerize for at least 20 min.
2. In the meantime, prepare ×1 running buffer by mixing ×10 running buffer with water in a ratio of 1:10.
3. After polymerization of the separating gel, pour off the isobutanol and rinse the surface of the gel once with water.
4. Prepare the stacking gel by mixing 3.05 ml water, 0.65 ml 30% acrylamide/bisacrylamide solution, 1.25 ml ×4 stacking buffer, 25 μl 10% APS solution, and 5 μl TEMED. Pour the gel and quickly insert the comb. Allow to polymerize for at least 20 min.

5. Once the gels are polymerized, carefully remove the comb, attach the gels in the electrophoresis chamber, fill the two chambers with running buffer, and wash the slots of the gels with buffer by means of a pipette.
6. Thaw the protein samples, denature for 3 min at 96°C (×6 His ladder only 1 min) and quickly chill on ice. Centrifuge for 1 min at maximum speed.
7. Load about 5 µl samples on the gel. As a standard load, 10 µl following molecular weight markers: prestained marker (NEB) for Coomassie staining, ×6 His ladder for western blot.
8. Run the gels at 200V until the dye front has nearly reached the lower edge.
9. In the meantime, cut a nitrocellulose membrane to a size that equals the (separating) gel. Additionally, cut eight pieces of Whatman paper to the same size.
10. After electrophoresis, remove and discard the stacking gels. Carefully remove the separating gels from the glass plates.
11. For Coomassie staining, incubate one gel twice for 10 min each in water with gentle shaking. Then incubate the gel for 1 h (or longer) in Coomassie dye with gentle shaking. To remove excessive dye, wash the stained gel several times in water. For permanent documentation, the stained gel can be air-dried and should be stored in the dark.
12. For western blot employing anti-penta His antibody, incubate the other gel, the Whatman papers, and the nitrocellulose membrane in western transfer buffer for 5 min and place them as a sandwich on the positive electrode of the semi-dry transfer apparatus in the following order: (a) four Whatman papers, (b) nitrocellulose membrane, (c) gel, and (d) four Whatman papers (*see* **note 7**).
13. Place the upper electrode of the transfer apparatus on top of the blot and run the transfer at 1.5 mA per cm^2 membrane for 25 min.
14. Once the transfer is complete, wash the membrane for 5 min with TBS buffer.
15. Incubate membrane for 1 h in blocking buffer at room temperature.
16. Wash membrane for 5 min in TBS–Tween/Triton buffer at room temperature.
17. Wash membrane for 5 min with TBS buffer at room temperature.
18. Incubate in anti-His HRP-conjugate solution (1:2000 dilution of conjugate stock solution in blocking buffer) at room temperature for 1 h.
19. Wash membrane twice for 5 min each time in TBS–Tween/Triton buffer at room temperature.
20. Wash for 5 min in TBS buffer.
21. For detection, mix equal volumes of the stable peroxide solution and the luminol/enhancer solution (SuperSignal West Pico Chemiluminescent Substrate) and spread about 0.125 ml solution per cm^2 of membrane.
22. Incubate blot with the solution for 5 min at room temperature with gentle shaking.
23. Remove blot from the solution and place it in a plastic membrane protector (e.g., a plastic wrap may be used). Remove excessive liquid and air bubbles between the blot and the plastic membrane.

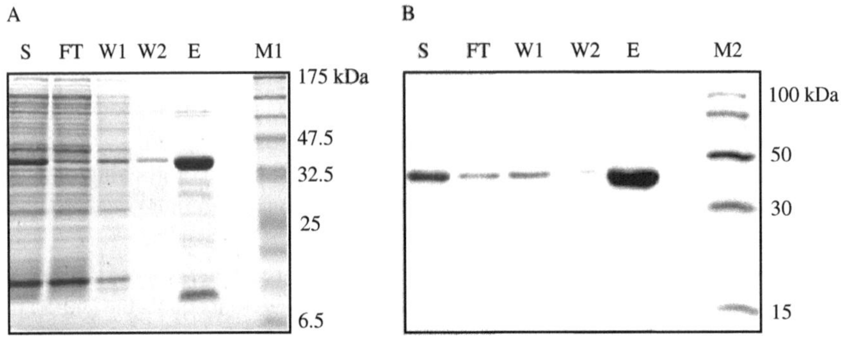

Fig. 2. Purification of recombinant His-TAT-NLS-Cre (HTNCre) from *Escherichia coli* as analyzed by Coomassie staining (a) and western blot employing an anti-penta His antibody (b). S, supernatant; FT, flow through; W1, washing fraction 1; W2, washing fraction 2; E, elution fraction; M1, prestained marker; M2, ×6 His ladder.

24. Place the protected membrane in an X-ray film cassette with the protein side facing up.
25. In the dark place, an X-ray film on top of the membrane and close the cassette. A recommended first exposure time is 60 s. The optimal exposure time must be determined empirically.
26. Develop the film using appropriate developing solution and fixative. An example result is shown in **Fig. 2**.

3.3. Cre Protein Transduction into ES Cells

3.3.1. Cre Protein Transduction into mES Cells

Cre protein transduction in mES cells results in highest recombination efficacy when cells are treated within 4–24 h after plating. Adding HTNCre to multilayer colonies strongly reduces the accessibility of the protein to inner cells, resulting in diminished overall recombination efficiency. The HTNCre concentration in this protocol is optimized for ES medium with 10–15% fetal calf serum (FCS). For serum-free ES medium, a 5- to 10-fold lower concentration of HTNCre is recommended because FCS inhibits the transduction. (*see* **note 8**)

1. Plate single-cell suspension of mES cells in normal ES medium.
2. Prepare ES medium containing 10 μM HTNCre protein by diluting an appropriate amount of HTNCre glycerol stock solution with normal ES medium. Sterilize medium with 0.22-μm filter.
3. After 5–6 h or when the cells are attached, wash carefully with PBS.

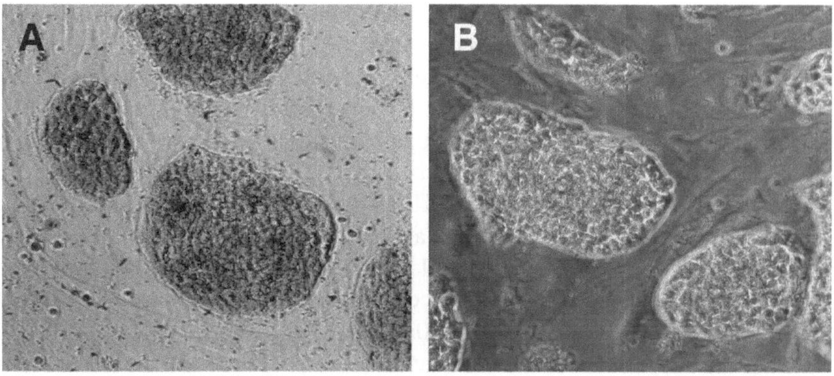

Fig. 3. Cre protein transduction into mouse embryonic stem (mES) Cre reporter MS4pAM ES cells *(26)*. His-TAT-NLS-Cre (HTNCre) induces highly efficient recombination in mES cells as analyzed by a Cre-inducible *LacZ* reporter gene. (a) Cells were incubated under serum-free conditions for 16 h with 2 μM HTNCre and analyzed 3 days after transduction. Cells were fixed and stained for β-galactosidase activity with X-Gal solution. (b) Untreated cells served as controls.

4. Incubate cells for 8–20 h with HTNCre-containing medium.
5. After protein transduction, wash once with PBS and add normal ES medium. An example employing a Cre-inducible LacZ ES cell line is shown in **Fig. 3**.

3.3.2. Cre Protein Transduction into Undifferentiated hES Cells

We routinely culture hES cells in KO/SR medium on tissue culture plates coated with mitotically inactivated mouse embryonic fibroblasts (MEFs). Cells are passaged by manual scraping or collagenase IV treatment. In our experience, Cre protein transduction into hES cells is most efficient if the colonies are very small, ideally clumps of few cells. Therefore, the hES cell colonies are dissociated using accutase II (PAA Laboratories) prior transduction. Accutase II has the advantage that it selectively detaches the hES cells from feeder cells if incubation does not exceed 30 min. The single-cell suspension is plated on MEFs again and allowed to adhere for 24 h. Within this time, very small hES cell colonies are formed, which have the optimal size for protein transduction.

Transduction efficiency is optimal for Cre concentrations of 6 μM and transduction times of 6 h. Longer incubation times do not significantly increase transduction efficiency. The optimal cell density for Cre protein transduction is about 50,000 hES cells per cm^2.

1. One day prior to protein transduction plate mitotically inactivated MEF cells on the desired number of tissue culture plates.
2. Remove the medium from the hES cells and add enough prewarmed accutase II solution to cover the colonies evenly. Incubate for 25 min at 37°C.
3. Detach the cells with KO/SR medium and transfer them to a 50-ml centrifugation tube. Take a sample of 15 μl and mix it with the equal volume of trypan blue solution to stain the cells. Take 15 μl stained cells, count the cell number in a Fuchs-Rosenthal chamber, and calculate the total number of hES cells. Calculate how many cells are required for the desired number of tissue culture plates.
4. Transfer the calculated volume of the cell suspension to 15- or 50-ml centrifugation tubes and centrifuge for 5 min at 1000 × *g* and 4°C.
5. Resuspend the pellet in KO/SR medium and dispense the cell suspension on the MEF cell-coated tissue culture plates. Allow the cells to adhere for 24 h at 37°C and 5% CO_2.
6. Dilute the Cre protein stock solution (*see* **section 3.1.3.**) in KO/SR medium to the desired concentration.
7. Remove the KO/SR medium from the plated hES cells and add enough Cre-containing KO/SR medium to cover the cells evenly.
8. Incubate for 1–6 h at 37°C and 5% CO_2.
9. Remove the Cre-containing medium, wash the cells once with knockout DMEM and add normal KO/SR medium again.
10. Prior further use of the Cre-transduced cells, allow the cells to recover for at least 24 h. An example of Cre protein transduction into hES cells is shown in **Fig. 4**.

3.4. Cre Protein Transduction in hES Cell-Derived Neural Precursors

Neural precursor cells can be prepared from hES cells according to a protocol published by Zhang et al. *(28)*. Embryoid bodies (EBs) prepared from hES cells are transferred to PO-coated cell culture plates and are propagated for 10 days in ITSFn medium. Within this time period, neural tube-like structures develop in the EB. These structures are isolated mechanically and are propagated as free-floating neurospheres in N2 medium for 2 weeks. Finally, neural precursor cells are isolated by enzymatic dissociation of these neurospheres, plating on PO/laminin-coated culture plates and further cultivation in the presence of FGF2 *(28)*. Alternatively, human neural precursors may be derived from fetal Central Nervous System (CNS) tissue as described *(29)*. Optimal parameters for Cre transduction in ES cell-derived neural precursors are a Cre concentration of 1 μM and a transduction time of 6 h. For maximum survival, it is crucial that the cells are confluent before transduction.

1. Dilute the Cre protein stock solution (*see* **section 3.1.3.**) in the medium used for the cultivation of neural precursors to a final concentration of 1 μM.

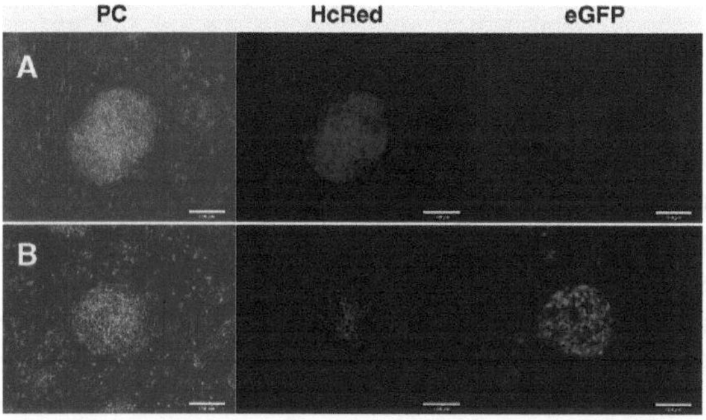

Fig. 4. Cre protein transduction into human embryonic stem (hES) cells. For Cre protein transduction, the double reporter hES cell line hES-FDR1 *(25)* was used, carrying a loxP-flanked *HcRed* gene and a downstream *eGFP* gene under control of a CAG promoter. Cre-mediated recombination induces a genetic switch from the expression of red to green fluorescent protein. (a) Untreated control cells exhibit homogeneous and intense red fluorescence. (b) Cells were treated with 6 μM His-TAT-NLS-Cre (HTNCre) for 6 h. After 24 h of transduction, the hES cells display markedly reduced HcRed expression but widespread green fluorescence indicative of highly efficient Cre recombination. Ph, phase contrast; HcRed, HcRed fluorescence; eGFP, eGFP fluorescence. Scale bar = 100 μM.

2. Remove the culture medium from the neural precursor cells and add the Cre protein solution. Incubate for 6 h at 37°C and 5% CO_2.
3. Wash the cells twice with DMEM : F12 medium, then add again the suitable culture medium.
4. Before further use, allow the cells to recover for at least 24 h.

Acknowledgments

We thank J. Itskovitz-Eldor (Rambam Medical Center, Haifa, Israel) for providing the hES cells. Special thanks go to Philipp Koch, Simone Haupt, F. Thomas Wunderlich, and Henrike Siemen, whose experimental work contributed significantly to the development of these protocols. We thank Corrinne G. Lobe (University of Toronto) and Michael Reth (Max-Planck Institute, Freiburg) for providing the Cre reporter ES cell lines and Katrin Hesse for text editing. Studies in our laboratories were supported by grants from the Stem Cell Network North Rhine Westphalia (400 004 03), the European Union (LSHB-CT-20003-503005; EUROSTEMCELL), the Volkswagen Foundation (Az I/77864), the DFG (BR 1337/3-2), and the Hertie Foundation.

Notes

1. Avoid excessive formation of foam when resuspending the bacteria pellets.
2. Make sure to resuspend the Ni-NTA agarose beads thoroughly before use.
3. For culture volumes larger than 4 l, distribute the Ni-NTA agarose cell lysate suspension on two columns.
4. Pipette the washing buffer carefully into the columns, avoid resuspension of the agarose beads layer.
5. At the beginning of protein elution, the elution buffer normally gets turbid because of the high protein concentration. The highly concentrated protein solution tends to precipitate. Therefore, fresh elution buffer should be added step by step to the protein solution immediately after the end of elution until the solution becomes clear.
6. A rapid and reliable assay for validation of the HTNCre transduction and recombination activity is the CV1-5B Cre reporter system ***(9)***. This cell line carries a Cre-inducible *LacZ* cassette that is readily detectable upon Cre recombination by X-Gal staining. Quantification of the recombination efficiency is achievable by either determining the percentage of β-galactosidase-positive cells or Southern blotting ***(21)***. For validation in mouse ES cells and/or in mice, a couple of reporter lines are published in the literature. These include classical reporter lines exhibiting Cre-inducible expression of reporter genes such as *LacZ* (***26, 30***). or *GFP* ***(31)***. More refined reporter systems contain double reporter cassettes, enabling a genetic switch from one reporter gene to another such as the Z/EG [*LacZ* to *GFP* ***(32)***] or fluorescience double reporter (FDR) cell line [red to green fluorescence ***(25)***].
7. It is critical that no air bubbles remain between the gel and the nitrocellulose membrane. To avoid this, gently roll a pipette over the sandwich.
8. For specific questions relating to the protocols presented in this chapter, please contact Frank Edenhofer, Stem Cell Engineering Group, Institute of Reconstructive Neurobiology, LIFE & BRAIN Center, University of Bonn, Sigmund-Freud-Strasse 25, D-53105 Bonn, Germany, Tel.: +49-228-6885-529, Fax: +49-228-6885-531, E-mail: f.edenhofer@uni-bonn.de, Web site: http://imbie.meb.uni-bonn.de/rnb/index.php.

References

1. Sauer, B. and Henderson, N. (1988) Site-specific DNA recombination in mammalian cells by the Cre recombinase of bacteriophage P1. *Proc. Natl. Acad. Sci. U. S. A.* **85**, 5166–5170.
2. Branda, C. S. and Dymecki, S. M. (2004) Talking about a revolution: the impact of site-specific recombinases on genetic analyses in mice. *Dev. Cell* **6**, 7–28.
3. Lewandoski, M. (2001) Conditional control of gene expression in the mouse. *Nat. Rev. Genet.* **2**, 743–755.
4. Kuhn, R. and Torres, R. M. (1997) *Laboratory Protocols for Conditional Gene Targeting*, Oxford University Press. (Oxford, New York, Tokyo).

5. Rohlmann, A., Gotthardt, M., Willnow, T. E., Hammer, R. E., and Herz, J. (1996) Sustained somatic gene inactivation by viral transfer of Cre recombinase. *Nat. Biotechnol.* **14**, 1562–1565.
6. Shibata, H., Toyama, K., Shioya, H., Ito, M., Hirota, M., Hasegawa, S., Matsumoto, H., Takano, H., Akiyama, T., Toyoshima, K., Kanamaru, R., Kanegae, Y., Saito, I., Nakamura, Y., Shiba, K., and Noda, T. (1997) Rapid colorectal adenoma formation initiated by conditional targeting of the Apc gene. *Science* **278**, 120–123.
7. Badorf, M., Edenhofer, F., Dries, V., Kochanek, S., and Schiedner, G. (2002) Efficient in vitro and in vivo excision of floxed sequences with a high-capacity adenoviral vector expressing Cre recombinase. *Genesis* **33**, 119–124.
8. Metzger, D., Clifford, J., Chiba, H., and Chambon, P. (1995) Conditional site-specific recombination in mammalian cells using a ligand-dependent chimeric Cre recombinase. *Proc. Natl. Acad. Sci. U. S. A.* **92**, 6991–6995.
9. Kellendonk, C., Tronche, F., Monaghan, A. P., Angrand, P. O., Stewart, F., and Schutz, G. (1996) Regulation of Cre recombinase activity by the synthetic steroid RU 486. *Nucleic Acids Res.* **24**, 1404–1411.
10. Wunderlich, F. T., Wildner, H., Rajewsky, K., and Edenhofer, F. (2001) New variants of inducible Cre recombinase: a novel mutant of Cre-PR fusion protein exhibits enhanced sensitivity and an expanded range of inducibility. *Nucleic Acids Res.* **29**, E47.
11. Brooks, H., Lebleu, B., and Vives, E. (2005) Tat peptide-mediated cellular delivery: back to basics. *Adv. Drug Deliv. Rev.* **57**, 559–577.
12. Dietz, G. P. and Bahr, M. (2004) Delivery of bioactive molecules into the cell: the Trojan horse approach. *Mol. Cell Neurosci.* **27**, 85–131.
13. Wadia, J. S. and Dowdy, S. F. (2003) Modulation of cellular function by TAT mediated transduction of full length proteins. *Curr. Protein Pept. Sci.* **4**, 97–104.
14. Schwarze, S. R., Ho, A., Vocero-Akbani, A., and Dowdy, S. F. (1999) In vivo protein transduction: delivery of a biologically active protein into the mouse. *Science* **285**, 1569–1572.
15. Caron, N. J., Torrente, Y., Camirand, G., Bujold, M., Chapdelaine, P., Leriche, K., Bresolin, N., and Tremblay, J. P. (2001) Intracellular delivery of a Tat-eGFP fusion protein into muscle cells. *Mol. Ther.* **3**, 310–318.
16. Takenobu, T., Tomizawa, K., Matsushita, M., Li, S. T., Moriwaki, A., Lu, Y. F., and Matsui, H. (2002) Development of p53 protein transduction therapy using membrane-permeable peptides and the application to oral cancer cells. *Mol. Cancer Ther.* **1**, 1043–1049.
17. Noguchi, H., Kaneto, H., Weir, G. C., and Bonner-Weir, S. (2003) PDX-1 protein containing its own antennapedia-like protein transduction domain can transduce pancreatic duct and islet cells. *Diabetes* **52**, 1732–1737.
18. Guelen, L., Paterson, H., Gaken, J., Meyers, M., Farzaneh, F., and Tavassoli, M. (2004) TAT-apoptin is efficiently delivered and induces apoptosis in cancer cells. *Oncogene* **23**, 1153–1165.

19. Patsch, C. and Edenhofer, F. (2007) Conditional mutagenesis by cell permeable proteins: potential, limitations and prospects, in *Handbook of Experimental Pharmacology*, 178,203–232.
20. Jo, D., Nashabi, A., Doxsee, C., Lin, Q., Unutmaz, D., Chen, J., and Ruley, H. E. (2001) Epigenetic regulation of gene structure and function with a cell-permeable Cre recombinase. *Nat. Biotechnol.* **19**, 929–933.
21. Peitz, M., Pfannkuche, K., Rajewsky, K., and Edenhofer, F. (2002) Ability of the hydrophobic FGF and basic TAT peptides to promote cellular uptake of recombinant Cre recombinase: a tool for efficient genetic engineering of mammalian genomes. *Proc. Natl. Acad. Sci. U. S. A.* **99**, 4489–4494.
22. Will, E., Klump, H., Heffner, N., Schwieger, M., Schiedlmeier, B., Ostertag, W., Baum, C., and Stocking, C. (2002) Unmodified Cre recombinase crosses the membrane. *Nucleic Acids Res.* **30**, e59.
23. Joshi, S. K., Hashimoto, K., and Koni, P. A. (2002) Induced DNA recombination by Cre recombinase protein transduction. *Genesis* **33**, 48–54.
24. Lin, Q., Daewoong, J., Grebre-Amlak, K. D., and Ruley, E. (2004) Enhanced cell-permeant Cre protein for site-specific recombination in cultured cells. *BMC Biotechnol.* **4**, 25.
25. Nolden, L., Edenhofer, F., Haupt, S., Koch, P., Wunderlich, F. T., Siemen, H., and Brustle, O. (2006) Site specific recombination in human embryonic stem cells induced by cell permeable Cre recombinase permits efficient conditional gene modification, *Nat. Methods* 3, 461–467.
26. Zhang, Y., Riesterer, C., Ayrall, A. M., Sablitzky, F., Littlewood, T. D., and Reth, M. (1996) Inducible site-directed recombination in mouse embryonic stem cells. *Nucleic Acids Res.* **24**, 543–548.
27. Utomo, A. R., Nikitin, A. Y., and Lee, W. H. (1999) Temporal, spatial, and cell type-specific control of Cre-mediated DNA recombination in transgenic mice. *Nat. Biotechnol.* **17**, 1091–1096.
28. Zhang, S. C., Wernig, M., Duncan, I. D., Brüstle, O., and Thomson, J. A. (2001) In vitro differentiation of transplantable neural precursors from human embryonic stem cells. *Nat. Biotechnol.* **19**, 1129–1133.
29. Brüstle, O., Choudhary, K., Karram, K., Hüttner, A., Murray, K., Dubois-Dalcq, M., and McKay, R. D. G. (1998) Chimeric brains generated by intraventricular transplantation of fetal human brain cells into embryonic rats. *Nat. Biotechnol.* **16**,1040–1044.
30. Soriano, P. (1999) Generalized lacZ expression with the ROSA26 Cre reporter strain. *Nat. Genet.* **21**, 70–71.
31. Mao, X., Fujiwara, Y., Chapdelaine, A., Yang, H., and Orkin, S. H. (2001) Activation of EGFP expression by Cre-mediated excision in a new ROSA26 reporter mouse strain. *Blood* **97**, 324–326.
32. Novak, A., Guo, C., Yang, W., Nagy, A., and Lobe, C. G. (2000) Z/EG, a double reporter mouse line that expresses enhanced green fluorescent protein upon Cre-mediated excision. *Genesis* **28**, 147–155.

3

Human Embryonic Stem Cells for Tissue Engineering

Daniel Kitsberg

Summary

Human embryonic stem cells (HESCs) are characterized by their ability to self-renew and capacity to differentiate into almost every cell type. As a result, they have enormous potential for use in tissue engineering and transplantation therapy. If these cells can be induced to differentiate into a particular cell type, they may provide an almost unlimited source of cells for transplantation for treating certain diseases where normal cell function is impaired. The challenge lies in the development of techniques to induce differentiation into a specific cell type, to enrich for that population, and to isolate it. It is essential that the starting material, the undifferentiated embryonic stem cell line, is growing under optimal conditions that preserve its pluripotent potential and maintain a stable karyotype. This review will discuss methods for the growth, maintenance, and spontaneous differentiation of HESCs and methods to genetically manipulate them.

Key Words: Human embryonic stem cells; Pluripotent; Differentiation; Embryoid bodies; Transfection; Feeder layer; Self-renewal; Transplantation.

1. Introduction

Tissue engineering is a field that aims to eventually replace damaged or dysfunctional tissue with tissue that is capable of restoring function. One approach is the use of organs from cadavers. However, such organs are in severely short supply. Another approach would be cell therapy, by which functional cells are grown and expanded in vitro and then transplanted into the patient. These cells might have the ability to restore function, for example, pancreatic β-cells restoring insulin function in diabetic patients. They may also help to some extent in providing signals that may enable the regeneration of

From: *Methods in Molecular Medicine, 2nd ed.: Tissue Engineering*
Edited by: H. Hauser and M. Fussenegger © Humana Press Inc., Totowa, NJ

a damaged tissue such as in the case of primary hepatocytes transplanted into patients with liver disease.

One problem associated with cell therapy is finding a source of cells for transplantation, particularly in view of the fact that large numbers of cells are required for successful transplantation therapy. From where can such large numbers of cells be obtained? Primary cells obtained from organs are in short supply because of the lack of available organs. Furthermore, primary cells often have a very short lifespan in culture. There has therefore been a lot of excitement about the potential of human embryonic stem cells (HESCs) as a source of cells for cell therapy.

Mouse embryonic stem (ES) cells were first derived in the early 1980s (*1*, *2*). However, it was not till the late 1990s that Thomson and his coworkers succeeded in deriving ES cells from human blastocysts (*3*). Since then, there has been much excitement about the potential for these cells to be used for transplantation for treating various diseases such as diabetes, Parkinson's disease, and muscular dystrophy.

ES cells are derived from the inner cell mass (ICM) of blastocyst-stage embryos. They are pluripotent cells that are characterized by their potential to differentiate into almost any cell type and by their capacity for self-renewal. These properties may make them an excellent starting material for the production of a specific cell type for cell therapy. If HESCs can be induced to differentiate into a specific cell type and that cell type can be isolated and further expanded, we might be able to have a virtually unlimited source of cells for transplantation. One of the challenges that we are facing today is to develop methodologies to induce differentiation of HESCs along a particular differentiation pathway and to enrich for a specific particular cell type. These cells will then need to be isolated and expanded to produce large numbers. Much effort is being made today to study and characterize different differentiation pathways and optimize conditions for identifying, isolating, and expanding a pure population of a specific cell type from HESCs.

To use HESCs for tissue engineering, there are several prerequisites that are essential for achieving any success. It is imperative that the starting material, the ES cell line, is in an undifferentiated, healthy, and proliferative state. This requires that the cells be of relatively low passage number and be maintained under conditions that allow them to preserve their pluripotent potential and karyotypic stability. HESCs have been shown to be capable of remaining karyotypically normal after multiple passages in culture (*3*, *4*). However, recently it was shown that when low passage HESCs were compared with later passage HESCs, various genetic alterations that are often seen in human cancers were observed *(5)*. This does not detract from the therapeutic potential of the HESCs but stresses the importance of continual monitoring for genomic mutations and using early passage cells.

There is much discussion about factors that affect karyotypic stability. One possible factor is the way that HESCs are passaged. HESCs can be passaged by mechanical dissociation (cut and paste) or enzymatic dissociation (e.g., collagenase or trypsin). Several laboratories that use the mechanical method report that in general, their cells maintain a stable karyotype (*6, 7*). However, it was demonstrated that when these cells are passaged enzymatically, karyotypic abnormalities accumulate after several passages. It has been argued that the enzymatic dissociation into single cells leads to aneuploidies. The passage of cells in clumps or clusters may be important to preserve integrity of the culture *(8)*. It is possible that the interaction of cells within the cluster may in some way prevent more uncontrolled division of the cells than might occur with individual cells where there is no cell-to-cell contact. In mechanically dissociated cells where there is massive cell-to-cell contact, we would expect significantly less divisions of the cells and this might explain the more stable karyotype. For large-scale culture, the mechanical methods are too cumbersome and time consuming. Thus if enzymatic dissociation techniques are utilized, it may be important to allow the cells to detach from the plate but to remain in clumps. It is possible that use of collagenase IV may be preferable over trypsin because it is less aggressive on the cells and enables them to remain in clusters. It has also been suggested that trypsin without ethylenediaminetetraacetic acid (EDTA) is less potent than trypsin–EDTA solution *(6)*.

The HESCs should also be subjected to minimal stress enabling them to maintain a stable normal karyotype. This includes growing the cells under optimal cell density so that there are sufficient cells to support each other but not too many cells that will deplete the important factors in the medium and oxygen (O_2) supply too quickly and thus compromise the quality of the entire culture. The cells should be cultured in optimal conditions with respect to the medium's composition, osmolarity, temperature, and O_2 saturation.

The undifferentiated state is evident when observing the cells both morphologically and on the molecular level, from expression of various undifferentiated cell markers including SSEA-3, SSEA-4, TRA-1-60, TRA-1-81, Rex1 and Oct4, and alkaline phosphatase, and they should display high telomerase activity (*3, 9*). These optimal conditions enable the cells not only to remain undifferentiated but also to preserve their pluripotent potential so that they can differentiate into the required cell type. It is the ability of the cells to differentiate into almost any cell type that defines them as a stem cell and makes them so useful for tissue engineering.

Until now, the successful derivation and propagation of ES cells have required the use of a feeder layer (typically mouse embryo fibroblasts) (***1***, ***3***), which provides support for cell growth and secretes factors necessary for maintaining the undifferentiated state. In the case of mouse embryonic stem cells (MESCs), it was shown that leukemia inhibitory factor (LIF), which acts through STAT3, can successfully replace the need for feeder layers to maintain this state and is presumably the major factor secreted by feeders required for self-renewal (***10***, ***11***). In the case of HESCs, it was shown that LIF is not able to support undifferentiated growth, and therefore there are presumably other factors secreted by feeders that maintain the pluripotent state (***12***, ***13***). The nature of these factors still remains elusive. One important factor is basic fibroblast growth factor (bFGF), which is normally added to the medium at a concentration of 4–8 ng/ml ***(4)***. Higher concentrations of bFGF have been shown to support HESC growth in the absence of mouse embryonic fibroblasts (MEFs) ***(14–18)***. However, these concentrations of bFGF are higher than those present in conditioned medium. It was demonstrated that bFGF is more stable in conditioned medium compared with unconditioned medium suggesting that MEFs presumably secrete factors that stabilize bFGF. In the absence of these factors, much higher levels of bFGF are required to attain the same self-renewal effect ***(18)***. The factors required for the self-renewal of HESCs therefore still need to be elucidated.

As discussed, the HESCs can be maintained in culture almost indefinitely in an undifferentiated pluripotent state if cultured under appropriate conditions. If they are grown either on feeder layers or in feeder-conditioned medium, and in the presence of bFGF, they preserve an undifferentiated state. However, when cultured in non-adherent Petri dishes in the absence of bFGF, the cells cluster together to form embryoid bodies (EBs). The cells in these EBs gradually differentiate into various cells representing all three germ layers ***(19)***. The addition of various growth factors can direct differentiation preferentially toward a particular cell type ***(20)***. Histological and immunohistochemical observations of the EBs over time show the development of a wide variety of morphologically and functionally different cell types including neurons, cartilage, bone, and pulsating cardiomyocytes. Microarray analyses of EBs at different stages compared with ES cells show that on the molecular level, markers of the undifferentiated cell are gradually turned off, whereas markers of differentiation are sequentially turned on. These markers include early differentiation markers such as LEFTYA, LEFTYB, and NODAL ***(21)*** and then more mature markers such as albumin, cardiac actin, and neurofilament protein. To some extent, what occurs in the EB recapitulates the events that occur in the early developing embryo but in a much more disorganized manner *(22)*. Thus, to perform tissue engineering using HESCs, the undifferentiated pluripotent cells need to be induced to differentiate, and this is typically achieved by the initial formation of EBs that contain all three germ layers. Because the EB contains all types of cells, it is important to develop methods to direct differentiation in a particular direction so as to enrich for a specific cell type. The addition of various growth factors can direct the EBs to differentiate preferentially for enriching one cell type over another *(20)*. At some stage of EB development, the EBs are dissociated into single cells that are then plated on adherent plates and are grown in medium containing specific growth factors that favor differentiation into a specific cell type. The challenge facing many stem cell researchers today is to optimize and define conditions that will direct the EBs towards the cell type in which they are interested.

Another advantage of HESCs is that they are capable of genetic manipulation. This opens up an exciting window of opportunity that enables us to perform various genetic changes in the HESCs such as knocking out a specific gene by homologous recombination (*23*, *24*) or knockdown of gene expression using siRNA *(25–27)* or overexpression of various genes under constitutive or tissue-specific promoters or expression of various reporter constructs that enable us to genetically mark cells (*28*, *29*).

The optimization of protocols for differentiation of HESCs into various tissue types requires a way to monitor the appearance of the cells of interest and methods to enrich and isolate those cells. One way to monitor these cells is to genetically modify the cells with a fluorescent tag for a specific cell type. For example, Lavon et al. *(29)* transfected HESCs with a reporter construct that expressed the enhanced green fluorescent protein (eGFP) under the albumin promoter. When these cells were allowed to differentiate into EBs, eGFP-expressing cells were observed that co-localized with albumin positive cells. Using fluorescence-activated cell sorting (FACS), it was possible to separate these green cells from the other cells in the EB and thus isolate a relatively pure population of HESC-derived hepatic-like cells.

Genetic manipulation of HESCs can be achieved by viral infection or transfection of DNA [reviewed in *(30)*]. Viral infection can be achieved by retroviruses that require cell division for chromosomal integration (*17*, *31*). Lentiviruses can also be used for viral infection. They were the first viral vectors used to genetically modify HESCs. They do not require cell division for transduction and are less susceptible to transcriptional silencing *(27, 32–35)*. However, they have a limit on the size of the transgene that can be used. Another system is adenovirus, which has been successfully used in transducing many cell types. They do not integrate and can introduce genes up to 30 kb *(36)*. They have been successfully used in MESCs, but to date there are few reports of their use in HESCs. Although viral vectors in general demonstrate high transfection efficiencies and can provide good expression levels, viral transduction has safety concerns and requires viral packaging and extraction which is time consuming.

It has been demonstrated that HESCs can be efficiently transfected by various methods including calcium phosphate, electroporation *(23)*, or using cationic reagents such as ExGen 500 and Trans-IT (*24*, *37*), or lipid reagents such as Fugene 6 (Roche), Lipofectamine 2000 (Gibco BRL), and Effectene (Qiagen) *(38)*. Recently, Siemen et al. *(38)* have customized conditions for the transfection of HESCs using Nucleofector™ technology *(39)*. They demonstrated that transfection efficiency using the optimized nucleofection protocol is 65% compared with 40% with the electroporation protocol of Zwaka and Thomson *(23)* and less than 30% with Fugene 6 or Lipofectamine 2000. Although cationic and lipid reagents may be effective for introducing DNA into HESCs, it is possible that physical methods such as electroporation may be required for achieving effective homologous recombination in HESCs (*23*, *40*).

This review will discuss conventional protocols for the growth, maintenance, cryopreservation, spontaneous differentiation, and genetic manipulation of HESCs. These protocols will offer a basis for the growth of HESCs in a healthy and consistent manner that will provide a suitable starting material for the future development of methods for the induced differentiation of HESCs toward a specific cell type.

It should be stressed that though HESC lines show many similarities and common properties, there are clear differences in morphology, growth, epigenetic modifications *(41)*, transcriptional profiles (*42*, *43*), methods of handling, and sensitivity to various conditions (be it to inductive growth factors or conditions of stress). Furthermore, differences in batches of serum replacement (SR) and other reagents and differing growth conditions in laboratories worldwide lead to differences in the same cell line that is grown in different laboratories that may also affect growth and karyotypic stability of the cells. This review therefore discusses protocols that have been successful in the growth of many different HESC lines in our laboratory in our hands. However, it is probable that the growth of other cell lines and even the growth of the same cell line in other laboratories using slightly different reagents will necessitate the recalibration or fine-tuning of optimal conditions by each individual laboratory for each different HESC line.

2. Materials

2.1. Derivation of MEFs

1. MEF medium: 500 ml Dulbecco's modified Eagle's medium (DMEM) for MEF cells with high glucose (4.5 g/l) and L-glutamine (Sigma D5796), 50 ml fetal calf serum (FCS) (Biological Industries 04-001-1), 5 ml penicillin (10,000 U/ml), and streptomycin (10 mg/ml) ×100 stock (Gibco BRL 15140-122 or Biological Industries 03-031-1B). Medium can be stored at 4°C for up to a month. Medium should be prewarmed in a 37°C water bath before use.
2. Dulbecco's phosphate-buffered Saline (DPBS; Sigma D8662) can be stored at room temperature.
3. 0.5 g gelatin (Sigma G1890) is added to 500 ml double-distilled water and is solubilized and sterilized in the autoclave. It can then be kept at room temperature for several weeks.
4. Mitomycin-C-containing medium: Prepare mitomycin-C stock solution by dissolving 1 mg mitomycin-C (Sigma M0903) in 1 ml medium or PBS. To make mitomycin-C-containing medium, add 120 μl stock solution to 12 ml MEF medium (final concentration of 10 μg/ml). We prefer to make the mitomycin-C stock solution fresh because we find that when it is stored at 4°C, the mitomycin-C comes out of solution. Some laboratories dilute 2 mg mitomycin-C in 250 ml DMEM, filter, and store it in aliquots at −20°C. Aliquots can be thawed and added to cells that need to be mitotically inactivated. Caution: Mitomycin-C is hazardous and should be handled with gloves and disposed of accordingly.

2.2. Growth of HESCs

1. ES medium (for undifferentiated growth): 500 ml Knockout™ DMEM-optimized DMEM for ES cells (Gibco BRL 10829-018) (*see* **note 1**), 6 ml non-essential amino acids ×100 (Gibco BRL 11140-035), 60 μl β-mercaptoethanol 1 M stock solution (final concentration 0.1 mM), 75 ml (*see* **note 2**) Knockout™ SR-serum-free formulation (Gibco BRL 10828-028) (*see* **note 3**), 6 ml L-glutamine ×100 (200 mM; Gibco BRL 25030-024 or Biological Industries 03-020-1; final concentration 2 mM), 1.2 ml human bFGF stock solution (final concentration: 4 ng/ml), 3 ml insulin-transferrin-selenium (ITS; Gibco BRL 41400-045) (*see* **note 4**). Optional: 3 ml penicillin (10,000 U/ml) and streptomycin (10 mg/ml; Gibco BRL 15140-122) (*see* **note 5**). The medium is prepared under sterile conditions in the hood. Because the Knockout™ SR-serum-free formulation is light sensitive, the bottle of medium should be wrapped in aluminum foil (*see* **note 6**).

2. Basic FGF stock solution (2 μg/ml) : 50 μg vial of bFGF (Peprotech 100-18B or Gibco BRL 13256-029) is dissolved in 25 ml PBS containing 0.1% bovine serum albumin (BSA), which has been filter sterilized. 1.2 ml is aliquotted into Eppendorf tubes and stored at −20°C until needed (*see* **note 7**). The BSA that we use is from Sigma (A4919).
3. β-Mercaptoethanol 1 M stock solution: 14 M β-mercaptoethanol (Sigma M7522) is diluted at 1:14 in double-distilled water to give a 1 M solution, which can be stored at 4°C in a dark bottle (β-mercaptoethanol is light sensitive).

2.3. Passaging HESCs Cells

1. Collagenase type IV: The collagenase type IV (Worthington LS004188 or Gibco BRL 17104-019) is prepared by using a solution of 1 mg/ml (*see* **note 8**) (ideally the collagenase activity should be >300 U/mg) in DMEM without any other additives (some laboratories dissolve in PBS). The solution is filtered through a 0.2-μm membrane and can be kept at 4°C for up to a week.
2. Trypsin–EDTA solution B: 0.25% trypsin and 0.05% EDTA in Puck's saline A (Biological Industries 03-052-1A or Gibco BRL 25200-072). Trypsin–EDTA solution can be aliquotted into 50-ml Falcon tubes and stored at −20°C. Trypsin can be stored for several days at 4°C without losing too much activity.
3. Trypsin solution (without EDTA): 0.25% trypsin in DPBS with calcium and magnesium (Biological Industries 03-045-1B or Gibco BRL 25050-014). Trypsin can be aliquotted into 50-ml Falcon tubes and stored at −20°C.

2.4. Cryopreservation

1. Freezing medium (with serum): 60% KO-DMEM (Gibco BRL 10829-018), 20% fetal bovine serum defined (FBSd; Hyclone), 20% dimethylsulfoxide (DMSO; Sigma D2650). The components of the freezing medium are combined and filter sterilized.
2. Freezing medium (without serum): 50% KO-DMEM (Gibco BRL 10829-018), 30% Knockout™ SR-serum-free formulation (Gibco BRL 10828-028), 20% DMSO (Sigma D2650). The components of the freezing medium are combined and filter sterilized.
3. Freezing medium (high serum): 90% FBSd (Hyclone), 10% DMSO (Sigma D2650). The components of the freezing media are combined and filter sterilized.

2.5. Differentiation of HESCs into EBs

1. EB medium: EB medium is identical to ES medium except no bFGF is added (*see* **note 9**). EBs are grown in non-adherent plastic 90-mm Petri dishes (Greiner 632 180 or Miniplast 20090-01) or alternatively in non-adherent 6-well plates (Greiner 60-657-102) that have been UV-irradiated in the tissue culture hood for half an hour.

2.6. Genetic Manipulation of HESCs

1. Geneticin (G418; Sigma G9516 or Gibco BRL 11811-031): Geneticin is dissolved in PBS at a concentration of 40 mg/ml and can be stored at −20°C in aliquots. Medium containing G418 can be stored at 4°C for up to 2 weeks. G418 selection is carried out at a concentration of 40–200 μg/ml.
2. Hygromycin B 50 mg/ml (Roche 843555): Hygromycin B is stored at 4°C and is light sensitive. Hygromycin B selection is carried out at a concentration of 40–200 μg/ml.
3. Puromycin (Sigma P8833): Stock solution is 25 mg/ml in double-distilled water, filter sterilized, and stored in aliquots at −20°C. Selection is carried out at about 300 ng/ml.
4. Trans-IT (Mirus MIR2300).
5. OptiMEM™ (Gibco BRL 31985-047).
6. ExGen 500 (Fermentas RO511): For ExGen 500 transfection, a 150-mM solution of NaCl is required and should be filter sterilized.

2.7. Consumables

1. Cryotube™ vials (Nunc 363401).
2. Falcon 100 × 20-mm tissue culture dish (Becton Dickinson 3003).
3. Greiner 90 × 20-mm tissue culture dish (Greiner 664160).
4. Greiner 140 × 20-mm tissue culture dish (Greiner 639160).
5. Greiner 6-well tissue culture dish (Greiner 657160).
6. Nunc 140-mm tissue culture dish (Nunc 168 381).
7. Nunc 90-mm tissue culture dish (Nunc 150 350).
8. Electroporation gap cuvette 0.4 mm (Biorad 165-2088 or Cell Projects EP-140).

2.8. Equipment

1. Nalgene cryo 1°C freezing container (Nalgene 5100-0001).

3. Methods

3.1. Derivation and Growth of MEF Feeders (see note 10)

Growth of HESCs on a feeder layer enables the cells to maintain their undifferentiated state. MEFs are prepared from mouse embryos. It is important to maintain a constant stock of MEFs. MEFs can be frozen untreated and are thawed when needed and treated with mitomycin-C to mitotically inactivate the cells (see section 3.1.2). However, it is recommended to freeze a stock of vials of mitomycin-C-treated MEFs that can be thawed and used immediately when required.

3.1.1. Derivation of MEFs

1. Male and female mice are set up to mate and are checked daily for plugs.
2. At 12.5 or 13.5 days post coitum (dpc), the females are killed. The abdomen is swabbed with 70% ethanol and is opened up, and the uterine horns containing the embryos are removed (*see* **note 11**) and placed in a Petri dish in about 15 ml PBS.
3. At this time, work is transferred to a laminar hood.
4. The membranes surrounding the embryos are cut with sterile fine scissors and released into the PBS and are washed with PBS several times to clear some of the blood and the debris.
5. The placental tissue and membranes are dissected away from the embryos. The head is held with forceps with one hand, and using the other hand, the innards (liver, lung, kidney and heart, etc.) are removed with tweezers, which are inserted into the abdomen (*see* **note 12**). The embryos are then decapitated and the remains of the embryo are washed several times in PBS.
6. The remaining embryo is placed in 3 ml trypsin–EDTA solution and cut into small pieces with forceps and scissors.
7. The cut up embryo is incubated at 37°C for 5 min in the trypsin–EDTA solution. The trypsinization is stopped by adding MEF medium containing serum and the cells are spun down in a centrifuge at 600 × *g* for 5 min, and the pellet is resuspended in medium by pipetting up and down in order to break up the cell clumps.
8. The cells are aliquotted to gelatinized tissue culture dishes. In general, one 140-mm gelatinized plate is sufficient for 1.5–2 embryos and 15 ml MEF medium is added to each plate.
9. The following day, a large proportion of cells will have attached. The cells are washed with PBS several times to remove the debris and fresh medium is added.
10. When the plates reach 80–90% confluency (typically 2–3 days after dissection), they are split 1:3 or 1:4 by trypsinization.
11. When these cells reach confluency, they are trypsinized. Trypsinization is stopped by adding MEF medium.
12. The cells are spun down in a 15-ml conical tube at 600 × *g* for 5 min and the pellet is resuspended in freezing medium and transferred to a Nunc cryotube (one 150-mm plate per cryotube) and they are stored in liquid nitrogen (*see* **note 13**).

3.1.2. Mitomycin-C Mitotic Inactivation of MEFs (see **note 14**)

1. One cryotube of MEFs (from step 12 in section 3.1.1) is thawed into MEF medium and spun down in a centrifuge at 600 × *g* for 5 min. The pellet is resuspended in MEF medium and the cell suspension is split between three 140-mm plates.
2. The cells are allowed to grow to 80–90% confluency.
3. The cells are split 1:3 twice, so that eventually 27 plates of near-confluent MEFs are obtained.

4. Cells are inactivated by adding mitomycin-C-containing medium (120 μl mitomycin-C stock solution is added to 12 ml MEF medium to give a final concentration of 10 μg/ml). Cells are incubated at 37°C for 3 h.
5. Mitomycin-C medium is aspirated and cells are washed three times with PBS.
6. Cells are then trypsinized and spun down in a 50-ml Falcon tube at 600 × *g* for 5 min.
7. The supernatant is aspirated and cells are counted in a hemocytometer.
8. Cells can be directly plated on gelatinized plates at a concentration of 2×10^6 cells per 10-cm plate.
9. The cells attach within a few hours. The plates should have a uniform layer of MEFs that is 80% confluent. If the plate is too sparse, it can be topped up with more MEFs. Cells can be used for up to 5 days after plating.
10. Mitomycin-C-treated cells from step 7 can also be frozen for later use. Normally, 6×10^6 cells are resuspended in 1 ml freezing medium and transferred to Nunc cryotubes and stored in liquid nitrogen. Each tube will have sufficient cells for three 10-cm plates (*see* **note 15**).

3.1.3. Gelatinization of Plates

1. To gelatinize plates, we added 5 ml 0.1% gelatin solution to a 100-mm tissue culture plate so that the whole area of the plate is covered. Plates are allowed to stand with gelatin for 1 h at room temperature (*see* **note 16**).
2. The gelatin solution is aspirated and medium is added to the plates, and appropriate cells (typically MEFs) are seeded on the plate in MEF medium.

3.2. Normal Growth and Maintenance of HESCs

1. The day before working with HESCs, tissue culture dishes should be plated with MEFs (*see* **note 17**). Plates are gelatinized (as in section 3.1.3) and a vial of MEFs is thawed at 37°C and added to 9 ml MEF medium in a 10-ml conical tube. The tube is spun for 5 min at 600 × *g*. The supernatant is aspirated and the pellet is resuspended in MEF medium and is split between three gelatinized 100-mm tissue culture plates containing MEF medium. The cells need 1–4 h to attach depending on the type of plate used (*see* **note 18**). When plating ES cells, MEF medium is replaced with ES medium.
2. Cells are maintained in ES medium in a 37°C incubator at 5% CO_2 (*see* **note 19**). Many researchers recommend changing medium daily (*see* **note 20**).
3. Work with ES cells should be performed according to standard operating procedures employed for normal tissue culture using a class II hood under sterile conditions.
4. HESCs seem to grow best when surrounded by other HESCs that probably secrete factors that have a positive effect on their own growth and the growth of neighboring

cells. Cells therefore do not grow well if the culture is too sparse. On the contrary, if plates are too crowded, they will deplete the nutrients and factors in the medium too quickly leading to more cell death, and cells begin to differentiate. As a result, cultures should be carefully monitored daily to assess how well the cells are growing and how healthy they look. It is important to ensure that the colonies do not overgrow because once they get too big, they begin to differentiate or become cystic. A "good" undifferentiated colony has smooth defined edge and has a uniform appearance throughout the colony (*see* **Fig. 1**). As the cells differentiate, the outside borders of the colony become less defined and cells of different morphology begin to appear. The inside of the colony begins to take on a cystic appearance or grow upwards.

5. It is also recommended to perform routine karyotypic analysis on the cell lines to ensure that karyotypic stability is maintained (*see* **note 21**).
6. Cells are usually split once every 3 days, but it obviously depends on how crowded the plate is and on the morphology of the colonies.

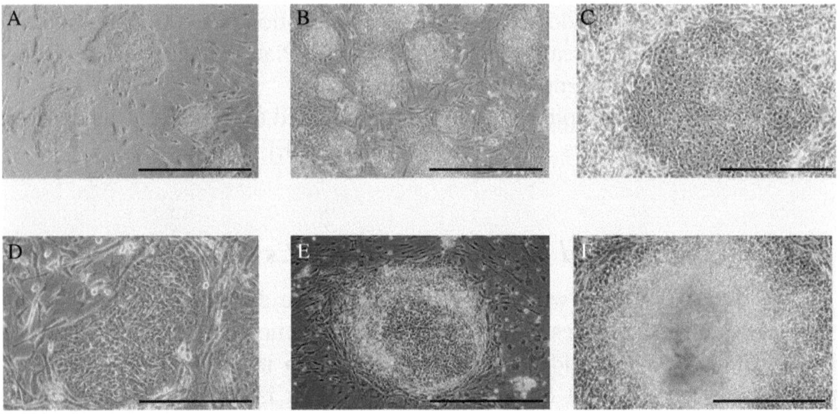

Fig. 1. Morphologies of various human embryonic stem cells (HESCs). (a) H9 HESC line—note the clear borders and uniform cells; (b) H9 HESC line—note the uniform appearance of the colonies; (c) I6 HESC line—note the uniform appearance of the cells within the colony; (d) I3 HESC line; (e) a differentiated I3 HESC colony. Note the differentiated cells that are beginning to spread out from the defined borders of the colony. (f) A differentiated I3 HESC colony. Note the differentiated cells that are beginning to develop in the center of the colony, which is becoming cystic. The colony is beginning to grow upwards as opposed to outwards. The scale line in a, b, and e represents 100 mM and in c, d, and f represents 200 mM.

3.3. Passaging ES Cells

As discussed, there are several methods that are used to passage the HESCs. The three major options are:

1. Collagenase type IV.
2. Trypsin.
3. Avoidance of enzymatic procedures and instead physical disruption (cut and paste) (*see* **note 22**).

3.3.1. Passaging Cells with Collagenase Type IV (see **note 23**)

1. In the case of a 10-cm plate of cells, the medium is aspirated, and the cells are washed twice with PBS and 2-ml collagenase type IV solution (1 mg/ml) is added. The cells are incubated for between 10 min and 2 h at 37°C. In the case of some ES cell lines, and for longer incubation times, colonies can be observed to detach from the feeder layer leaving the MEFs intact. Otherwise, the cells are dislodged by pipetting up and down with a 1-ml pipette or by the use of a sterile disposable rubber policeman (*see* **note 24**).
2. ES medium is added up to a volume of 10 ml. Cells are then spun down for 5 min at $600 \times g$ in a 10-ml conical tube.
3. The medium is aspirated and the pellet is resuspended in 1 ml ES medium by pipetting up and down several times in such a way to break up the cell aggregates into smaller clumps of cells or individual cells.
4. The cells are then plated on plates with fresh feeder cells with or without conditioned medium. Typically, a relatively confluent plate is split 1:3.

3.3.2. Passage of HESCs by Trypsinization (see **note 25**)

1. Medium is aspirated from the plate and the cells are washed twice with warm PBS, and prewarmed 0.05% trypsin–EDTA solution is added to the cells.
2. The plate is left at room temperature and the cells are observed under the microscope until the MEFs begin to retract and the ES cell colonies become more rounded and rough on the edges.
3. Cells are then resuspended in MEF medium (which contains FCS, which will help inactivate the trypsin) by pipetting up and down and dislodging the MEFs and HESCs from the plate.
4. The cells are transferred to a 10- or 15-ml conical bottomed tube and the plate washed with another 1 ml medium to remove the remaining cells.
5. The tube is spun in a centrifuge at $600 \times g$ for 5 min at room temperature. The medium is aspirated and the pellet is resuspended with a 1-ml pipette in ES medium and plated on fresh MEFs (*see* **note 26**).

3.4. Cryopreservation of HESCs (see note 27)

3.4.1. Freezing HESCs

1. A healthy looking culture is detached from the plate by collagenase or trypsin treatment (as in section 3.3).
2. Cells are spun down in a centrifuge at 600 × *g* for 5 min.
3. The supernatant is removed and the cell pellet is resuspended in 1 ml medium pipetting up and down to break up the large cell clumps and to generate a homogenous cell suspension. 1 ml freezing medium (with serum or with SR) is added dropwise while tapping the bottom of the tube to ensure continual mixing.
4. The cells are transferred to a cryotube and are allowed to freeze slowly at −80°C using a Nalgene Cryo 1°C freezing container.
5. After 2 or 3 days at −80°C, the cells are transferred to liquid nitrogen for long-term storage.

3.4.2. Thawing HESCs

1. Cryovials are removed from liquid nitrogen storage and thawed quickly by putting in a 37°C waterbath.
2. Thawed cells are diluted in prewarmed medium and spun down at 600 × *g* for 5 min.
3. The supernatant is aspirated and the pellet is carefully resuspended in prewarmed medium to break up the clumps.
4. The cells are transferred to an appropriate sized plate of MEFs with ES medium.
5. In the case of some cell lines, the addition of conditioned medium to the ES medium in a 1:1 ratio may aid the survival and initial growth of the cells.
6. The following day, the cells should be monitored. Although there is always some cell death, there should also be at least 60% of the cells that attach, divide and survive. The medium is changed and if there is a lot of cell debris, the plate is washed with PBS one or two times before adding fresh medium.

3.5. Differentiation of HESCs into EBs

There are two common techniques for preparing EBs (*see* **note 28**):

1. EB formation in suspension culture.
2. EB formation in hanging drop culture.

3.5.1. EB Formation in Suspension (see **note 29**)

1. A relatively confluent 10-cm plate of cells is washed three times with PBS and treated with 2 ml collagen type IV (1 mg/ml in DMEM) for 30 min. The cells are harvested by pipetting up and down or by using a cell scraper and then they are transferred to a 15-ml conical tube and centrifuged at 600 × *g* for 5 min (*see* **note 30**).

2. The supernatant is aspirated and the pellet is resuspended in EB medium by pipetting up and down with a 1-ml pipette.
3. Cells should be counted with a hemocytometer.
4. About 4×10^6 cells from the cell suspension are then transferred to a sterile (UV-irradiated) non-adherent 100-mm plastic Petri dish containing 15 ml EB medium (*see* **note 31**).
5. The plate is incubated at 37°C at 5% CO_2 for up to a month (*see* **note 32**).
6. The development of EBs is monitored (*see* **Fig. 2**). It is recommended that for the first 2 days at least, the plate is not moved at all or as little as possible.
7. Every second day, half of the medium is carefully removed from the plate in such a way as to not take up the larger EBs, and fresh EB medium is added. This can be achieved by tilting the plate at an angle. The EBs normally move to the bottom and the medium above is relatively free of EBs. Fresh EB medium is then added to the plate.

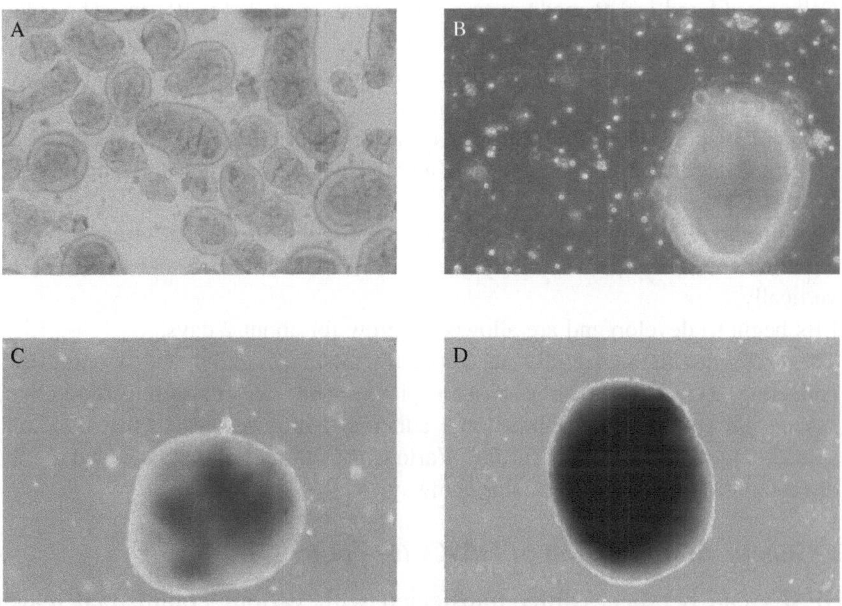

Fig. 2. Generation of embryoid bodies (EBs) in suspension culture. (a) Day 3.5; (b) Day 8; (c) Day 12; (d) Day 30. Over time, the cells within the EB divide and the EBs become larger. By day 30, they become very cystic. At the same time, histologically, the cells within the EB differentiate into a variety of cell types and structures (data not shown).

8. At various stages, EBs can be removed for histology by fixation, sectioning and staining or for immunohistochemistry, or for DNA, RNA and protein extraction for molecular analysis or can be broken up by trypsinization and analyzed by FACS.

An alternative method for producing EBs is by a method called the hanging drop method *(44)*. This method produces much smaller numbers of EBs but, on the contrary, produces EBs in a more controlled and uniform manner with a defined initial number of cells.

3.5.2. Embryoid Body Formation in Hanging Drops

1. A plate of cells is washed three times with PBS and treated with 2 ml collagenase type IV (1 mg/ml in DMEM) for 30 min or with trypsin for 5–10 min. The cells are harvested by pipetting up and down or using a cell scraper, and then they are transferred to a 15-ml conical tube and centrifuged at 600 × *g* for 5 min.
2. The supernatant is aspirated and the pellet is resuspended in EB medium by pipetting up and down with a 1-ml pipette.
3. The cells are counted in a hemocytometer.
4. Cells are diluted in EB medium to a concentration of 2–4 × 10^5/ml. The cells are then placed in autoclave-sterilized trough.
5. A lid of a 14-cm tissue culture dish is turned upside down, and using a multichannel pipette, we placed 25-μl drops on the lid at intervals in an orderly array. The cell suspension in the trough is occasionally pipetted up and down so as to prevent the cells settling and to ensure a uniform solution. Each drop should contain 5000–10,000 cells.
6. Medium or PBS is put in the bottom of the tissue culture dish to ensure a moist environment. The lid is carefully placed back onto the dish and the drops hang vertically.
7. EBs begin to develop and are allowed to grow for about 2 days.
8. The drops are then carefully harvested by washing the lid with medium or by collecting each drop with a Pasteur pipette and are either transferred to EB suspension culture or are plated onto adherent culture dishes so that they attach, spread out, and differentiate further. Various growth factors can be added to induce differentiation toward a particular pathway.

3.6. Genetic Manipulation of HESCs (see note 33)

HESCs can be successfully transfected with various commercial reagents including ExGen 500, Trans-IT, Fugene, Lipofectamine 2000, and Effectene (*37*, *38*, *45*) essentially following the manufacturers' protocols. However, they can also be efficiently transfected by calcium phosphate, electroporation *(23)*, and nucleofection *(38)*. It should be stressed that there is variation in transfection efficiencies between different HESC lines using different reagents. It is recommended that for each cell line, transfection efficiencies using different parameters are tested by performing a control transfection using a ubiquitously expressed EGFP (e.g., pEGFP-N1 from Clontech) and observing the number of cells that are green 1 day and 2 days post transfection.

3.6.1. Preparation of DNA for Transfection

1. Before electroporation, DNA must be prepared. Pure DNA preps can be prepared using CsCl or using a commercial maxiprep kit (such as Qiagen).
2. If stable transfections are to be performed, the plasmid DNA is linearized by a restriction enzyme that cuts in the vector sequence and not in the construct itself. Complete digestion is checked by running a small sample on an agarose gel, and the DNA is then ethanol precipitated by adding 1/10th volume of 3 M sodium acetate and two volumes of ethanol (*see* **note 34**). The DNA is washed with 70% ethanol and air-dried and resuspended in sterile double-distilled water. The concentration is calculated using a spectrophotometer or a nanodrop.

3.6.2. Transfection by Trans-IT

Trans-IT is a low-toxicity transfection reagent that takes advantage of the intrinsic properties of histone HI and a polyamine to deliver DNA.

1. HESCs are plated in a gelatinized 6-well dish containing antibiotic resistant MEFs (*see* **note 35**), so that the following day, there are a large number of small colonies or clumps plated uniformly over the well (typically 20–40% confluency).
2. The following day, 7.5 μl Trans-IT reagent is diluted in 250 μl OptiMEM™ in an Eppendorf tube and is mixed and allowed to stand at room temperature for 10 min.
3. 5 μg DNA is added to the diluted reagent and mixed by tapping the tube, and the tube is allowed to stand at room temperature for 20–30 min.
4. The contents of the tube are added dropwise to the well of cells in the 6-well plate, and the plate is rocked to and fro to mix in the Trans-IT and DNA uniformly.
5. The following day (1 day post transfection), the medium is removed and fresh ES medium is added.
6. If stable transfection is being performed, then the following day (2 days post transfection), selection is begun (G418, hygromycin or puromycin) (*see* **note 36**).
7. Depending on the extent of cell death, selection medium is changed every 2 days.
8. Over the next few days, there is massive cell death and small resistant colonies begin to appear which gradually grow larger.
9. At about days 10–12 after selection begins, colonies are large enough to pick.
10. Medium is removed from the plate, the cells are washed with PBS, and 2 ml collagenase IV is added. Cells are incubated for 15 min at 37%. This should ensure that on one hand, the cells still remain attached to the plate, but on the other hand, it will be easier to detach them from the MEFs with a pipette.

11. 8 ml medium is added to the plate and under an inverted light microscope, colonies are picked using a p200 Gilson pipette and sterile tips or using a mouth pipette attached to drawn-out Pasteur pipettes (*see* **note 37**).
12. Colonies are typically transferred to a well in a 12- or 24-well plate containing MEFs in HESC medium with selection (*see* **note 38**) and are pipetted up and down in an effort to break the colony up.
13. When the colonies have expanded enough in the 12- or 24-well dish, they are transferred to a 6-well plate containing MEFs.
14. They are then transferred to 2 wells of a 6-well dish. One well can be cryopreserved for later use and the other well can either be expanded or used to prepare DNA, RNA, or protein for molecular analysis.

3.6.3. Transfection by ExGen 500

ExGen 500 is a polyethylenimine with a high cationic charge that enables it to condense and complex with the DNA and delivers it to the nucleus. It also acts as a "proton sponge" that allows for endosomal buffering and protects the DNA from lysosomal degradation.

1. HESCs are plated in a gelatinized 6-well dish containing antibiotic resistant MEFs, so that the following day, there are a large number of small colonies or clusters plated uniformly over the well (typically 20–40% confluency).
2. One hour prior to transfection, change medium to 1 ml fresh medium per well.
3. For each well of a 6-well tissue culture dish, prepare an Eppendorf tube containing 4 μg DNA in 100 μl 150 mM NaCl and vortex briefly and spin down.
4. Add 13 μl ExGen 500 (not reverse order) vortex immediately for 10 s.
5. Allow to stand for 10 min at room temperature.
6. Add 100 μl ExGen 500/DNA mixture dropwise to each well.
7. Gently rock the plate to and fro to equally distribute the complexes on the cells.
8. Centrifuge culture trays in a swinging bucket centrifuge immediately for 5 min at 280 × *g*.
9. Incubate at 37°C at 5% CO_2 for 30 min.
10. Wash twice with PBS, add ES medium, and return to incubator.
11. Two days later, selection can be initiated (*see* **section 3.6.2.**).

3.6.4. Electroporation (Essentially According to the Protocol of Zwaka and Thomson) (23) (see note 39)

1. Cells are grown in a 10-cm plate until the plate is greater than 70% confluent. The cells should look healthy and undifferentiated (*see* **note 40**).
2. Cells are detached from the plate by collagenase IV (or trypsin) treatment (*see* **note 41**).
3. Cells are spun down in a centrifuge at 600 × *g* for 5 min. The supernatant is aspirated and the pellet is resuspended in 500 μl ES medium ($1.5–3 \times 10^7$ cells). In a separate

Eppendorf tube, 20–30 μg linearized DNA is diluted in PBS to a final volume of 300 μl.

4. The DNA solution is combined with the cell suspension and transferred to a 4-mm gap electroporation cuvette.
5. The electroporation is performed in a BioRad Gene Pulser giving a single pulse of 320 volts and 250 μF (*see* **note 42**).
6. After electroporation, the cells are allowed to stand in the cuvette for 10 min and then the contents of the cuvette are transferred to two 10-cm plates containing MEFs in fresh HESC medium.
7. The following day (day 1 post transfection), the cells are washed twice with PBS to remove cell debris and fresh ES medium is added.
8. If stable transfection is being performed, then the following day (day 2 post transfection), selection is begun (G418, hygromycin, or puromycin) (*see* **section 3.6.2.**).

3.7. Future Challenges

The successful derivation, propagation, and expansion of various HESC lines has provided scientists with an exciting and unique tool that may have enormous potential and promise for tissue engineering and transplantation therapy. However, many challenges that still lay ahead need to be addressed. It will be essential to develop technologies to direct differentiation of the HESCs toward a specific cell type so as to enrich that cell type. Methods will need to be developed to scale up production and isolation of the specific differentiated cell type that will provide sufficient quantities of a pure population of cells for transplantation.

Another issue that will also need to be addressed is the immunogenicity of the transplanted cells. It has been shown that undifferentiated HESCs express very low levels of MHC1, but expression is increased on differentiation to EBs or teratomas and it is further induced by interferons ***(46)***. Therefore, it will be important to generate a non-immunogenic cell that will not be subject to graft rejection by the immune system (***47***, ***48***) or that will have significantly reduced immunogenicity that will enable the administration of lower doses of immunosuppressant drugs. It will also be important to ensure that the cells do not form tumors like teratomas that develop when undifferentiated cells are transplanted under the kidney capsule in nude mice. This will require that we ensure that transplanted differentiated cells do not form tumors and perhaps will require the incorporation of a safety mechanism in the form of a suicide gene ***(49)*** in the event that tumors do develop.

It is obvious that for HESCs to be eventually useful for tissue engineering and transplantation, the cells will need to be derived, expanded, propagated, and induced to differentiate under conditions that are totally animal-free. It was shown that human ES cells grown on animal feeders or grown in medium with animal-derived products express Neu5Gc, a non-human sialic acid that could potentially be immunogenic for cells used for transplantation ***(50)***. Furthermore, mouse-derived MEFs may contain animal pathogens that also would be problematic for transplantation. As a result, much research is being done to find conditions to grow cells under either animal-free or feeder-free and serum-free conditions. One approach has been to grow cells on human-derived feeders ***(51–54)***. It has been shown that human-derived feeders are efficient alternatives to MEFs and will eventually lead to xeno-free clinically compliant HESCs. The growth of HESCs on feeders still requires the derivation and growth of feeders. It would therefore be preferable to develop systems that enable feeder-independent growth. It was shown that HESCs can grow efficiently under feeder-free conditions on human fibronectin-coated plates using medium supplemented with 15% SR, transforming growth factor (TGF)β1, and bFGF ***(55)***.

Ludwig et al. ***(56)*** have reported the definition of a serum-free, animal-free medium that supports feeder independent growth. They define physiochemical culture conditions such as pH 7.2, osmolarity of 350 mOsMol, and 10% CO_2/5% O_2. They also defined medium conditions and tested these conditions on several cell lines. Furthermore, they showed that they were able to derive two new HESCs using these conditions. This is an excellent starting point for further optimization of conditions and in the future will enable a universally acceptable standard operating procedure that will reduce variability and enable the derivation of new HESCs that are animal-free and feeder-free.

Despite the hurdles ahead, there is enormous potential for using HESCs for tissue engineering and cell therapy in the future. The further development of protocols for the growth, expansion, and directed differentiation of these cells under xeno-free conditions will ultimately bring us closer to realizing their potential.

Notes

1. Some laboratories use DMEM-F12 1:1 (Gibco BRL 31330-038)instead.
2. Some laboratories use up to 20% SR, but we have found that it is not necessary. In fact, it is possible that too much SR might adversely affect the osmolarity of the medium. Koivisto et al. ***(57)*** compared growth of an ES cell line on human foreskin fibroblasts under various conditions including SR at 10, 15, or 20% or human serum. They found that the cells had a highest growth rate in 20% SR but grew well in 15% SR. However, they grew significantly less in human serum. It should however be stressed that this important study was performed on one cell

line that was grown from the beginning on human fibroblast feeders and it will be important to perform similar studies on other cell lines.

3. Although the knockout serum-free formulation should be relatively standard, we have found differences between batches and there are some batches that support HESC growth better than others. It is recommended that the researcher tests several batches and monitors proliferation, maintenance of the undifferentiated state, and ability to differentiate into EBs to select for the preferred batch for the particular HESC lines that are being grown.
4. Other laboratories do not add ITS to the medium. It is possible that it is not necessary because the SR presumably contains enough of these supplements ***(57)***.
5. There are some laboratories that supplement the ES medium with 3 ml penicillin–streptomycin. We prefer not to add antibiotics if possible because it is preferable to work with cultures that are free of bacteria, and the antibiotics may mask their presence.
6. Some laboratories go to the extent of performing all tissue culture work with this SR-containing medium in a tissue culture hood without the light on.
7. There is much discussion as to how stable bFGF is in solution ***(18)***. There are some laboratories that add bFGF from a frozen stock to the medium on the day of use.
8. For better and faster results, some laboratories use a stock solution of 2 mg/ml collagenase type IV.
9. Some laboratories grow EBs in medium containing 20% FBSd instead of SR. This medium also lacks bFGF.
10. There are two common types of feeder layers: mouse embryo fibroblasts and STO cells. STO cells have the advantage that being a cell line, they are easy to culture and can grow almost indefinitely. On the contrary, MEFs being primary cells have a limited lifespan and reach senescence after about 15 divisions, although MEFs are not normally used after passage 4 because they are less effective. This means that MEFs need to be continually derived and expanded. Despite these disadvantages, it seems that MEFs are more commonly used than STO cells.
11. Some laboratories dip the uterine horns very briefly in 70% ethanol before transferring to PBS.
12. Some laboratories perform this step under a stereomicroscope, but we find that the innards, which are red-colored, are clearly visible and can be removed with the naked eye.
13. The successful growth of HESCs on MEFs is very dependent on MEF cell density. If the MEFs are too sparse, then the MEFs do not provide sufficient support to the HESCs and there will not be sufficient secreted factors for maintaining the undifferentiated state and for promoting self-renewal. On the contrary, if MEFs are too crowded, they physically hinder the growth and attachment of the HESCs and deplete the nutrients and oxygen in the medium that the HESCs require. Finding an optimal MEF density is therefore very important. A preliminary study showed that using H9 cells, a density of 20,000 cells/cm^2 is optimal while 32,000 cell/cm^2

has a strong negative effect on growth *(58)*. However, it was also reported that other HESC lines such as the Harvard University embryonic stem cell (HUES) cells developed at Harvard University or the Embryonic stem cell International (ESI) cell lines grow on feeders with a density of 60–75,000 cells/cm^2. We find that 30,000 cells/cm^2 (2×10^6 per 10-cm plate) is optimal. It is recommended that each laboratory test the optimal feeder density for supporting growth.

14. An alternative to mitomycin-C treatment of MEFs is γ-irradiation. This requires the availability of a γ-irradiation source. The effect of this source on the cells needs to be first calibrated. This is done by exposing 50-ml Falcon tubes (in ice) containing a fixed number of cells to the gamma source (typically 3000 rads) for different time intervals. The cells are then plated on gelatinized plates to assess at what time interval they completely stop cell division. If such a γ-irradiation source is available and the conditions have been calibrated, this approach is easier, faster, and cheaper than mitomycin-C treatment. However, both methods achieve the final aim and we have not observed any clear differences between MEFs inactivated by either method. What is most important is that the cells are mitotically inactivated completely.
15. It is interesting to note that it was reported that when using MEF-conditioned medium on HESCs growing on Matrigel, the quality was an important factor *(59)*. It was found that passage 2 or passage 3 MEFs provided optimal conditions suggesting that after this passage, there is a significant decline in the secretion of factors necessary for HESC maintenance and growth.
16. Some laboratories expose plates to gelatin for only 5–10 min. They argue that this is sufficient because the gelatin solution is concentrated and therefore saturates the plate quickly. When coating plates with Matrigel or fibronectin solutions, which are less concentrated, longer incubation time of an hour or more is required to effectively coat the plate.
17. Some laboratories find that MEFs can be plated 3–4 h before use. It is important to check that the MEFs have attached and spread out on the plate. In the case of MEFs that are plated on the day before use, it is possible to replace MEF medium with ES medium several hours after plating so as to provide conditioned medium for the following day.
18. We have found that there is much variability between different brands of tissue culture dishes. This can be seen in both how quickly feeders attach to the plates and how the cells grow and spread out on the plate. We have found Falcon plates and Greiner plates to be excellent for maintaining growth under the conditions we use. Nunc plates also support growth well but cells take longer to attach.
19. Ezashi et al. *(60)* have argued that early stage mammalian embryos develop under hypoxic conditions. They studied the effect of changing conditions for HESC growth from the conventional 21% O_2 to 1–4% O_2. They demonstrated that not only do the cells grow as efficiently in hypoxic conditions but they exhibited

significantly less differentiation as assessed morphologically, biochemically, and immunohistologically.

20. Some laboratories replace only 50% of the medium daily with fresh medium. In a 10-cm plate containing 10 ml, 5 ml medium is removed every day and 5 ml fresh medium is added so that the cells are growing in partially "conditioned" medium. Other laboratories routinely change medium with MEF-conditioned medium (or at least MEF-conditioned medium and fresh medium in a 1:1 ratio).
21. Routine karyotype analysis is performed by preparing G-banded chromosome spreads and counting chromosomes and monitoring for any gross abnormalities. Recently, spectral karyotype analysis (SKY) has become very popular (***61***, ***62***) and can provide more detailed information. It should be stressed that these analyses will detect trisomies or chromosome losses or relatively large translocations or deletions, but smaller mutations and chromosomal aberrations will not be detected. Even if the gross karyotype is normal, it is recommended to use a relatively low passage number. It is also recommended to monitor cell growth and when cells seem to be growing at a faster rate; this might indicate that they have undergone some mutation that might promote growth.
22. The cut and paste method to split colonies is performed by using Pasteur pipettes that has been put in a Bunsen and drawn out and broken so that the edge is thin and sharp. Using these sterilized Pasteur pipettes, relatively large colonies are cut into eight or so equal-sized sections, leaving the central section (which has often begun to become cystic). Cutting is performed by forming a grid or radial spokes from the middle. The cut up colonies are peeled off the plate or aspirated into a 20-μl pipette and are added dropwise to the medium carefully dispersing over the plate uniformly (for more information, see http://www.escellinternational.com/pdfs/stemcellproducts/hES_Culture_Methodology_Manual.pdf).
23. The original paper on HESCs used collagenase to split cells. Collagenase type IV action is believed to be less aggressive than trypsin and thus may preserve the karyotypic stability of cells. Sjogren-Jansson et al. ***(59)*** compared various methods for the transfer of HESCs to Matrigel. They demonstrated, using their specific cell lines, that mechanical dissociation had higher survival rates over collagenase. However, using the mechanical method, they found that colonies were larger and fewer as compared with collagenase-treated cultures. Some cell lines were derived and propagated in the absence of trypsin and collagenase and as a result have not been adapted to these agents. Often cell lines that have not been adapted to trypsin will die or differentiate when exposed to trypsin. These cells can only therefore be cultured by physical dissociation (i.e., cut and paste). Although this method probably is least disruptive on the cells, it is a very cumbersome, time consuming, and demanding technique and is not suitable for large-scale culture. Furthermore, the nature of the technique is such that the resultant cells after splitting are relatively large colonies which are therefore limited in the extent of their growth as compared with single cells or small clumps of cells that are produced by collagenization or

trypsinization. As discussed, it may be preferable to passage the cells in small clumps as opposed to single cells because it has been suggested that this may help to preserve a stable karyotype.

24. Different HESCs behave differently with collagenase treatment. Batches of collagenase differ greatly. Collagenase is sold by weight but the number of units per milligram may vary significantly. We typically aim to buy batches that are >300 U/mg. For each cell line that is used, it is recommended that the time of collagenase treatment be calibrated. If collagenase treatment is too short, then it is difficult to dislodge cells from the plate; cell clumps are not sufficiently dissociated, and after plating, colonies are too large and begin to differentiate. If collagenase treatment is too long, there is limited cell survival and significant cell death on the following day. The cells are also present as single cells as opposed to clumps, which may be less desirable for survival and stability. We have found that in the case of some HESC lines after incubation for an hour in collagenase, the cells detach nicely from the MEFs leaving a "hole" in the MEF layer and enabling us to harvest them by just adding a few milliliters of medium and taking up the medium from the plate. Other cells do not detach so easily from the MEFs and need to be dislodged physically by pipetting up and down sometimes quite forcefully or using a sterile rubber policeman.
25. Various laboratories believe that a short treatment of the cells with trypsin does not have adverse effects on the ES cells. The laboratory of Dr. Melton at Harvard University has developed several new HESC lines that were adapted from an early stage to trypsin and that they routinely passage with trypsin ***(63)*** (http://www.mcb.harvard.edu/melton/hues/HUES_manual.pdf).
26. As mentioned, it may be preferable not to dissociate the HESCs into single cells but rather into clumps. This necessitates the use of milder procedures for detaching the cell colonies from the plates that do not break the cells up into single cells. Some laboratories have suggested using dilute trypsin by diluting the trypsin solution 1:3 in PBS. Although the cells may take longer to detach, this treatment is considered to be less stressful on the cells. We have found that use of a trypsin solution that does not contain EDTA is very effective for passaging HESCs. The trypsin solution needs to be added to the cells for 10–20 min and incubated at 37°C. After about 10 min, the cells are periodically monitored to see when they detach. Under these conditions, the MEFs remain firmly attached to the plate but the ES cells detach from the plates in clumps (leaving holes in the MEF layer where the cells were attached). The cells are spun down and resuspended in ES medium and plated on MEFs. We have found that there is greater survival, and though the colonies may be bigger than that seen when using trypsin with EDTA, they look healthier and grow better. We have yet performed a comprehensive study to follow karyotype over time after passaging with various reagents.
27. There are various techniques for freezing cells. The two most commonly used techniques for most somatic cells are the (a) conventional slow stepwise freezing

method using cryovials and storage in liquid nitrogen and (b) nap freezing vitrification using an open pulled straw and storage in liquid nitrogen ***(64)***. It has been shown by some laboratories in the case of HESCs that the controlled rate freezing method results in low thaw survival rates and low plating efficiencies. Other laboratories have reported high efficiencies ***(59)***. There are reports that storage in liquid nitrogen can lead to contaminations between vials due to seepage of liquid nitrogen from cryovials and various viruses have been shown to survive well in liquid nitrogen ***(65)***. It has therefore been argued that the vitrification method consistently yields higher plating efficiency (***9***, ***51***, ***64***, ***66***). However, this method involves direct contact of the HESCs with the liquid nitrogen at the open end of the pulled straw, and this can result in contamination. Some techniques make use of FCS for cryopreservation, which exposes the HESCs to animal products. Richards et al. ***(67)*** have thus optimized techniques for cryopreservation of HESCs in a xeno-free manner. They performed vitrification in closed sealed straws using human serum albumin in place of FCS as the major cryoprotectant. They also perform the long-term storage in the vapor phase of liquid nitrogen. This technique shows high thaw-survival rates and low differentiation. Some techniques though very effective are long and cumbersome or might require specialized equipment. In our experience, HESCs freeze and thaw relatively well using standard techniques. Although there might be some inevitable cell death, there is a high enough survival rate to make these techniques sufficient. It is possible that the surviving cells may have some growth advantage that leads to selection of a specific population, but we have not found evidence to suggest that these cells have any disadvantages such as abnormal karyotype, inability to self-renew, or impaired pluripotent potential.

28. Another simple technique has been suggested for EB formation using a 1.5 ml polypropylene conical tube ***(68)***. It has been successfully used for MESCs.
29. Although this protocol is the most commonly used, there are several limitations. It was demonstrated with MESCs that EB agglomeration has a gradually more negative effect on cell proliferation and differentiation ***(69)***. With time, EBs in static culture in flasks demonstrate extensive cell death and large necrotic areas appear. This is due to mass transport limitations. Furthermore in the case of static culture, it is difficult to scale up to obtain sufficient cells that might be required for transplantation. Methods have been developed that address this problem. One approach has been to grow cells in dynamic culture in spinner flasks ***(70)***. Another approach has been to use slow rotating bioreactors ***(71)***. This allows four-fold more EB formation over static culture but controls agglomeration thus enhancing efficiency of viable EB formation and differentiation with minimal cell death.
30. It is sometimes recommended to remove as much of the feeder layer as possible before starting differentiation. This can be achieved by transferring the cells once or twice to gelatinized plates lacking MEFs and growing the cells in MEF-conditioned medium. It is also possible to reduce the number of MEFs by treating the cells with a more dilute trypsin or collagenase. If the cells are carefully monitored,

one will observe the HESCs detaching from the MEFs and going into suspension before the feeders begin to detach. In such a way, one can obtain a culture that has significantly reduced amount of feeders.

31. Putting more than 4×10^6 cells per 100-mm dish can compromise the quality of the EBs. For routine EB production, it is recommended to calibrate the optimal cell number to produce EBs for each cell line. It is also important that the cells are growing well and are looking healthy and undifferentiated. Unhealthy cultures do not produce good EBs.
32. After about 4 days in culture, the EBs can be transferred back to adherent plates and grown on a substrate such as gelatin, collagen, fibronectin, or Matrigel. The EBs start to outgrow onto the substrate and various cell phenotypes can be seen to appear. These phenotypes can be further enriched or directed by the addition of various growth factors.
33. Eiges et al. ***(37)*** first demonstrated that HESCs were capable of transfection using various reagents albeit with different efficiencies. At the time they found that ExGen 500 was the most effective reagent for transfection, but since then it has been found that other techniques including calcium phosphate and electroporation are also effective. We have successfully used commercial reagents such as ExGen 500, Fugene, Trans-IT, and Lipofectamine 2000. In another comparative study, it was shown that Effectene gave high efficiencies that improved in combination with silica beads ***(45)***. We have found that the HESCs may be more sensitive than other cell lines to some of these reagents and when the manufacturer recommends incubation with a certain reagent for between a few hours to overnight; we prefer to incubate the cells for a shorter period than overnight.
34. Some people prefer to purify the DNA further, by performing a phenol extraction, then a phenol/chloroform extraction, and then a chloroform extraction before ethanol precipitation. We have found that the DNA without the phenol and chloroform extractions seems to be clean enough to give reasonable transfection efficiencies.
35. It should be remembered that the MEFs should be resistant to the same antibiotic that is used for selection. MEFs need to be derived from transgenic mice containing either a neomycin resistance gene or a different antibiotic resistance gene. Mice should be homozygous for the transgene so that when mice are bred, all progeny are resistant to the antibiotic. Tucker et al. ***(72)*** have created the DR4 mice that contain resistance genes for neomycin puromycin, hygromycin, and 6TG. The MEFs derived from these mice will be suitable for most selection applications. These mice are available from Jackson Laboratories (003208).
36. It should be noted that different HESC lines vary significantly in their intrinsic resistance to antibiotic. It is important that for each cell line used, a killing curve is set up to determine at which concentration of antibiotic all non-transfected cells on a plate die. We have found that the optimal concentration for G418 selection varies between 50 and 200 μg/ml depending on the HESC line. The optimal concentration

for hygromycin B selection is 50–200 μg/ml and the optimal concentration for puromycin selection for the cell lines that we have used is 300 ng/ml.

37. The method of picking is very much dependent on the cell line. Some cell lines are strongly attached to the MEFs and trying to pick colonies without use of collagenase is difficult because it is hard to detach them from the MEF layer. If collagenase is not used, it is possible to cut around the colony with a sterile tip or with a drawn out Pasteur pipette and thus also separate the MEFs supporting the colony from surrounding MEFs. Many laboratories pick colonies without using collagenase and they find that the colonies detach easily. Some laboratories transfer the detached whole colony to a 96-well plate containing trypsin or collagenase to break up the colony into smaller clumps so that when it is placed in the 24 well, it disperses over the well. Obviously, the transfer of a single large colony which has not be broken up into a 24 well will mean that it will be hard for it to expand and furthermore it will be prone to differentiation.
38. After picking colonies, it is recommended to maintain the cells in selection medium as this ensures that the integrated DNA does not get lost during passaging.
39. Zwaka and Thomson ***(23)*** developed parameters for the electroporation of HESCs. They found that using standard protocols for mouse ES cells (220 V and 960 μF in PBS) they obtained a low stable transfection rate of HESCs of about 10^{-7}. They came to the conclusion that HESCs are significantly bigger than mouse ES cells (~14 versus ~8 μm) and they therefore tested parameters used for larger cells. They therefore electroporated clumps as opposed to single cells and plated the cells at high density to promote cell survival and performed the electroporation in an isotonic protein-rich solution (as opposed to PBS) at room temperature. They were able to increase transfection efficiencies significantly to 2×10^{-5}. With these conditions, they were able to obtain homologous recombinants.
40. In the original protocol, 1 week before electroporation, cells were plated on Matrigel and cultured in fibroblast-conditioned medium. However, we find that as long as we grow the HESCs on feeders that will also be resistant to the antibiotic with which we select, the electroporation works just as efficiently.
41. Zwaka and Thomson ***(23)*** have argued that the efficiency of electroporation and cell survival after the procedure is significantly increased if it is performed on clumps of cells as opposed to single cells. As a result, they recommend treating with collagenase for a minimal amount of time so that they detach from the feeder layer but do not break up into single cells.
42. There are many different electroporators on the market. The conditions for electroporation need to be calibrated for each type of electroporator for obtaining the highest transfection efficiency. Too harsh conditions will kill too many cells and will leave very few transfectants. If the conditions are too mild, there will be few cells into which the DNA inserts. Zwaka and Thomson ***(23)*** used a BioRad Gene Pulser II and gave a pulse of 320 V and 200 μF.

Acknowledgments

The ES cell work was performed in collaboration with the laboratory of Prof. Nissim Benvenisty at the Hebrew University of Jerusalem, Israel and the laboratory of Prof. Joseph Itskovitz-Eldor at the Technion Institute of Science, Haifa, Israel. I thank Prof. Benvenisty and the members of his laboratory together with the members of the Itskovitz laboratory for helpful discussions in preparing this manuscript.

References

1. Evans, M. J. and Kaufman, M. H. (1981) Establishment in culture of pluripotential cells from mouse embryos. *Nature* **292,** 154–156.
2. Martin, G. R. (1981) Isolation of a pluripotent cell line from early mouse embryos cultured in medium conditioned by teratocarcinoma stem cells. *Proc. Natl. Acad. Sci. U. S. A.* **78**, 7634–7638.
3. Thomson, J. A., Itskovitz-Eldor, J., Shapiro, S. S, Waknitz, M. A., Swiergiel, J. J., Marshall, V. S., and Jones, J. M. (1998) Embryonic stem cell lines derived from human blastocysts. *Science* **282**, 1145–1147.
4. Amit, M., Carpenter, M. K., Inokuma, M. S., Chiu, C. P., Harris, C. P., Waknitz, M. A., Itskovitz-Eldor, J., and Thomson, J. A. (2000) Clonally derived human embryonic stem cell lines maintain pluripotency and proliferative potential for prolonged periods of culture. *Dev. Biol.* **227**, 271–278.
5. Maitra, A., Arking, D. E., Shivapurkar, N., Ikeda, M., Stastny, V., Kassauei, K., Sui, G., Cutler, D. J, Liu, Y, Brimble, S. N., Noaksson, K., Hyllner, J., Schulz, T. C., Zeng, X., Freed, W. J., Crook, J., Abraham, S., Colman, A., Sartipy, P., Matsui, S., Carpenter, M., Gazdar, A. F., Rao, M., and Chakravarti, A. (2005) Genomic alterations in cultured human embryonic stem cells. *Nat. Genet.* **37**, 1099–1103.
6. Brimble, S. N., Zeng, X., Weiler, D. A., Luo, Y., Liu, Y., Lyons, I. G., Freed, W. J., Robins, A. J., Rao, M. S., and Schulz, T. C. (2004) Karyotypic stability, genotyping, differentiation, feeder-free maintenance, and gene expression sampling in three human embryonic stem cell lines derived prior to August 9, 2001. *Stem Cells Dev.* **13**, 585–597.
7. Buzzard, J. J., Gough, N. M., Crook, J. M., and Colman, A. (2004) Karyotype of human ES cells during extended culture. *Nat. Biotechnol.* **22**, 381–382; author reply 382.
8. Hoffman, L. M. and Carpenter, M. K. (2005) Characterization and culture of human embryonic stem cells. *Nat. Biotechnol.* **23**, 699–708.
9. Reubinoff, B. E., Pera, M. F., Fong, C. Y., Trounson, A., and Bongso, A. (2000) Embryonic stem cell lines from human blastocysts: somatic differentiation in vitro. *Nat. Biotechnol.* **18**, 399–404.

10. Smith, A. G., Heath, J. K., Donaldson, D. D., Wong, G. G., Moreau, J., Stahl, M., and Rogers, D. (1988) Inhibition of pluripotential embryonic stem cell differentiation by purified polypeptides. *Nature* **336**, 688–690.
11. Williams, R. L., Hilton, D. J., Pease, S., Willson, T. A., Stewart, C. L., Gearing, D. P., Wagner, E. F., Metcalf, D., Nicola, N. A., and Gough, N. M. (1988) Myeloid leukaemia inhibitory factor maintains the developmental potential of embryonic stem cells. *Nature* **336**, 684–687.
12. Humphrey, R. K., Beattie, G. M., Lopez, A. D., Bucay, N., King, C. C., Firpo, M. T., Rose-John, S., and Hayek, A. (2004) Maintenance of pluripotency in human embryonic stem cells is STAT3 independent. *Stem Cells* **22**, 522–530.
13. Daheron, L., Opitz, S. L., Zaehres, H., Lensch, W. M., Andrews, P. W., Itskovitz-Eldor, J., and Daley, G. Q. (2004) LIF/STAT3 signaling fails to maintain self-renewal of human embryonic stem cells. *Stem Cells* **22**, 770–778.
14. Xu, R. H., Peck, R. M., Li, D. S., Feng, X., Ludwig, T., and Thomson, J. A. (2005) Basic FGF and suppression of BMP signaling sustain undifferentiated proliferation of human ES cells. *Nat. Methods* **2**, 185–190.
15. Klimanskaya, I., Chung, Y., Meisner, L., Johnson, J., West, M. D., and Lanza, R. (2005) Human embryonic stem cells derived without feeder cells. *Lancet* **365**, 1636–1641.
16. Wang, L., Li, L., Menendez, P., Cerdan, C., and Bhatia, M. (2005) Human embryonic stem cells maintained in the absence of mouse embryonic fibroblasts or conditioned medium are capable of hematopoietic development. *Blood* **105**, 4598–4603.
17. Wang, G., Zhang, H., Zhao, Y., Li, J., Cai, J., Wang, P., Meng, S., Feng, J., Miao, C., Ding, M., Li, D., and Deng, H. (2005) Noggin and bFGF cooperate to maintain the pluripotency of human embryonic stem cells in the absence of feeder layers. *Biochem. Biophys. Res. Commun.* **330**, 934–942.
18. Levenstein, M. E., Ludwig, T. E., Xu, R. H., Llanas, R. A., Vandenheuvel-Kramer, K., Manning, D., and Thomson, J. A. (2006) Basic FGF support of human embryonic stem cell self-renewal. *Stem Cells* **24**, 568–574.
19. Itskovitz-Eldor, J., Schuldiner, M., Karsenti, D., Eden, A., Yanuka, O., Amit, M., Soreq, H., and Benvenisty, N. (2000) Differentiation of human embryonic stem cells into embryoid bodies comprising the three embryonic germ layers. *Mol. Med.* **6**, 88–95.
20. Schuldiner, M., Yanuka, O., Itskovitz-Eldor, J., Melton, D. A., and Benvenisty, N. (2000) Effects of eight growth factors on the differentiation of cells derived from human embryonic stem cells. *Proc. Natl. Acad. Sci. U. S. A.* **97**, 11307–11312.
21. Dvash, T., Mayshar, Y., Darr, H., McElhaney, M., Barker, D., Yanuka, O., Kotkow, K. J., Rubin, L. L., Benvenisty, N., and Eiges, R. (2004) Temporal gene expression during differentiation of human embryonic stem cells and embryoid bodies. *Hum. Reprod.* **19**, 2875–2883.

22. Dvash, T. and Benvenisty, N. (2004) Human embryonic stem cells as a model for early human development. *Best Pract. Res. Clin. Obstet. Gynaecol.* **18**, 929–940.
23. Zwaka, T. P. and Thomson, J. A. (2003) Homologous recombination in human embryonic stem cells. *Nat. Biotechnol.* **21**, 319–321.
24. Urbach, A., Schuldiner, M., and Benvenisty, N. (2004) Modeling for Lesch-Nyhan disease by gene targeting in human embryonic stem cells. *Stem Cells* **22**, 635–641.
25. Matin, M. M., Walsh, J. R., Gokhale, P. J., Draper, J. S., Bahrami, A. R., Morton, I., Moore, H. D., and Andrews, P. W. (2004) Specific knockdown of Oct4 and beta2-microglobulin expression by RNA interference in human embryonic stem cells and embryonic carcinoma cells. *Stem Cells* **22**, 659–668.
26. Vallier, L., Rugg-Gunn, P. J., Bouhon, I. A., Andersson, F. K., Sadler, A. J., and Pedersen, R. A. (2004) Enhancing and diminishing gene function in human embryonic stem cells. *Stem Cells* **22**, 2–11.
27. Zaehres, H., Lensch, M. W., Daheron, L., Stewart, S. A., Itskovitz-Eldor, J., and Daley, G. Q. (2005) High-efficiency RNA interference in human embryonic stem cells. *Stem Cells* **23**, 299–305.
28. Soria, B., Roche, E., Berna, G., Leon-Quinto, T., Reig, J. A., and Martin, F. (2000) Insulin-secreting cells derived from embryonic stem cells normalize glycemia in streptozotocin-induced diabetic mice. *Diabetes* **49**, 157–162.
29. Lavon, N., Yanuka, O., and Benvenisty, N. (2004) Differentiation and isolation of hepatic-like cells from human embryonic stem cells. *Differentiation* **72**, 230–238.
30. Menendez, P., Wang, L., and Bhatia, M. (2005) Genetic manipulation of human embryonic stem cells: a system to study early human development and potential therapeutic applications. *Curr. Gene Ther.* **5**, 375–385.
31. Robertson, E., Bradley, A., Kuehn, M., and Evans, M. (1986) Germ-line transmission of genes introduced into cultured pluripotential cells by retroviral vector. *Nature* **323**, 445–448.
32. Pfeifer, A., Ikawa, M., Dayn, Y., and Verma, I. M. (2002) Transgenesis by lentiviral vectors: lack of gene silencing in mammalian embryonic stem cells and preimplantation embryos. *Proc. Natl. Acad. Sci. U. S. A.* **99**, 2140–2145.
33. Gropp, M., Itsykson, P., Singer, O., Ben-Hur, T., Reinhartz, E., Galun, E., and Reubinoff, B. E. (2003) Stable genetic modification of human embryonic stem cells by lentiviral vectors. *Mol. Ther.* **7**, 281–287.
34. Ma, Y., Ramezani, A., Lewis, R., Hawley, R. G., and Thomson, J. A. (2003) High-level sustained transgene expression in human embryonic stem cells using lentiviral vectors. *Stem Cells* **21**, 111–117.
35. Xiong, C., Tang, D. Q., Xie, C. Q., Zhang, L., Xu, K. F., Thompson, W. E., Chou, W., Gibbons, G. H., Chang, L. J., Yang, L. J., and Chen, Y. E. (2005) Genetic engineering of human embryonic stem cells with lentiviral vectors. *Stem Cells Dev.* **14**, 367–377.
36. Volpers, C. and Kochanek, S. (2004) Adenoviral vectors for gene transfer and therapy. *J. Gene Med.* **6** (Suppl. 1), S164–S171.

37. Eiges, R., Schuldiner, M., Drukker, M., Yanuka, O., Itskovitz-Eldor, J., and Benvenisty, N. (2001) Establishment of human embryonic stem cell-transfected clones carrying a marker for undifferentiated cells. *Curr. Biol.* **11**, 514–518.
38. Siemen, H., Nix, M., Endl, E., Koch, P., Itskovitz-Eldor, J., and Brustle, O. (2005) Nucleofection of human embryonic stem cells. *Stem Cells Dev.* **14**, 378–383.
39. Lakshmipathy, U., Pelacho, B., Sudo, K., Linehan, J. L., Coucouvanis, E., Kaufman, D. S., and Verfaillie, C. M. (2004) Efficient transfection of embryonic and adult stem cells. *Stem Cells* **22**, 531–543.
40. Vasquez, K. M., Marburger, K., Intody, Z., and Wilson, J. H. (2001) Manipulating the mammalian genome by homologous recombination. *Proc. Natl. Acad. Sci. U. S. A.* **98**, 8403–8410.
41. Hoffman, L. M., Hall, L., Batten, J. L., Young, H., Pardasani, D., Baetge, E. E., Lawrence, J., and Carpenter, M. K. (2005) X-Inactivation status varies in human embryonic stem cell lines. *Stem Cells* **23**, 1468–1478.
42. Abeyta, M. J., Clark, A. T., Rodriguez, R. T., Bodnar, M. S., Pera, R. A., and Firpo, M. T. (2004) Unique gene expression signatures of independently-derived human embryonic stem cell lines. *Hum. Mol. Genet.* **13**, 601–608.
43. Bhattacharya, B., Miura, T., Brandenberger, R., Mejido, J., Luo, Y., Yang, A. X., Joshi, B. H., Ginis, I., Thies, R. S., Amit, M., Lyons, I., Condie, B. G., Itskovitz-Eldor, J., Rao, M. S., and Puri, R. K. (2004) Gene expression in human embryonic stem cell lines: unique molecular signature. *Blood* **103**, 2956–2964.
44. Mountford, P., Nichols, J., Zevnik, B., O'Brien, C., and Smith, A. (1998) Maintenance of pluripotential embryonic stem cells by stem cell selection. *Reprod. Fertil. Dev.* **10**, 527–533.
45. Tan, S. M. and Droge, P. (2005) Comparative analysis of sequence-specific DNA recombination systems in human embryonic stem cells. *Stem Cells* **23**, 868–873.
46. Drukker, M., Katz, G., Urbach, A., Schuldiner, M., Markel, G., Itskovitz-Eldor, J., Reubinoff, B., Mandelboim, O., and Benvenisty, N. (2002) Characterization of the expression of MHC proteins in human embryonic stem cells. *Proc. Natl. Acad. Sci. U. S. A.* **99**, 9864–9869.
47. Drukker, M. (2004) Immunogenicity of human embryonic stem cells: can we achieve tolerance? *Springer Semin. Immunopathol.* **26**, 201–213.
48. Drukker, M. and Benvenisty, N. (2004) The immunogenicity of human embryonic stem-derived cells. *Trends Biotechnol.* **22**, 136–141.
49. Schuldiner, M., Itskovitz-Eldor, J., and Benvenisty, N. (2003) Selective ablation of human embryonic stem cells expressing a "suicide" gene. *Stem Cells* **21**, 257–265.
50. Martin, M. J., Muotri, A., Gage, F., and Varki, A. (2005) Human embryonic stem cells express an immunogenic nonhuman sialic acid. *Nat. Med.* **11**, 228–232.
51. Richards, M., Fong, C. Y., Chan, W. K., Wong, P. C., and Bongso, A. (2002) Human feeders support prolonged undifferentiated growth of human inner cell masses and embryonic stem cells. *Nat. Biotechnol.* **20**, 933–936.

52. Richards, M., Tan, S., Fong, C. Y., Biswas, A., Chan, W. K., and Bongso, A. (2003) Comparative evaluation of various human feeders for prolonged undifferentiated growth of human embryonic stem cells. *Stem Cells* **21**, 546–556.
53. Amit, M., Margulets, V., Segev, H., Shariki, K., Laevsky, I., Coleman, R., and Itskovitz-Eldor, J. (2003) Human feeder layers for human embryonic stem cells. *Biol. Reprod.* **68**, 2150–2156.
54. Hovatta, O., Mikkola, M., Gertow, K., Stromberg, A. M., Inzunza, J., Hreinsson, J., Rozell, B., Blennow, E., Andang, M., and Ahrlund-Richter, L. (2003) A culture system using human foreskin fibroblasts as feeder cells allows production of human embryonic stem cells. *Hum. Reprod.* **18**, 1404–1409.
55. Amit, M., Shariki, C., Margulets, V., and Itskovitz-Eldor, J. (2004) Feeder layer- and serum-free culture of human embryonic stem cells. *Biol. Reprod.* **70**, 837–845.
56. Ludwig, T. E., Levenstein, M. E., Jones, J. M., Berggren, W. T., Mitchen, E. R., Frane, J. L., Crandall, L. J., Daigh, C. A., Conard, K. R., Piekarczyk, M. S., Llanas, R. A., and Thomson, J. A. (2006) Derivation of human embryonic stem cells in defined conditions. *Nat. Biotechnol.* **24**, 185–187.
57. Koivisto, H., Hyvarinen, M., Stromberg, A. M., Inzunza, J., Matilainen, E., Mikkola, M., Hovatta, O., and Teerijoki, H. (2004) Cultures of human embryonic stem cells: serum replacement medium or serum-containing medium and the effect of basic fibroblast growth factor. *Reprod. Biomed. Online* **9**, 330–337.
58. Heng, B. C., Liu, H., and Cao, T. (2004) Feeder cell density–a key parameter in multicolor FISH probe sets and their applications human embryonic stem cell culture. *In Vitro Cell Dev. Biol. Anim.* **40**, 255–257.
59. Sjogren-Jansson, E., Zetterstrom, M., Moya, K., Lindqvist, J., Strehl, R., and Eriksson, P. S. (2005) Large-scale propagation of four undifferentiated human embryonic stem cell lines in a feeder-free culture system. *Dev. Dyn.* **233**, 1304–1314.
60. Ezashi, T., Das, P., and Roberts, R. M. (2005) Low O_2 tensions and the prevention of differentiation of hES cells. *Proc. Natl. Acad. Sci. U. S. A.* **102**, 4783–4788.
61. Bayani, J. and Squire, J. A. (2002) Spectral karyotyping. *Methods Mol. Biol.* **204**, 85–104.
62. Liehr, T., Starke, H., Weise, A., Lehrer, H., and Claussen, U. (2004) Multicolor FISH probe sets and their applications. *Histol. Histopathol.* **19**, 229–237.
63. Cowan, C. A., Klimanskaya, I., McMahon, J., Atienza, J., Witmyer, J., Zucker, J. P., Wang, S., Morton, C. C., McMahon, A. P., Powers, D., and Melton, D. A. (2004) Derivation of embryonic stem-cell lines from human blastocysts. *N. Engl. J. Med.* **350**, 1353–1356.
64. Reubinoff, B. E., Pera, M. F., Vajta, G., and Trounson, A. O. (2001) Effective cryopreservation of human embryonic stem cells by the open pulled straw vitrification method. *Hum. Reprod.* **16**, 2187–2194.
65. Tedder, R. S., Zuckerman, M. A., Goldstone, A. H., Hawkins, A. E., Fielding, A., Briggs, E. M., Irwin, D., Blair, S., Gorman, A. M., Patterson, K. G., Linch, D.,

Heptonstall, J., and Brink, N. (1995) Hepatitis B transmission from contaminated cryopreservation tank. *Lancet* **346**, 137–140.

66. Hovatta, O. (2003) Cryobiology of ovarian and testicular tissue. *Best Pract. Res. Clin. Obstet. Gynaecol.* **17**, 331–342.
67. Richards, M., Fong, C. Y., Tan, S., Chan, W. K., and Bongso, A. (2004) An efficient and safe xeno-free cryopreservation method for the storage of human embryonic stem cells. *Stem Cells* **22**, 779–789.
68. Kurosawa, H., Imamura, T., Koike, M., Sasaki, K., and Amano, Y. (2003) A simple method for forming embryoid body from mouse embryonic stem cells. *J. Biosci. Bioeng.* **96**, 409–411.
69. Dang, S. M., Kyba, M., Perlingeiro, R., Daley, G. Q., and Zandstra, P. W. (2002) Efficiency of embryoid body formation and hematopoietic development from embryonic stem cells in different culture systems. *Biotechnol. Bioeng.* **78**, 442–453.
70. Wartenberg, M., Donmez, F., Ling, F. C., Acker, H., Hescheler, J., and Sauer, H. (2001) Tumor-induced angiogenesis studied in confrontation cultures of multicellular tumor spheroids and embryoid bodies grown from pluripotent embryonic stem cells. *FASEB J.* **15**, 995–1005.
71. Gerecht-Nir, S., Cohen, S., and Itskovitz-Eldor, J. (2004) Bioreactor cultivation enhances the efficiency of human embryoid body (hEB) formation and differentiation. *Biotechnol. Bioeng.* **86**, 493–502.
72. Tucker, K. L., Wang, Y., Dausman, J., and Jaenisch, R. (1997) A transgenic mouse strain expressing four drug-selectable marker genes. *Nucleic Acids Res.* **25**, 3745–3746.

4

Culture and Characterization of Human Bone Marrow Mesenchymal Stem Cells

Bruno Delorme and Pierre Charbord

Summary

Bone marrow (BM) mesenchymal stem cells (MSCs) are non-hematopoietic cells capable of generating colonies of plastic-adherent marrow mesenchymal cells, each derived from a single cell termed a colony-forming unit fibroblasts (CFU-Fs). In addition to their role in establishing the marrow microenvironment, these cells have been shown to differentiate into several types of mesenchymal and non-mesenchymal lineages. Because of their multipotency, MSCs represent an attractive cellular source in the promising field of cellular therapy. In this chapter, we will focus on culture conditions for human BM MSC expansion and CFU-F assays. We also describe the methodologies to analyze the primary cultures obtained, both at the phenotypic and at the functional levels. Phenotypically, MSCs can be defined with a minimal set of markers as CD31-, CD34-, and CD45-negative cells and CD13-, CD29-, CD73-, CD90-, CD105-, and CD166-positive cells. Functionally, we describe the culture conditions (specific media and cellular preparations) for in vitro differentiation of MSCs into the adipogenic, osteogenic, and chondrogenic lineages. The corresponding colorimetric assays (oil red O, Von Kossa and alizarin red S, and safranin O and alcian blue stains, respectively) are also described.

Key Words: Bone marrow; Mesenchymal stem cells; Culture; CFU-F; Phenotype; Adipocytes; Osteoblasts; Chondrocytes; Differentiation.

1. Introduction

Friedenstein et al. *(1)* were the first to isolate mesenchymal stem cells (MSCs) from the bone marrow (BM). The isolation method was based on the adherence of marrow-derived cells to the plastic of the cell-culture plate. To date, Friedenstein's method represents a standard protocol to isolate BM MSCs. When BM cells are cultured in vitro, the plastic-adherent cellular population

From: *Methods in Molecular Medicine, 2nd ed.: Tissue Engineering*
Edited by: H. Hauser and M. Fussenegger © Humana Press Inc., Totowa, NJ

forms compact colonies of rapidly growing spindle-shaped cells, morphologically similar to fibroblasts and termed colony-forming unit fibroblast (CFU-F) *(1)*. These cells currently referred to as MSCs are also known as marrow stromal cells.

MSCs have attracted interest because these non-hematopoietic cells exhibit multilineage differentiation capacity. Many reports have shown that they are capable of giving rise to diverse mesenchymal derivatives such as bone, cartilage, and adipose tissues [for reviews on the biological and phenotypic characteristics of MSCs, see *(2–7)*]. In addition, they have been shown to be able to differentiate into other non-mesenchymal derivatives such as hepatic, cardiac, or even neural cells. They represent therefore an attractive and promising field in tissue regeneration for a wide range of diseases or traumas. In addition, these progenitors play a central role in establishing the marrow microenvironment both in vitro and in vivo.

Here we describe a culture protocol developed in the laboratory to expand MSCs from human BM samples, which give rise to a homogeneous cellular population. We also describe the methodologies to characterize these primary cultures, both at the phenotypic level (flow cytometric analysis of MSC membrane antigens) and at the functional level (histochemistry). In this protocol, plastic-adherent BM MSCs are defined as cells with the following characteristics:

1. Negative for CD31 (PECAM), CD34, and CD45, and positive for CD13 (ANPEP), CD29 (β1 integrin), CD73 (NTSE), CD90 (THY1), CD105 (Endoglin), and CD166 (Alcam).
2. Able to differentiate (when cultured in appropriate media) into the adipogenic, osteogenic, and chondrogenic lineages (revealed by specific colorimetric assays).

2. Materials

2.1. Cell Culture

1. Proliferation medium composed of minimum essential medium alpha (α-MEM) with L-glutamine but without ribonucleosides or deoxyribonucleosides (Gibco/BRL; cat. no. 22561-021), supplemented with 10% fetal bovine serum (FBS; Hyclone) (*see* **notes 1 and 2**), L-glutamine (Gibco/BRL; cat. no. 25030024) 2 mM, amphotericin B (Fungizone; Bristol-Myers Squibb) 0.0025 mg/ml, penicillin G 100 U/ml, and streptomycin sulfate (Gibco/BRL; cat. no. 15140122) 0.1 mg/ml. Filter the medium on a sterile filter unit with 0.22-μm pores.
2. Basic fibroblast growth factor (bFGF/FGF2; AbCys; cat. no. P100-18B) dissolved at 50 μg/ml and stored in single-use aliquots at −20°C.
3. 0.5% trypan blue solution (Biochrom; cat. no. L6323) ready to use.
4. Acetic acid (Sigma; cat. no. A6283).

5. Phosphate-buffered saline (PBS) ×10 without calcium or magnesium (Gibco/BRL; cat. no. 14200059).
6. Solution of trypsin (0.25% v/v) and ethylenediaminetetraacetic acid (EDTA; 1 mM; Gibco/BRL; cat. no. 25300054).
7. Dimethyl sulfoxide (DMSO), tissue-culture grade (Sigma; cat. no. D4540).
8. Sterile 250- or 500-ml filter units with 0.22-μm pores (Millipore; cat. no. SCGPU02RE or SCGPU05RE, respectively).
9. Cryofreezing container (Nalgene; cat. no. 22432) containing 100% isopropanol (Aldrich; cat. no. 19,076-4).
10. Cryovials (NUNC; cat. no. 377267).
11. Sterile tissue-culture flasks: T25, T75, or T150 (Falcon; cat. no. 353108, 353136, or 355001, respectively).
12. Labtek (NUNC; cat. no. 177445).
13. Tubes 15 or 50 ml (Falcon; cat. no. 352096 or 352070, respectively).
14. Deionized (DI) water (Gibco; cat. no. 15230-089).
15. Transfer bag containing acid citrate dextcose (ACD) (Macopharma; cat. no. MSE 0100Q).

2.2. CFU-F Assays

1. May-Grunwald-Giemsa (VWR; cat. no. 1014241000).
2. Methanol (Sigma; cat. no. 270474).

2.3. Differentiation Assays

2.3.1. Prepare Stock Solutions

1. Dexamethasone (Dex; Sigma; cat. no. D8893) 1 mM in ethyl alcohol 100% (*see* **note 3**).
2. Isobutylmethylxantine (IBMX; Sigma; cat. no. I5879) 45 mM in ethyl alcohol 100%.
3. Indomethacin (Sigma; cat. no. I7378) 50 mM in ethyl alcohol 100%.
4. L-Ascorbic acid (Sigma; cat. no. 25,556-4) 25 mg/ml in distilled water.
5. Sodium phosphate monobasic (NaH_2PO_4; Sigma; cat. no. S0751) 3 M in Dulbecco's minimal essential medium (DMEM).
6. Sodium pyruvate 0.1 M (Sigma; cat. no. S8636; solution ready to use).
7. L-Ascorbic acid-2-phosphate (Sigma; cat. no. A8960) 17 mM in distilled water.
8. Proline (Sigma; cat. no. P-5607) 35 mM in distilled water.
9. Insulin–transferin–selenium (ITS) ×500 (Cambrex; cat. no. 17-838Z; solution ready to use). Sterilize the prepared solutions by filtration on 0.22-μm pores and store aliquots at −20°C.

2.3.2. Prepare Adipogenic Differentiation Medium

Low-glucose DMEM (Gibco/BRL; cat. no. 31885023) supplemented with 10% serum, 10^{-6} M Dex, 0.5 mM IBMX, and 60 μM indomethacin. Keep the medium at 4°C. The same medium can be used for the total period of the differentiation experiment.

2.3.3. Prepare Osteogenic Differentiation Medium

High-glucose DMEM (Gibco/BRL; cat. no. 41965039) supplemented with 10% serum, 10^{-7} M Dex, 25 μg/ml L-ascorbic acid, and 3 mM NaH_2PO_4 (*see* **note 4**). Keep the medium at 4°C. The same medium can be used for the total period of the differentiation experiment.

2.3.4. Prepare Chondrogenic Differentiation Medium

The incomplete chondrogenic medium is composed of high-glucose DMEM supplemented with Dex 10^{-7} M, sodium pyruvate 1 mM, L-ascorbic acid-2-phosphate 0.17 mM, proline 0.35 mM, and ×1 ITS (e.g., add 200 μl ITS to 100 ml solution). The complete chondrogenic medium is composed of the incomplete medium supplemented with human transforming growth factor-β1 (TGF-β1; AbCys; cat. no. P100-21) at a final concentration of 10 ng/ml. Prepare complete medium on the day of use.

2.4. Colorimetric Assays

1. Prepare 4% (v/v) formaldehyde solution (Sigma; cat. no. F1635) in PBS ×1.
2. Prepare stock solution of 1 mg/ml nile red O (Sigma; cat. no. 73189) in DMSO; store aliquots at −20°C.
3. Prepare 5% (w/v) silver nitrate solution ($AgNO_3$; Sigma; cat. no. S8157) in distilled water; light sensitive.
4. Prepare 5% (w/v) sodium thiosulfate (Sigma; cat. no. S7026) solution in distilled water.
5. Prepare 2% (w/v) alizarin red S (Sigma; cat. no. A5533) solution in distilled water, mix the solution, and adjust the pH to 4.1–4.3 (using 0.5% ammonium hydroxide). The pH is critical.
6. Prepare 0.02% fast green (w/v; Sigma; cat. no. F7258) solution in distilled water.
7. Prepare 0.1% (w/v) safranin O (Sigma; cat. no. S2255) solution in distilled water.
8. Prepare 3% (w/v) alcian blue (Sigma; cat. no. A5268) solution in distilled water.
9. Prepare 3% methanol (Sigma; cat. no. 27,0474-4) solution in distilled water.
10. Hematoxylin Harris (Sigma; cat. no. HHS16).
11. Ethanol (Prolabo; cat. no. 20821).
12. Isopropanol (Aldrich; cat. no. 19,076-4).
13. Vectashield with 4', 6-Diamidino-z-phenylindole dihydrochloride (DAPI) (AbCys; cat. no. H1200).
14. Xylene (Aldrich; cat. no. 29,588-4).
15. Paraffin DiARATH (Hicrom microtech France; cat. no. 030980).

2.5. Flow Cytometric Analysis

1. PBS ×10 without calcium or magnesium (Gibco/BRL; cat. no. 14200059).
2. DI water 500 ml (Gibco; cat. no. 15230-089).
3. Solution of trypsin (0.25%) and EDTA (1 mM; Gibco/BRL; cat. no. 25300054).
4. Phycoerythrin (PE)-conjugated antibodies (Abs): IgG1-PE (clone MOPC-2; BD Biosciences; cat. no. BD 555 749), CD13-PE (clone WM15; BD Biosciences; cat. no. BD 555 394), CD29-PE (clone MAR4; BD Biosciences; cat. no. BD 555 443), CD31-PE (clone WM59; BD Biosciences; cat. no. BD 555 446), CD34-PE (clone 8G12; BD Biosciences; cat. no. BD 345 802), CD45-PE (clone HI30; BD Biosciences; cat. no. BD 555 483), CD73-PE (clone AD2; BD Biosciences; cat. no. BD 550 257), CD90-PE (clone 5E10; BD Biosciences; cat. no. BD 555 596), CD105-PE (clone SN6; Caltag), and CD166 (clone 3A6; BD Biosciences; cat. no. BD 559 263).

2.6. Freezing Procedure

Prepare fresh freezing medium consisting of 90% serum and 10% DMSO (Sigma; cat. no. D4540).

3. Methods

3.1. Human MSC Cultures

All reagents and materials must be sterile, and the cell culture must be performed under a laminar vertical flow tissue culture hood. Because all samples should be assumed to contain harmful agents (viruses, etc.), take the necessary precautions: work in L2 laboratory room and wear laboratory coats and gloves.

1. Primary human BM MSCs are obtained from BM aspirates (iliac crest) of patients undergoing hip replacement surgery, after informed consent, using transfer bag containing ACD. 10 ml ACD is aspirated to rinse carefully the syringe. Keep 1 or 2 ml ACD in the syringe and perform a BM aspirate of 9 or 18 ml, respectively (final concentration of 10% ACD in the syringe). In this way, the BM aspirate will be well homogenized.
2. Transport the BM sample to the laboratory in laminar airflow cabinet.
3. Rinse cell suspension in a solution of PBS ×1/2 mM EDTA medium and centrifuge at 350 × *g* for 10 min at room temperature. After centrifugation, resuspend the cellular pellets in 20 ml PBS/2 mM EDTA.
4. Count the number of mononuclear cells (MNCs) present in the cell suspension. It is generally necessary to dilute 1/10 the cell suspension in PBS. Incubate 20 μl 1/10 diluted cellular sample for 2–3 min with 180 μl solution of 10% acetic acid for red cell lysis. In parallel, a second aliquot of 20 μl cellular suspension is mixed with an equal volume of trypan blue solution to evaluate the proportion of dead cells. Count the number of dead cells (stained with trypan blue) and viable cells

(non-stained). Calculate the percentage of viability. The number of MNCs is estimated using a hemocytometer.

5. Prepare the proliferation medium. Just before use, add 1 ng/ml sterile bFGF to the proliferation medium (*see* **note 5**). Pre-warm at 37°C the volume of medium necessary.
6. Plate MNCs in the proliferation medium at cell density of 50, 000 cells/cm^2 (i.e., 1.25×10^6 cells per 25-cm^2 flask; 3.75×10^6 cells per 75-cm^2 flask; or 7.5×10^6 cells per 150-cm^2 flask) (*see* **note 6**). NB: Keep an aliquot of the cell suspension for the CFU-F assays (*see* **section 3.2.**).
7. The cells are maintained under standard cell-culture conditions (humidified atmosphere, 5% CO_2, 37°C). The first medium change is performed 3 days after plating of MNCs.
8. Thereafter, the medium is changed twice a week.
9. When cultures reach confluency (between days 14 and 18), remove supernatant and trypsinize cells. In a T75 flask, wash cells with 5 ml pre-warmed PBS ×1, then add 2 ml 0.25% (v/v) trypsin/1 mM EDTA for 2–3 min (incubation at 37°C), check on the invertoscope that cells are detached, add 8-ml medium to neutralize the trypsin/EDTA, centrifuge (300 × *g*, 5 min), resuspend the pellets in 5-ml medium, and count the viable cells (trypan blue negative cells). NB: If the total number of viable cells obtained is sufficient, dedicate a part of the cell culture to prepare a stock of passage 0 (P0) frozen cells (*see* **section 2.6.**) and, eventually, for cytometric analysis of the cells (*see* **section 3.5**) and differentiation assays (*see* **section 3.3**). If CD45-positive cells are detected in the cultures as shown by flow cytometry, or if the number of cells obtained is too low to perform the experiments, continue the amplification procedure.
10. For cellular expansion, replate the cells in the proliferation medium at a concentration of 1000 cells/cm^2 (P1). Every 3 days, remove medium with gentle aspiration with vacuum suction and replace it with the same volume of freshly prepared proliferation medium. Monitor cell morphology by phase microscopy (*see* **note 7**).
11. When cells reach confluence (usually 14–16 days later), remove medium, wash cells with PBS, and trypsinize cells as previously described.
12. At this stage, a homogeneous population of human BM MSCs is microscopically observed. A stock of P1 frozen cells, differentiation assays, and cytometric analysis can be performed (*see* **sections 3.6, 3.3,** and **3.5,** respectively).
13. For additional cellular expansion (P2), repeat steps 10–12 (*see* **note 8**).

3.2. CFU-F Assays

Assays for CFU-Fs provide a measure to estimate the number of clonogenic MSCs present in the cultures and consequently represent a kind of "quality control assay." CFU-F assays can be performed at the beginning of the culture (from freshly obtained BM samples) or after each cell passage to verify the maintenance of the clonogenic capacity of the cells.

1. Count cells with hemocytometer (from fresh samples, count MNCs after red cell lysis with acetic acid as previously described).
2. For fresh BM samples, seed MNCs at three different concentrations: 1, 5×10^5, and 1×10^6 cells in 5 ml proliferation medium, each concentration being plated in a 25-cm^2 flask. For successive passages (P1 and eventually P2), seed cells at the three following concentrations: 1, 2, and 5×10^2 cells in 5-ml proliferation medium, each concentration being plated in a 25-cm^2 flask (*see* **note 9**).
3. Renew the medium 3 and 7 days later.
4. On day 10, remove the medium, wash twice in PBS ×1, fix the cells in methanol 100% for 10 min at room temperature.
5. Wash the flask with PBS ×1 and perform a May-Grunwald-Giemsa staining.
6. Pipette enough May-Grunwald to cover the surface of the membrane; incubate at 37°C for 2 min.
7. Rinse twice with distilled water.
8. Pipette enough Giemsa (1/10) to cover the surface of the flask. Incubate at 37°C for 10 min.
9. Rinse twice with distilled water.
10. Dry the flask at 37°C overnight.
11. Examine flasks with the stained colonies under an inverted microscope and count colonies with ≥ 50 cells. Record the number of colonies counted divided by cells originally plated ×100 as "% CFU-Fs" (cloning efficiency) (*see* **note 10**).

3.3. Differentiation of the Human MSCs into the Adipogenic, Osteogenic, and Chondrogenic Lineages

Split cells in three aliquots and culture in three specific media.

3.3.1. Adipogenic Differentiation

1. Trypsinize cells and centrifuge them at $350 \times g$ for 5 min to remove culture medium.
2. Resuspend cells in proliferation medium and plate at cell density of 20,000–30, 000 cells/cm^2 for 24–48 h (until complete confluency is reached) (*see* **note 11**).
3. At complete confluency, switch to adipogenic differentiation medium.
4. Change the medium with the adipogenic medium every 2 days until day 14.

The adipogenic differentiation of the human MSCs is characterized by the appearance of numerous very refringent lipid droplets in the cytoplasm of the cells (*see* **Fig. 1,** panel c). These lipid droplets stain positively with nil red O fluorescent marker (*see* **section 4.1.**). Such lipid droplets appear at about day 3 and/or 4 and become bigger each following day.

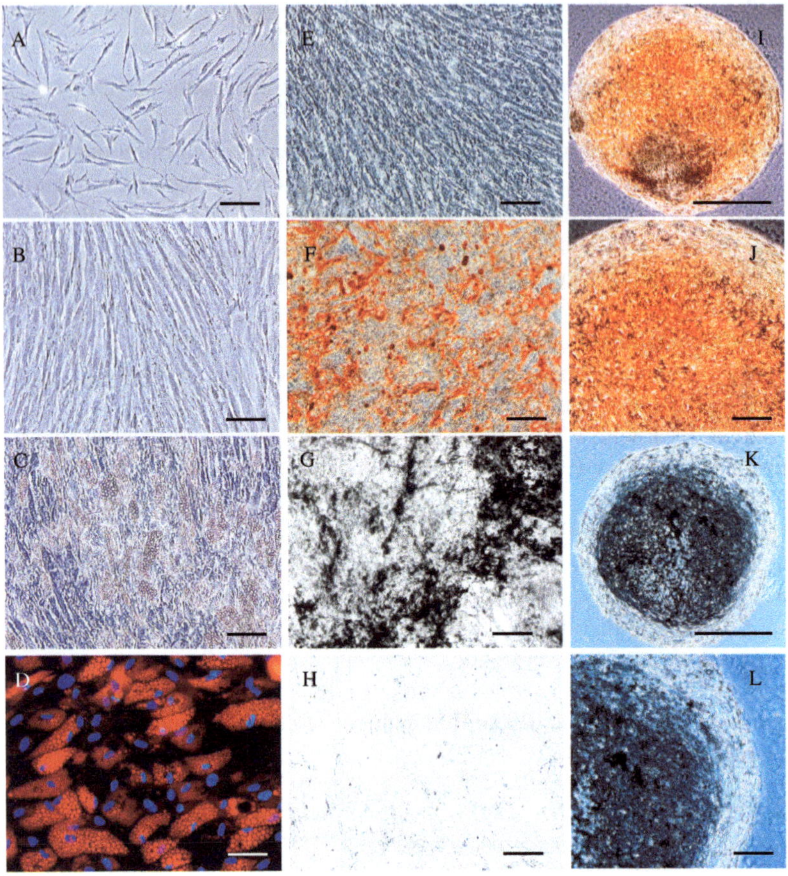

Fig. 1. Colorimetric assays of human bone marrow (BM) mesenchymal stem cells (MSCs) differentiated into adipogenic, osteogenic, and chondrogenic lineages. Phase contrast of (a) subconfluent and (b) confluent MSCs cultured in proliferation medium [cells at passage 1 (P1)]; (c) phase contrast of MSCs after 14 days of culture in adipogenic medium. Note the appearance of refringent lipid droplets in the cytoplasm of numerous cells; (d) oil red O staining of the lipid droplets (day 14 of adipogenic differentiation); (e) phase contrast of MSCs after 21 days of culture in osteogenic medium; (f) alizarin red S staining and (g) Von Kossa staining of MSCs after 21 days of culture in osteogenic medium. Note the presence of numerous mineralized orange-red and black positive areas, respectively; (h) as a negative control of mineralization, Von Kossa staining of MSCs after 21 days of culture in osteogenic medium without the addition of inorganic phosphate (NaH_2PO_4). Note the absence of black positive areas; (i and j) safranin O and (k and h) alcian blue staining of paraffin sections of MSC

3.3.2. Osteogenic Differentiation

1. Trypsinize P1 cells and centrifuge them at 350 × *g* for 5 min to remove culture medium.
2. Resuspend cells in proliferation medium and plate them at a cellular density of 20,000–30,000 cells/cm^2 for 24–48 h (until complete confluency is reached).
3. At complete confluency, switch to the osteogenic differentiation medium.
4. Change the medium with the osteogenic medium every 2 days until day 21.

3.3.3. Chondrogenic Differentiation

Chondrogenic differentiation is performed by cultures in micropellets in the presence of TGF-β1.

1. Trypsinize cells and centrifuge them at 350 × *g* for 5 min to remove culture medium.
2. Resuspend cells in incomplete medium for chondrogenesis and centrifuge at 350 × *g* for 5 min to wash the cells.
3. Resuspend cells in complete medium for chondrogenesis (containing TGF-β1 and without serum) and count them; adjust the cell concentration to 5×10^5 cells/ml. Plate 500 μl cell suspension (2.5×10^5 cells/tube) in 15-ml polypropylene tubes (Falcon).
4. Centrifuge at 350 × *g* for 5 min to pellet the cells.
5. Bring the tubes to the incubator carefully, incubate pellets at 37°C in 5% CO_2, and unscrew caps to permit gas exchange during culture.
6. At day 1, detach slowly the pellets from the bottom by flicking the tubes.
7. At day 2, replace the medium carefully with 500 μl complete medium (be careful not to touch the pellets).
8. Replace the medium with 500 μl complete medium every 2 days until day 21.

3.4. Colorimetric Analysis of the Differentiated Cultures (Adipogenic, Osteogenic, and Chondrogenic)

3.4.1. Adipogenic Differentiation: Nil Red O Staining

Nil red O is a lysochrome (fat-soluble dye) predominantly used for demonstrating the presence of triglycerides.

1. Remove culture supernatant; wash once with PBS.
2. Fix cells with 4% formaldehyde in PBS for 10 min at room temperature.
3. Remove fixative and wash twice with PBS.

Fig. 1. micropellets cultured 21 days in chondrogenic medium. Note in both cases, a gradient of proteoglycan and glycosaminoglycan (GAG) staining from the center to the periphery of the micropellets. Bars: 50 μm in (a); 10 μm in (b–h); 500 μm in (i and k); 200 μm in (j and l).

4. Add PBS with 1 μg/ml freshly diluted nil red O.
5. Incubate for 30 min at 4°C in the dark.
6. Wash twice in PBS.
7. Mount in vectashield with DAPI to visualize the nuclei.
8. Analyze by fluorescence at 580 or 520 nm.

MSCs differentiated into adipogenic cells can be easily visualized with numerous red lipid droplets in the cytoplasm of the cells as shown in **Fig. 1,** panel d.

3.4.2. Osteogenic Differentiation: Von Kossa and Alizarin Red S Stains

3.4.2.1. Von Kossa Staining

The Von Kossa staining indicates the mineralization of the bone matrix. It reveals calcium salts (phosphate, carbonate, sulfate, and oxalate) by substituting to a metallic cation $AgNO_3$ that will be visualized after reduction into metallic silver, leading to a black staining of the cell cultures (*see* **Fig. 1,** panel g).

1. Fix cell layer with 4% formaldehyde in PBS ×1 for 10 min at 4°C.
2. Rinse with distilled water.
3. Add $AgNO_3$ for 30 min at room temperature in the dark.
4. Rinse with distilled water.
5. Cover with distilled water and expose to ultraviolet light for 10 min to 1 h (monitor appearance of black areas under phase microscope).
6. Rinse with water.
7. Treat with 5% thiosulfate solution for 2 min to stop the reaction.
8. Rinse with water.

3.4.2.2. Alizarin Red S Staining

Alizarin red S staining is a second technique to identify calcium deposits during the osteogenic differentiation process. Calcium forms an alizarin red S–calcium complex in a chelation process.

1. Fix cells in 4% formaldehyde (in PBS ×1) for 10 min at room temperature.
2. Rinse twice with distilled water.
3. Add alizarin red S solution on the cells for 30 s to 5 min.
4. Follow the reaction under microscope until the appearance of an orange-red coloration.
5. Stop the reaction by rinsing the wells several times with distilled water.

Calcium salt present in the cellular culture should be revealed in orange-red as shown in **Fig. 1,** panel f.

3.4.3. Chondrogenic Differentiation: Safranin O and Alcian Blue Stains

Pellets are recovered in an Eppendorf tube and washed twice in cold PBS, fixed with 4% formaldehyde in PBS for ≥ 24 h at 4°C, rinsed twice for 1 h in cold PBS, dehydrated, and then embedded in paraffin (standard methods).

Paraffin sections are stained with alcian blue or safranin O.

3.4.3.1. Safranin O Staining

Safranin O is a cationic stain that binds to cartilage proteoglycans and glycosaminoglycans (GAGs) such as chondroitin and keratan sulfate *(8)*. Staining of the pellets (*see* **Fig. 1**, panels i and j) demonstrates a positive safranin O reaction indicating synthesis of GAGs.

This protocol is for paraffin-embedded sections of tissues on slides.

1. Place paraffin slides in xylene for 10 min.
2. Place the slides in 100% ethanol for 10 min.
3. Place the slides in 90% ethanol for 10 min.
4. Place the slides in 70% ethanol for 10 min.
5. Place the slides in distilled water.
6. Place the slides in 0.02% fast green for 3 min.
7. Place the slides in 1% acetic acid for 30 s.
8. Place the slides in 0.1% safranin O for 5–15 min.
9. Rinse the slides in water (agitate slowly).
10. Place the slides in 70% ethanol for 5 min.
11. Place the slides in 90% ethanol for 5 min.
12. Place the slides in 100% ethanol for 5 min.
13. Place the slides in xylene for 3 min.
14. Mount the slides with coverslip.

3.4.3.2. Alcian Blue Staining

Alcian blue is one of the most widely used cationic (it has many positive charges on the molecule) dye for the demonstration of GAGs. It is thought to work by forming reversible electrostatic bonds between the cationic dye and the negative (anionic) sites on the polysaccharide. This protocol is for paraffin-embedded sections of tissues on slides.

1. Place paraffin slides in xylene for 10 min.
2. Place the slides in 100% ethanol for 10 min.
3. Place the slides in 90% ethanol for 10 min.
4. Place the slides in 70% ethanol for 10 min.
5. Place the slides in distilled water.
6. Hematoxylin Harris for 20 min.

7. Wash with tap water.
8. 1% HCl in 70% ethanol.
9. Wash with tap water.
10. Mix 1:1 v/v 3% alcian blue solution and 3% methanol solution; immerse the slides for 10 min.
11. Wash with tap water.
12. Place the slides in 70% ethanol for 5 min.
13. Place the slides in 90% ethanol for 5 min.
14. Place the slides in 100% ethanol for 5 min.
15. Place the slides in xylene for 3 min.
16. Mount the slides with coverslip.

Staining of the pellets in blue (*see* **Fig. 1**, panels k and l) demonstrates a positive alcian blue reaction indicating synthesis of GAGs.

3.5. Flow Cytometric Analysis of Membrane Antigens (PE-Conjugated Antibodies)

MSCs are negative for the endothelial and/or hematopoietic markers CD31, CD34, and CD45 and are highly positive for CD13, CD29, CD73, CD90, CD105, and CD166. This minimal phenotype enables to confirm the identity of the expanded cells, the absence of hematopoietic contamination, and the homogeneity of the cellular population (unimodal distribution of each marker) (*see* **Fig. 2**).

1. Trypsinize cells and centrifuge them at 350 × *g* for 5 min.
2. Wash trypsinized cells with cold (4°C) PBS and centrifuge them (350 × *g*, 5 min).
3. Discard supernatant and resuspend the pellets in cold (4°C) PBS.
4. Count cells and quantify cell viability by trypan blue exclusion.
5. Put 10^5 viable cells/tube in 200 μl cold PBS.
6. Add directly PE-conjugated monoclonal Ab (mAb) at saturating concentration (usually 10 μl Ab is sufficient for 10^5 viable cells).
7. Incubate for 30 min, at 4°C, in the dark.
8. Wash twice with cold PBS by centrifugation (350 × *g*, 5 min, 4°C) and resuspend the pellets with cold PBS (200 μl).
9. Proceed immediately with flow cytometric analysis.

3.6. Freezing and Thawing Procedures

3.6.1. Freezing

Proceed as quickly as possible.

1. Prepare fresh freezing medium and keep it at 4°C for at least 30 min.
2. Prepare freezing device (cold cryobox + isopropanol) and keep it at 4°C for at least 30 min.

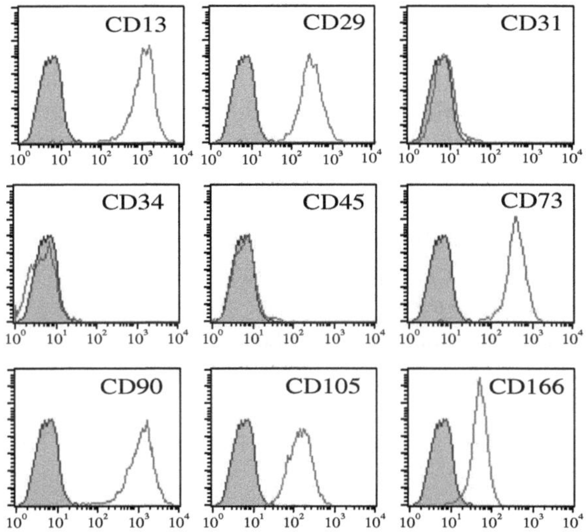

Fig. 2. Flow cytometric analysis of membrane antigen characteristics of human bone marrow (BM) mesenchymal stem cells (MSCs). Fluorescence intensity histograms with phycoerythrin (PE)-conjugated specific antibodies (Abs) for membrane antigens (red line) and irrelevant isotype-matched (IgG1 PE-conjugated) Ab as negative control (grey area). Immunophenotyping was performed on a FACSCalibur flow cytometer, and at least 10,000 events were collected for each sample. Human MSCs were found homogeneously positive for CD13, CD29, CD73, CD105, and CD166. Cells are negative for the hematopoietic and/or endothelial markers CD31, CD34, and CD45.

3. Prepare adequate number of 1.8-ml polypropylene cryotubes and identify each of them.
4. Trypsinize cells and centrifuge them at 350 × *g* for 5 min at 4°C. Discard supernatant.
5. Adjust cell suspension to 1×10^6/ml with cold freezing medium. (If the number of cells obtained is low, adjust the cell suspension to final concentrations ranging from 0.5 to 3×10^6/ml.)
6. Distribute the mixture in 1.8-ml cryotubes and place the tubes in cold (4°C) cryobox.
7. Put cryobox at −80°C.
8. After 24–48 h, store cell aliquots in liquid nitrogen containers.

3.6.2. Thawing

Proceed as quickly as possible.

1. Heat waterbath at 37°C.
2. The thawing medium is the classical proliferation medium used for MSC cultures.

3. Take the frozen sample (in polypropylene cryotubes) out of the liquid nitrogen container.
4. Thaw the sample quickly (< 2 min) in the 37°C waterbath under constant agitation.
5. Transfer immediately cell suspension in a 50-ml tube containing 20 ml thawing medium (for up to 1.8-ml cell suspension).
6. Centrifuge at $350 \times g$ for 10 min at 20°C, and then discard supernatant.
7. Count cells and evaluate cell mortality using a trypan blue (*see* **note 12**).
8. Plate cells in the proliferation medium at the desired cellular density (1000 cells/cm^2 for a large expansion of the cells).
9. Change the medium next day.
10. Thereafter, change medium every 3–4 days.

Acknowledgments

Work was supported by a grant from the European Union (Integrated project Gensotem no. 503161).

Notes

1. Differences in FBS batch strongly influence cell growth. Consequently, different lots of FBS must be tested before use in culture to determine the FBS that gives the best rate of growth, highest CFU-F activity, and best cellular morphology. This is a critical step for human MSC cultures.
2. Decomplementation of the serum is not necessary.
3. For aliquots prepared in ethanol, be careful to have caps well closed to avoid evaporation and consequently change in concentrations of the stock solutions.
4. It is recommended by people working in the field of bone formation not to use NaH_2PO_4 for theoretical reasons that remain unclear.
5. Addition of bFGF at 1 ng/ml is optional. However, bFGF was found to be a potent mitogen for CFU-Fs from multiple species and increased growth while maintaining multilineage differentiation potential of MSCs (*9*, *10*). Add bFGF in the culture medium just before use. No other cytokines are required.
6. Coating of flasks with fibronectin or collagen and Hypaque–Ficoll (or other density gradient separation) is not mandatory.
7. With practice you will be able to identify cultures that have (i) highest CFU-F activity (i.e., the number of CFU-Fs in the cultures), (ii) best cellular morphology (*see* **Fig. 1**; spindle cells with the presence of numerous filopods; high number of vacuoles indicates senescence), and (iii) rapid expansion.
8. If possible (sufficient number of cells obtained), stop the expansion of the culture at the end of P1. With passages, we noticed the appearance in the cultures of larger elongated cells that proliferate more slowly and have a more limited potential to differentiate. The presence of larger cells in culture indicates that cells are progressing toward senescence. With experience, the proportion of larger cells can be judged by phase-contrast microscopic observation of the cultures.

9. Be careful to dissociate well the cells before counting and plating.
10. The number of CFU-Fs is an important characteristic of these cells and should be recorded for each BM sample. Important variations between samples have been noticed. The number of CFU-Fs obtained from fresh samples is about 30 per [6] MNCs plated. The percentage of CFU-Fs obtained from P1 cells varies from 30 to 50%.
11. We noticed that high confluency of the cell layers is an important step for good adipogenic differentiation.
12. The expected yield of dead cells after thawing is between 2 and 10%.

References

1. Friedenstein, A. J., Gorskaja, J. F., and Kulagina, N. N. (1976) Fibroblast precursors in normal and irradiated mouse hematopoietic organs. *Exp. Hematol.* **4**, 267–274.
2. Caplan, A. I. (1991) Mesenchymal stem cells. *J. Orthop. Res.* **9**, 641–650.
3. Prockop, D. J. (1997) Marrow stromal cells as stem cells for nonhematopoietic tissues. *Science* **276**, 71–74.
4. Conget, P. A. and Minguell, J. J. (1999) Phenotypical and functional properties of human bone marrow mesenchymal progenitor cells. *J. Cell Physiol.* **181**, 67–73.
5. Pittenger, M. F., Mackay, A. M., Beck, S. C., Jaiswal, R. K., Douglas, R., Mosca, J. D., Moorman, M. A., Simonetti, D. W., Craig, S., and Marshak, D. R. (1999) Multilineage potential of adult human mesenchymal stem cells. *Science* **284**, 143–147.
6. Bianco, P., Riminucci, M., Gronthos, S., and Robey, P. G. (2001) Bone marrow stromal stem cells: nature, biology, and potential applications. *Stem Cells* **19**, 180–192.
7. Dennis, J. E. and Charbord, P. (2002) Origin and differentiation of human and murine stroma. *Stem Cells* **20**, 205–214.
8. Lammi, M. and Tammi, M. (1988) Densitometric assay of nanogram quantities of proteoglycans precipitated on nitrocellulose membrane with Safranin O. *Anal. Biochem.* **168**, 352–357.
9. Bianchi, G., Banfi, A., Mastrogiacomo, M., Notaro, R., Luzzatto, L., Cancedda, R., and Quarto, R. (2003) Ex vivo enrichment of mesenchymal cell progenitors by fibroblast growth factor 2.*Exp. Cell Res.* **287**, 98–105.
10. Tsutsumi, S., Shimazu, A., Miyazaki, K., Pan, H., Koike, C., Yoshida, E., Takagishi, K., and Kato, Y. (2001) Retention of multilineage differentiation potential of mesenchymal cells during proliferation in response to FGF. *Biochem. Biophys. Res. Commun.* **288**, 413–419.

5

Skeletal ("Mesenchymal") Stem Cells for Tissue Engineering

Pamela Gehron Robey, Sergei A. Kuznetsov, Mara Riminucci, and Paolo Bianco

Summary

Skeletal stem cells (SSCs, commonly referred to as "mesenchymal" stem cells) are found in the bone marrow stromal cell (BMSC) fraction of post-natal bone marrow. They can be isolated in culture as adherent, clonogenic cells endowed with the ability to grow and differentiate into multiple lineages, all of which correspond to tissues that are integral parts of the skeleton. The multipotency of SSCs is probed by in vivo transplantation assays. The ability of SSCs to generate a cell strain competent to form significant amounts of bone in vivo has led to the formulation of preclinical models of bone repair.

Key Words: Skeletal stem cells; Mesenchymal stem cells; Osteogenesis; Bone repair; Tissue engineering; Bone; Cartilage.

1. Introduction

The bone marrow stroma includes progenitor cells of all tissues found in the skeleton (bone, cartilage, fibrous tissue, fat, and hematopoiesis-supporting stroma). These cells, originally called "bone marrow stromal stem cells" *(1)*, were later renamed "mesenchymal stem cells." Throughout this chapter, the term skeletal stem cells (SSCs) is used to mark the fact that their differentiation potential includes all different skeletal lineages, as proven by reproducible in vivo assays, whereas their putative ability to differentiate toward other "mesenchymal" lineages (such as skeletal or cardiac muscle), or even non-mesodermal lineages (such as neurons) in vivo and in the absence of specific inductive cues, has not been unequivocally proven *(2)*. At the same time, the use of the term "SSCs" restricts the discussion to cells isolated from the bone

From: *Methods in Molecular Medicine, 2nd ed.: Tissue Engineering*
Edited by: H. Hauser and M. Fussenegger © Humana Press Inc., Totowa, NJ

marrow stroma. Claims of the existence of cells with similar biological properties in other tissues (fat, cord blood, etc.) have not been validated by the use of the same in vivo assays based on which marrow-derived SSCs are identified and probed.

Stromal nature, clonogenicity, phenotype, multipotency, and self-renewal are the five defining characteristics of SSCs *(2)*. Their stromal nature, originally surmised by their ability to adhere to a plastic substrate and lack of phagocytic properties, has been confirmed by subsequent phenotypic analyses. SSCs are found in the meshwork of cells making the scaffold upon which hematopoietic cells proliferate and mature in the bone marrow, but they are non-hematopoietic in origin, nature, and phenotype *(3, 4)*.

Only a small fraction of marrow stromal cells are able to establish, when plated at clonal density (< 1.6 nucleated, non-hematopoietic cell/cm^2), a discrete single-cell-derived colony of cells. This reflects the ability of a fraction of stromal cells for density-insensitive growth, taken as evidence for an inherent ability for significant growth, although that is not always the case. It has been proven that the multipotent progenitors of skeletal tissues are specifically found among stromal clonogenic cells. The progeny of a clonogenic cell is fibroblastic in habit, hence the classical designation of the stromal clonogenic cell as the colony-forming unit fibroblastic (CFU-F) *(5)*. CFU-Fs are known to be functionally heterogeneous *(6)*. The progeny of a single CFU-F can be expanded in culture and transplanted in vivo. This has proven that a fraction (10–20%) of CFU-Fs can give rise, in vivo, to all skeletal tissues, whereas others give rise to a restricted range of tissues *(7)*. On this basis, it is postulated that only a fraction of CFU-Fs correspond to multipotent SSCs, whereas others correspond to differentiation-restricted, committed bone marrow stromal cells (BMSCs).

Whether clonogenic cells exhibit a specific phenotype that could be used to purify them from a marrow cell suspension has been the subject of extensive investigation and significant effort. Although many markers define the clonogenic fraction of stromal cells, and several can be used to significantly enrich clonogenic cells from a marrow cell suspension, all proposed markers define the whole of the CFU-F population and none, to date, distinguish the multipotent CFU-Fs from the others. Antibodies to at least four markers actually enrich CFU-Fs in a marrow cell suspension when used for magnetic-activated, or fluorescence-activated, cell sorting (STRO-1, CD49a, CD146, and CD63); whether they are all in fact co-expressed in a homogeneous population of clonogenic cells is less clear *(2)*. Extensive work has also been conducted to define the phenotype of SSCs (actually, of SSC-derived cell strains) in culture. Although significantly affected by the extensive modulation of most investigated markers in vitro, these have outlined a somewhat consistent phenotype of SSC-derived cultures: $CD34^-CD45^-CD14^-CD13^+CD29^+CD44^+CD49a^+CD63^+CD90^+CD105^+CD106^+CD146^+CD166^+$ [*(8–14)*, reviewed in *(2)*]. In the same cultures, additional markers that do not necessarily fall into the category of surface epitopes are expressed. These include, most notably, the pivotal transcription factor and master gene of osteogenesis, Runx2/Cbfa1, although the influence of culture conditions on its expression is not known ***(15)***. BMSC cultures also express PPAR-gamma ***(16)***, a major regulator of adipogenesis, but do not express significant levels of other lineage-dictating transcription factors, such as the chondrogenic inducer Sox-9. They also variably express connective tissue proteins such as osteopontin, bone sialoprotein, and osteocalcin ***(17)***.

The multipotency of SSCs can only be probed in the progeny of a single CFU-F. All claims of dealing with SSCs (or MSCs) based on the use of differentiation assays as applied to non-clonal cultures are null and void. Most commonly, multipotency is probed in vitro, by exposing cultures to specific differentiation-inducing cocktails [reviewed in ***(18)***]. Although empirically valuable, these assays are affected by the artificial nature of the culture conditions employed. More importantly, they may not necessarily predict the outcome of more stringent, in vivo transplantation assays, which must be taken into account whenever in vitro assays are used alone, or a novel test population is being probed. In vivo transplantation assays of clonal strains of SSCs directly prove their differentiation potential ***(7)***. Although in vivo assays can prove multipotency, self-renewal is more elusive. Strictly speaking, self-renewal is proven by serial transplantation assays as has been demonstrated for hematopoietic stem cells ***(19)***, but these assays are hard to conduct and impractical in most circumstances is solid tissues.

The potential use of SSCs for tissue engineering is predicted by their differentiation potential and significant ability for growth exhibited by their progeny. However, the use of SSCs for tissue engineering (and also for cell therapy and stem cell-based gene therapy) that is envisioned as a possibility close to reality is mostly related to bones (***18***, ***20***). When cell strains derived from explanted BMSCs are expanded in culture, combined with an osteoconductive carrier (i.e., a mineral-based phase apt to facilitate the unfolding of the inherent osteogenic property of the cells, without exerting any inductive influence on their differentiation) and transplanted in vivo at an ectopic site, they generate significant quantities of histology-proven new bone ***(21)***. Whereas this approach was originally evolved as an assay designed to probe the biological function of SSCs within the BMSC population, the earliest attempts toward proving the efficacy of SSCs for bone repair in animal models were but a direct extension of the same frame of thinking. In the published preclinical models of SSC-based bone tissue engineering, cell-biomaterial constructs similar in design to those employed for ectopic transplantation in the immunocompromised mice were used for orthotopic transplantation, for example, at the site of a segmental resection in dogs, or sheep, or in craniotomy defects created in mice or dogs (*see* **Fig. 1**) ***(22–25)***. It is important to realize that the application of this principle to multiple specific clinical scenarios will call for significant modulations of the original approach. For example, new and different carriers may be required for sustained bone formation in individual clinical applications (e.g., spinal fusion), consequently no single method for bone tissue engineering can be outlined or sought. Significant advances are expected from the development of newer materials, including biodegradable collagen-based materials, and "smart" materials able to deliver bioactive factors in a controlled way ***(26)***.

On the contrary, the use of SSCs for repairing other skeletal tissues such as articular cartilage still require substantial experimentation, not only in the area of materials, but also in the relevant cell biology. One specific problem arises, for example, from the difficulty in obtaining hyaline cartilage tissue that remains stable over time, that is, without entering the hypertrophy–mineralization cascade that characterizes growth plates [reviewed in ***(27)***].

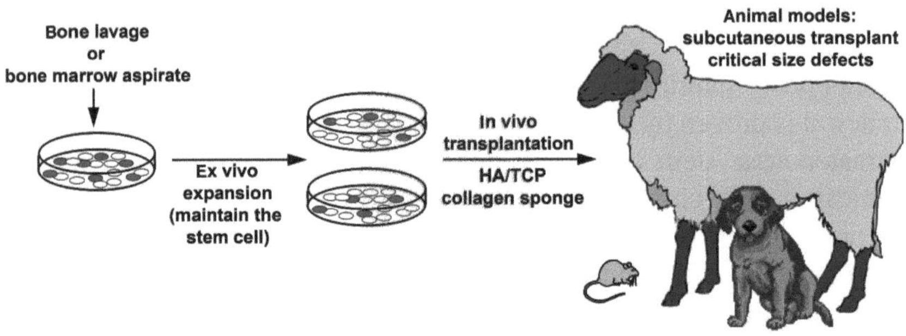

Fig. 1. Use of skeletal stem cells (SSCs) in bone regeneration. Populations of bone marrow stromal cells (BMSCs) that contain SSCs are initiated by single-cell suspensions of bone marrow obtained by washing surgical bone fragments and gently scraping with a scalpel or from a 2–5 cc bone marrow aspirate. The non-hematopoietic BMSCs rapidly adhere, thereby separating them from hematopoietic cells. These populations can be expanded through approximately 10 passages; however, it must be kept in mind that each passage dilutes the number of SSCs because of asymmetric division. These ex vivo populations can be used with hydroxyapatite/tricalcium phosphate (HA/TCP) particles (mouse and human BMSCs) or collagen sponges (mouse) to generate bone in various animal models, including the formation of subcutaneous ossicles, and in critical-size defects that would not heal under normal circumstances.

2. Materials

2.1. Murine Specimens

1. Long bones from any mouse strain cleaned of any connective tissue.
2. Institutionally approved protocol for the use of animals in research.

2.2. Human Specimens

1. Bone marrow aspirates (2–5 ml) (*see* **note 1**), combined with alpha-MEM containing heparin (100 U/ml; Sigma, St. Louis, MO), or bone specimens obtained as surgical waste, or from a core biopsy, stored in with alpha-MEM, containing 20% lot-selected fetal bovine serum (FBS) (*see* **note 2**), glutamine (2 mM), penicillin (100 U/ml), and streptomycin (100 μg/ml). Tissue culture reagents are not vendor specific.
2. Institutionally approved protocol or exemption (surgical waste, no patient identifiers) for the use of human subjects in research.

2.3. Murine and Human BMSC Cultures

1. Sterilized dissecting equipment (forceps, scissors, and scalpels).
2. Standard tissue culture supplies (pipettes, tubes, Petri dishes, and flasks).
3. Alpha-MEM (referred to as nutrient medium), and most times containing 20% lot-selected FBS (*see* **note 2**), glutamine (2 mM), penicillin (100 U/ml), and streptomycin (100 μg/ml) (referred to as serum-containing nutrient medium).
4. Cell strainers (Becton Dickinson, Franklin Lakes, NJ).
5. Hanks' balanced salt solution (HBSS).
6. Chondroitinase ABC (20 mU/ml) dissolved in HBSS, aliquoted, and stored at −20°C (Seikaguaku America, East Falmouth, MA) (*see* **note 3**).
7. Trypsin (0.05%)/ethylenediaminetetraacetic acid (EDTA; 0.53 mM) solution.
8. Hemocytometer.

2.4. Hydroxyapatite/Tricalcium Phosphate Transplants

1. Hydroxyapatite/tricalcium phosphate (HA/TCP), particle size 0.5–1 mm (Zimmer, Warsaw, IN) (*see* **note 4**), sterilized by placing the particles in a glass bottle, sealed with foil, and heated at 220°C for a minimum of 8 h. Particles are then aliquoted in a hood using sterile equipment at 40 mg/sterile 2 ml round-bottomed centrifuge tube.
2. Fibrinogen (Sigma), species-specific depending on the recipient (e.g., mouse fibrinogen if transplant is into a mouse) reconstituted in sterile phosphate-buffered saline (PBS) at 3.2 mg/ml, aliquoted, and stored at −80°C.
3. Thrombin (Sigma), species-specific depending on the recipient (e.g., mouse thrombin if transplant is into a mouse) reconstituted in sterile calcium chloride at 25 U/ml and stored at −80°C.

2.5. Collagen Sponge Transplants

1. Sterile Gelfoam™ (Upjohn, Kalamazoo, MI) cut into cubes or shapes of the desired size.

2.6. Immunocompromised Mouse Models

1. Female Bg Nu/Nu-Xid mice, 6 weeks to 6 months of age (Harlan, Indianapolis, IN).
2. 70% alcohol.
3. Betadine.
4. Sterile surgical equipment (forceps, scissors, scalpels, syringes, and drapes).
5. Anesthesia formulated by combining 225 μl ketamine with 69 μl acepromazine and 231 μl sterile water (total volume = 600 μl; Sigma). A 25-g mouse receives 100 μl, and less should be used for smaller mice.
6. A Dremell hand drill with sterile 5-mm burr (for calvarial defect model).
7. Autoclips or suture material (5-0 Vicryl, Johnson and Johnson, Cincinnati, OH).
8. Buprenorphine (0.05–0.2 mg/kg; Sigma) for post-surgical pain control.

2.7. Cartilage Pellet Culture

1. Coon's modified Ham's F12 medium supplemented with 10^{-6} M bovine insulin, 8×10^{-8} M human apo-transferrin, 8×10^{-8} M bovine serum albumin, 4×10^{-6} M linoleic acid (ITS^+, Gibco, Carlsbad, CA), 10^{-3} M sodium pyruvate, 10 ng/ml rhTGF-β (Austral Biologicals, San Ramon, CA), 10^{-7} M dexamethasone, and 2.5×10^{-4} M ascorbic acid.
2. 15-ml sterile polypropylene conical tubes.

3. Methods

Establishment of BMSC cultures (which contain SSCs), from single-cell suspensions of bone marrow, relies on the rapid adherence of BMSCs, whereas hematopoietic cells are non-adherent. As mentioned above, currently, there is no proven strategy to "purify" SSCs from more committed cell types in the BMSC population. However, although the adherent population is initially heterogeneous in nature, with passage, more differentiated cells are lost, and the population becomes less heterogeneous as determined by the list of markers described above. Although these BMSC populations are "unpurified," they are tried and true with respect to their ability to completely reform a bone/marrow organ upon in vivo transplantation (*see* **Figs 2–4**) *(22–25)*.

Murine BMSCs are commonly used in the development of different models for tissue engineering. However, they are not as easy to establish in culture as their human counterparts. First, cultures of rodent BMSCs are highly contaminated with macrophages, which adhere and very much resemble rodent BMSCs in their morphology, whereas macrophages do not adhere in cultures of human

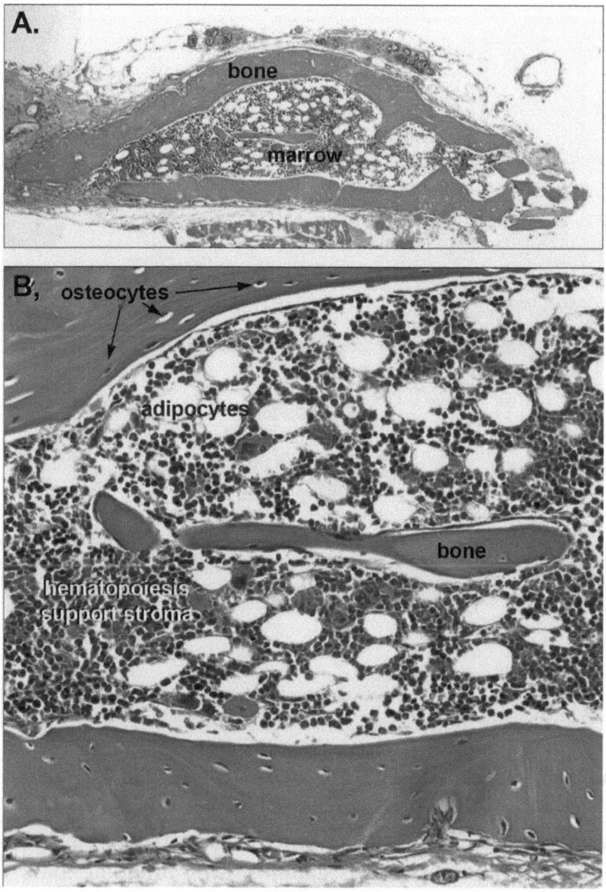

Fig. 2. Generation of a bone/marrow organ by murine bone marrow stromal cells (BMSCs) in collagen sponges. (a) Murine BMSCs are loaded onto collagen sponges as described under **sections 2 and 3** and placed into a subcutaneous pocket in immunocompromised mice. After 6 weeks, a thick cortical bone surrounds a fully formed hematopoietic marrow. (b) Closer examination reveals osteocytes of donor origin within the cortical bone, extensive hematopoiesis, and formation of marrow adipocytes.

marrow *(28)*. Without cell sorting, macrophages can only be eliminated by passage and are generally reduced to approximately 1% by the third passage. Although macrophage contamination does not influence the ability of murine BMSCs to form bone upon in vivo transplantation, it must be taken into consideration in other studies aimed to show that BMSCs can form cell types other than osteoblasts, chondrocytes, stroma, and adipocytes. Murine studies suggesting

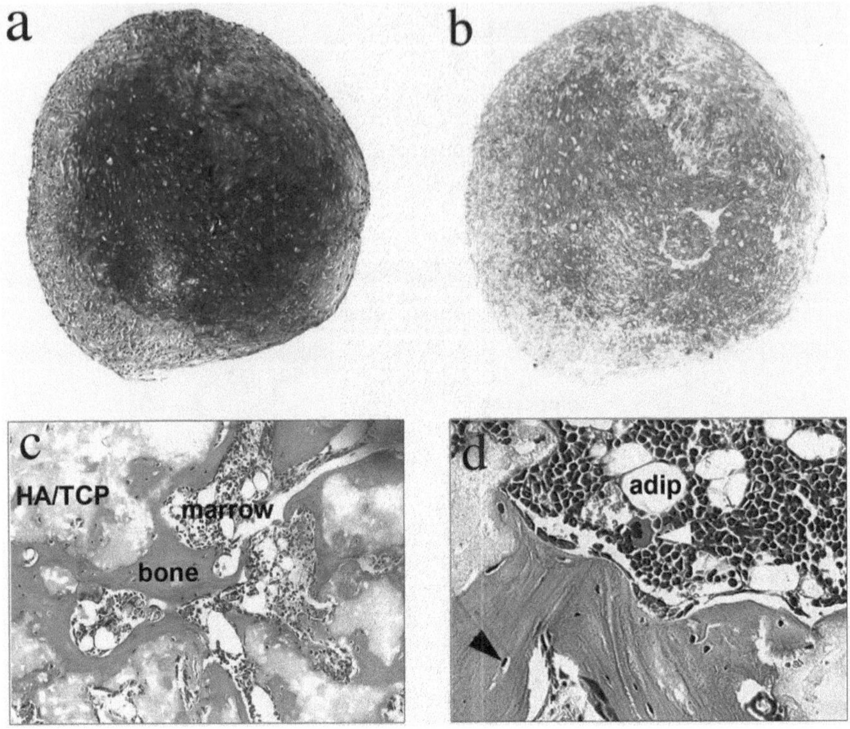

Fig. 3. Formation of cartilage and a bone/marrow organ by human bone marrow stromal cells (BMSCs). When placed into pellet-type cultures, which are relatively anerobic, as described under **sections 2 and 3**, early passage human BMSCs form cartilage that is metachromatic upon staining with toluidine blue (a) and stains positively for type II collagen by immunohistochemistry (b). When transplanted subcutaneously into immunocompromised mice along with hydroxyapatite/tricalcium phosphate (HA/TCP) particles, there is extensive bone formation and support of hematopoiesis (c). Osteocytes (black arrow head) of donor origin are found in bone that is primarily lamellar in nature, and the marrow contains adipocytes (adip), along with all subsets of hematopoietic cells, including megakaryocytes (white arrow head) (d).

that BMSCs infused into the circulation engraft into many tissues are most likely reflective of engraftment of contaminating macrophages, which is not unexpected. Second, the proliferation rate of murine BMSCs dramatically slows after the fourth passage. These cultures can stay quiescent for months but then occasionally begin to proliferate, that is, they spontaneously immortalize, as do many rodent cells ***(29)***. However, it is not a given that such immortalized cells are representative of primary murine MSCs, and all subsequent studies

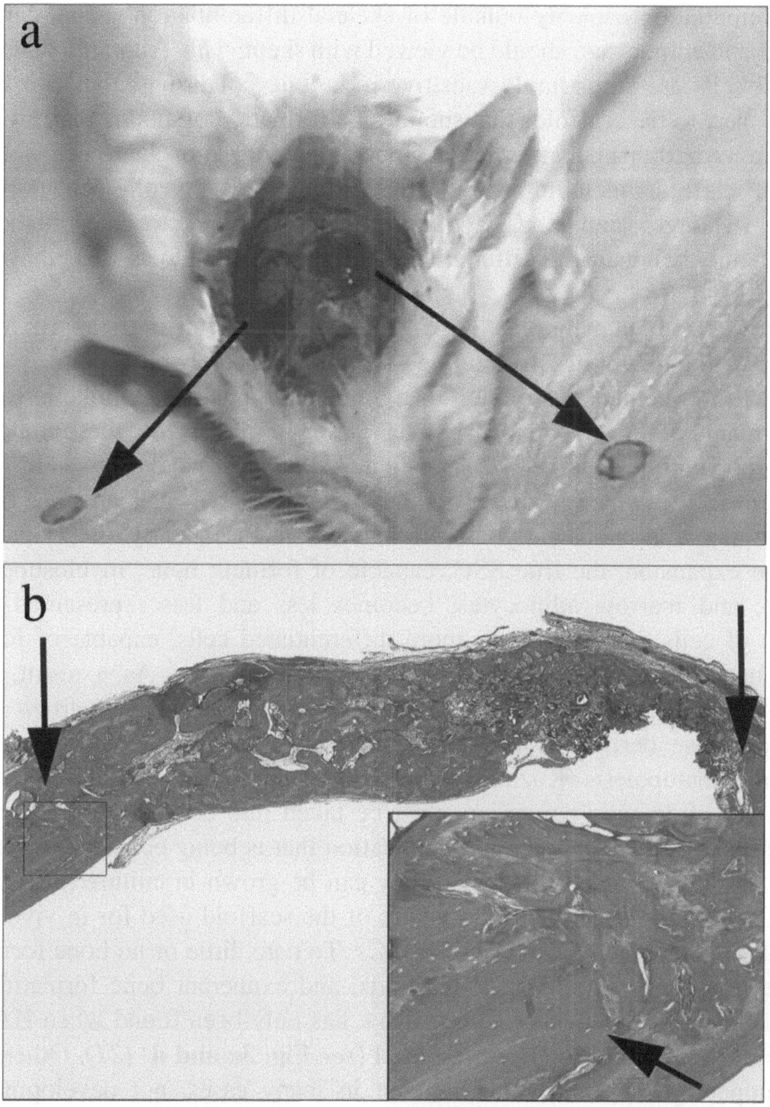

Fig. 4. Use of bone marrow stromal cells (BMSCs) for closure of a critical-size cranial defect in mice. In this instance, murine BMSCs contained within collagen sponges were placed into bilateral 5-mm cranial defects created as described under **sections 2 and 3** (a). The defect is completely regenerated with new bone of donor origin (b), which is well integrated at the margins of the defect (arrows and inset).

on differentiation capacity outside of skeletal differentiation (bone, cartilage, stroma, and adipocytes) should be viewed with skepticism. Although the murine stromal cells are surprisingly sensitive to culture conditions, they are less so with respect to the scaffolds that support the formation of a bone marrow organ upon in vivo transplantation ***(21)***. Good bone formation has been noted on various scaffolds, including poly-lactic acid and poly-glycolic acid polymers, collagen sponges, and HA/TCP (*see* **Fig. 2**). However, to date, methods of reproducibly being able to differentiate murine BMSCs into cartilage have been lacking.

BMSCs from humans can be ex vivo expanded up until approximately 25 population doublings (roughly 9–10 passages), at which point, proliferation ceases. Spontaneous immortalization is extremely rare, unlike in rodents. The amount of ex vivo expansion has a major impact on the composition of the population at large. It is postulated that post-natal stem cells divide asymmetrically, whereby one daughter remains a stem cell and the other a more committed, transiently amplifying cell. Consequently, with increased ex vivo expansion, the true SSC, capable of forming bone, myelosupportive stroma, and marrow adipocytes, becomes less and less represented in an aliquot of cells compared with more differentiated cells, capable of forming bone, but not capable of supporting marrow formation. As a result, transplants derived from early passage cells form a complete bone/marrow organ, whereas those derived from late passage form only bone and very rarely support hematopoiesis (Kuznetsov and Robey, unpublished results). Therefore, the extent of ex vivo expansion must be taken into consideration depending on the type of tissue engineering application that is being considered. Interestingly, despite the ease with which they can be grown in culture ***(28)***, human BMSCs are very sensitive to the nature of the scaffold used for in vivo transplantation compared with murine BMSCs. To date, little or no bone formation has been seen in non-HA/TCP scaffolds, and exuberant bone formation and establishment of a hematopoietic marrow has only been found when HA/TCP of 65%/35% composition has been used (*see* **Fig. 3c and d**) ***(21)***. Other ratios can support good bone formation, but in many cases, not development of hematopoiesis. Using a method similar to that developed for rabbit ***(30)***, human BMSCs, at least in early passages, are also able to make cartilage in high-density cultures that are relatively anerobic compared with standard monolayer cultures (*see* **Fig. 3a and b**) ***(31)***. However, they may lose this capacity with prolonged expansion. Furthermore, micromass cultures go on to hypertrophy, and the factors that prevent this stage of maturation are not well understood to date ***(20)***.

3.1. Establishment of Murine BMSC Cultures

1. Mice are euthanized appropriately, and the femora, tibiae, and humeri are dissected under aseptic conditions and thoroughly cleaned of all soft connective tissue.
2. After removing the epiphyses, the marrow is flushed from the medullary cavity with nutrient medium, and the samples from all bones are combined.
3. The cell suspension is prepared by pipetting up and down several times to break up aggregates, but unlike human cells, the suspension is not passaged through needles of decreasing gauge.
4. Nucleated cell number is determined with a hemocytometer.
5. Approximately 6–8 × 10^7 nucleated cells are plated into 75-cm^2 flask.
6. Cells are cultured in a humidified atmosphere of 5% CO_2 with air at 37°C.
7. Medium is replaced at day 7, and passage 1 is on days 12–14, when the cultures are approximately 70% confluent.
8. To passage, cultures are washed twice with HBSS and the incubated with chondroitinase ABC for 25–35 min at room temperature (*see* **note 3**).
9. The cells are then treated with trypsin/EDTA for 25–30 min at room temperature, cells are removed, and the remaining attached cells are removed by another trypsin/EDTA treatment for 25–30 min at 37°C, followed by a wash with nutrient medium.
10. As fractions are collected, FBS is added to a final concentration of 1% to inhibit trypsin activity. All fractions are combined.
11. The suspension is vigorously pipetted to break up cell aggregates and centrifuged at 135 × *g* for 10 min.
12. The resulting pellet is resuspended in fresh medium.
13. After counting with a hemocytometer, 2–10 × 10^6 cells are plated into 75-cm^2 flask, depending on the level of macrophage contamination.
14. Medium is replaced at 2- to 3-day intervals, and cells are passaged again when the cultures reach approximately 70% confluency.

3.2. Establishment of Human BMSC Cultures

1. Fragments of human trabecular bone are scraped with a scalpel and washed with nutrient medium until the bone is marrow free (as can be visually ascertained by the presence of blood).
2. Or bone marrow aspirates are obtained and immediately mixed in 5 ml nutrient medium with heparin.
3. A single-cell suspension from fragments and from aspirates is prepared by pipetting up and down several times, passaged through 16 20-G needles, and then through a cell strainer, to remove aggregates.
4. Nucleated cell number is determined with a hemocytometer.
5. Suspensions from bone fragments are plated at 5 × 10^6 to 5 × 10^7 nucleated cells per 75-cm^2 flask with 30 ml serum-containing nutrient medium.

6. Or, because of the presence of contaminating peripheral blood in bone marrow aspirates (*see* **section 4**), between 5×10^6 (for small samples) and 20×10^7 (for large samples) nucleated cells are plated in a 75-cm^2 flask along with 30 ml serum-containing nutrient medium.
7. Cultures are fed on days 3 and 7 with all non-attached cells removed at the first medium replacement and passaged between days 12 and 14 when the cells are approximately 70% confluent.
8. Cultures are washed with HBSS, followed by two consecutive treatments with trypsin/EDTA for 10–15 min each at room temperature, followed by a wash with nutrient medium.
9. As fractions are collected, FBS is added to a final concentration of 1% to inhibit enzyme activity. All fractions are combined.
10. The combined suspensions are vigorously pipetted to break up aggregates and centrifuged at $135 \times g$ for 10 min, and the cell pellet is resuspended in fresh medium.
11. Cells are plated at 2×10^6 cells per 75-cm^2 flask with 30 ml serum-containing nutrient medium.
12. Medium is replaced at 2- to 3-day intervals, and cells are again passaged when the cultures reach approximately 70% confluency.

3.3. Preparation HA/TCP Transplants

1. Passaged BMSCs (murine and human) are resuspended to a concentration of $1–2 \times 10^6$ cells/ml in serum-containing nutrient medium.
2. The HA/TCP particles are washed twice with serum-containing nutrient medium.
3. 1 ml cell suspension is added to each HA/TCP-containing tube and gently rotated (25 rpm) at 37°C for 70–100 min.
4. The particles with attached cells are centrifuged at $135 \times g$ for 1 min, and the supernatant is gently aspirated.
5. The particles are held together by gently mixing with mouse fibrinogen (15 μl) and mouse thrombin (15 μl). The tube is tightly sealed to prevent desiccation.
6. After incubating at room temperature for several minutes, a clot is formed that can be easily removed from the tube with a sterile spatula.

3.4. Preparation of Collagen Sponge Transplants

1. Sterile collagen sponges are cut into the desired size and shape with sterile scissors. For $1–2 \times 10^6$ cells, cubes of approximately 5 mm × 5 mm × 5 mm.
2. The sponge is hydrated in serum-containing nutrient medium and allowed to expand to their full size. Air bubbles are removed by squeezing the sponge with sterile forceps and allowing them to re-expand.

3. Murine BMSCs prepared as described above are resuspended at a concentration of 1–2 × 10^6 cells/ml serum-containing nutrient medium, and 1 ml is aliquoted into tubes and centrifuged at 135 × *g* for 10 min. The supernatant is aspirated until approximately 50 μl remains, and the cells are resuspended in that volume.
4. The sponges are blotted between two pieces of sterile filter paper to remove the medium and then placed immediately into the tubes with the resuspended cells. As the sponge expands, it draws in the cell suspension. The tube is tightly sealed to prevent desiccation.

3.5. In Vivo Subcutaneous Transplantation of BMSC Constructs

1. Mice are anesthetized and placed on sterile surgical drapes on a heating pad, and the skin is cleaned first with betadine and then with 70% alcohol. Immunocompromised mice have little or no fur, and consequently, shaving is not necessary. For syngeneic transplants in fur-bearing mice, shaving is performed.
2. With the mouse lying on its stomach, a single longitudinal incision (3 mm) is made along the dorsal surface.
3. Using sterile scissors, a pocket for the transplants is made by inserting the tips of the scissors subcutaneously and then opening the scissors by 1 cm.
4. HA/TCP transplants are inserted by using a sterile spatula, and collagen sponge transplants are put into place with forceps. Up to six transplants can be placed (three on each side of the incision).
5. The incision is closed with several autoclips or with suturing material. Although AAALAC regulations require removal of autoclips within 10 days, immunocompromised mice are exempt from this due to the fact that their removal causes excessive bleeding.
6. The animals are treated with buprenorphine for pain relief.
7. After full recovery, animals are returned to their cages and euthanized at various times for histological analyses.

3.6. BMSC Transplants in Critical-Size Calvarial Defects

1. Mice are anesthetized and prepped as described above and placed on their stomachs.
2. A 1-cm midline incision is made through the skin over the cranial vault. The skin and the periosteum are then separated from the skull by blunt dissection and retracted.
3. A full-thickness defect is created using a Dremell hand piece fitted with a 5-mm trephine burr. Care is taken not to damage the dura mater covering of the brain.
4. Either HA/TCP constructs or collagen sponge constructs cut to fit the size of the defects are placed in the defect.
5. After repositioning the skin, the incision is closed with suture material.
6. The animals are treated for pain with buprenorphine and, after full recovery, are returned to their cages.
7. Animals are euthanized at various time intervals for analyses (*see* **Fig. 4**).

3.7. Pellet Cultures

1. Human BMSCs are resuspended in chondrogenic medium following passage at a concentration of 5×10^4 cells per milliliter.
2. 5 ml cell suspension is placed in 15-ml polypropylene conical tubes and centrifuged at $500 \times g$ for 5 min.
3. The tubes are incubated with the lids partially unsealed for 3 weeks at 37°C in 5% CO_2, 95% air.
4. Medium is changed at 2- to 3-day intervals, and pellets should remain free-floating within the tube.
5. Pellets are harvested at 3 weeks for histological analyses.

Notes

1. It is necessary to keep the volume of a bone aspirate low (should not exceed 5 ml). Drawing larger amounts, even after repositioning the needle, results in substantial contamination of the marrow with peripheral blood. Peripheral blood in excess has a negative influence on the growth of human BMSCs.
2. Serum requirements vary from one animal species to another. Serum must be lot-selected for each animal species used in order to guarantee optimal growth *(28)*.
3. In primary mouse BMSC cultures, the adherent cells often secrete copious amounts of a gel-like substance (undoubtedly composed of hyaluronic acid and other matrix components). Large amounts of this material interfere with the enzymatic removal of the cells from the surface by standard trypsin treatment alone, and chondroitinase ABC pre-treatment improves release of the cells. However, the amount of this material appears to vary from one mouse strain to another; consequently, chondroitinase ABC digestion may not always be necessary.
4. There are many scaffolds that are commercially available; however, all are not equivalent. The ratio of HA to TCP determines the amount of bone formation (the higher the HA, the more vigorous the bone formation, the higher the TCP, the faster the removal of the scaffold). The optimal composition has yet to be determined. The composition of the HA/TCP used in the studies presented here is 65% HA and 35% TCP.

Acknowledgments

Work covered in this chapter was supported by grants from Telethon Fondazione Onlus (grant GGP04263), MIUR, and the EU (GENOSTEM) to MR and PB, and by the Division of Intramural Research, NIDCR, Intramural Research Program, NIH, DHHS (PGR and SAK). The authors thank Dr. Paul H. Krebsbach, University of Michigan, for providing photographs of calvarial defects in mice.

References

1. Owen, M. and Friedenstein, A. J. (1988) Stromal stem cells: marrow-derived osteogenic precursors. *Ciba Found. Symp.* **136**, 42–60.
2. Bianco, P. and Robey, P. G. (2004) Skeletal stem cells, in *Handbook of Adult and Fetal Stem Cells* (Lanza, R. P., ed.), Academic Press, San Diego, CA, pp. 415–424.
3. Friedenstein, A. J., Chailakhyan, R. K., Latsinik, N. V., Panasyuk, A. F., and Keiliss-Borok, I. V. (1974) Stromal cells responsible for transferring the microenvironment of the hemopoietic tissues. Cloning in vitro and retransplantation in vivo. *Transplantation* **17**, 331–340.
4. Friedenstein, A. J., Deriglasova, U. F., Kulagina, N. N., Panasuk, A. F., Rudakowa, S. F., Luria, E. A., and Ruadkow, I. A. (1974) Precursors for fibroblasts in different populations of hematopoietic cells as detected by the in vitro colony assay method. *Exp. Hematol.* **2**, 83–92.
5. Friedenstein, A. J. (1976) Precursor cells of mechanocytes. *Int. Rev. Cytol.* **47**, 327–359.
6. Bianco, P., Riminucci, M., Kuznetsov, S., and Robey, P. G. (1999) Multipotential cells in the bone marrow stroma: regulation in the context of organ physiology. *Crit. Rev. Eukaryot. Gene Expr.* **9**, 159–173.
7. Kuznetsov, S. A., Krebsbach, P. H., Satomura, K., Kerr, J., Riminucci, M., Benayahu, D., and Robey, P. G. (1997) Single-colony derived strains of human marrow stromal fibroblasts form bone after transplantation in vivo. *J. Bone Miner. Res.* **12**, 1335–1347.
8. Barry, F. P., Boynton, R. E., Haynesworth, S., Murphy, J. M., and Zaia, J. (1999) The monoclonal antibody SH-2, raised against human mesenchymal stem cells, recognizes an epitope on endoglin (CD105). *Biochem. Biophys. Res. Commun.* **265**, 134–139.
9. Deschaseaux, F. and Charbord, P. (2000) Human marrow stromal precursors are alpha 1 integrin subunit-positive. *J. Cell Physiol.* **184**, 319–325.
10. Filshie, R. J., Zannettino, A. C., Makrynikola, V., Gronthos, S., Henniker, A. J., Bendall, L. J., Gottlieb, D. J., Simmons, P. J., and Bradstock, K. F. (1998) MUC18, a member of the immunoglobulin superfamily, is expressed on bone marrow fibroblasts and a subset of hematological malignancies. *Leukemia* **12**, 414–421.
11. Gronthos, S., Zannettino, A. C., Graves, S. E., Ohta, S., Hay, S. J., and Simmons, P. J. (1999) Differential cell surface expression of the STRO-1 and alkaline phosphatase antigens on discrete developmental stages in primary cultures of human bone cells. *J. Bone Miner. Res.* **14**, 47–56.
12. Stewart, K., Monk, P., Walsh, S., Jefferiss, C. M., Letchford, J., and Beresford, J. N. (2003) STRO-1, HOP-26 (CD63), CD49a and SB-10 (CD166) as markers of primitive human marrow stromal cells and their more differentiated progeny: a comparative investigation in vitro. *Cell Tissue Res.* **313**, 281–290.

13. Vogel, W., Grunebach, F., Messam, C. A., Kanz, L., Brugger, W., and Buhring, H. J. (2003) Heterogeneity among human bone marrow-derived mesenchymal stem cells and neural progenitor cells. *Haematologica* **88**, 126–133.
14. Zannettino, A. C., Harrison, K., Joyner, C. J., Triffitt, J. T., and Simmons, P. J. (2003) Molecular cloning of the cell surface antigen identified by the osteoprogenitor-specific monoclonal antibody, HOP-26. *J. Cell Biochem.* **89**, 56–66.
15. Satomura, K., Krebsbach, P., Bianco, P., and Robey, P. G. (2000) Osteogenic imprinting upstream of marrow stromal cell differentiation. *J. Cell Biochem.* **78**, 391–403.
16. Gimble, J. M., Robinson, C. E., Wu, X., and Kelly, K. A. (1996) The function of adipocytes in the bone marrow stroma: an update. *Bone* **19**, 421–428.
17. Kuznetsov, S. A., Mankani, M. H., Gronthos, S., Satomura, K., Bianco, P., and Robey, P. G. (2001) Circulating skeletal stem cells. *J. Cell Biol.* **153**, 1133–1140.
18. Bianco, P., Kuznetsov, S. A., Riminucci, M., and Robey, P. G. (in press) Post-natal skeletal stem cells, in *Methods in Enzymology: Stem Cells*, vol. 2 (Lanza, R. P., ed.), Elsevier/Academic Press, San Diego, CA.
19. Harrison, D. E. and Astle, C. M. (1982) Loss of stem cell repopulating ability upon transplantation. Effects of donor age, cell number, and transplantation procedure. *J. Exp. Med.* **156**, 1767–1779.
20. Robey, P. G. and Bianco, P. (2004) Stem cells in tissue engineering, in *Handbook of Adult and Fetal Stem Cells* (Lanza, R. P., ed.), Academic Press, San Diego, CA, pp. 785–792.
21. Krebsbach, P. H., Kuznetsov, S. A., Satomura, K., Emmons, R. V., Rowe, D. W., and Robey, P. G. (1997) Bone formation in vivo: comparison of osteogenesis by transplanted mouse and human marrow stromal fibroblasts. *Transplantation* **63**, 1059–1069.
22. Krebsbach, P. H., Mankani, M. H., Satomura, K., Kuznetsov, S. A., and Robey, P. G. (1998) Repair of craniotomy defects using bone marrow stromal cells. *Transplantation* **66**, 1272–1278.
23. Kon, E., Muraglia, A., Corsi, A., Bianco, P., Marcacci, M., Martin, I., Boyde, A., Ruspantini, I., Chistolini, P., Rocca, M., Giardino, R., Cancedda, R., and Quarto, R. (2000) Autologous bone marrow stromal cells loaded onto porous hydroxyapatite ceramic accelerate bone repair in critical-size defects of sheep long bones. *J. Biomed. Mater. Res.* **49**, 328–337.
24. Marcacci, M., Kon, E., Zaffagnini, S., Giardino, R., Rocca, M., Corsi, A., Benvenuti, A., Bianco, P., Quarto, R., Martin, I., Muraglia, A., and Cancedda, R. (1999) Reconstruction of extensive long-bone defects in sheep using porous hydroxyapatite sponges. *Calcif. Tissue Int.* **64**, 83–90.
25. Mankani, M. H., Kuznetsov, S. A., Shannon, B., Nalla, R. K., Ritchie, R. O., Qin, Y., and Robey, P. G. (2006) Canine cranial reconstruction using autologous bone marrow stromal cells. *Am. J. Pathol.* **168**, 542–550.

26. Sittinger, M., Hutmacher, D. W., and Risbud, M. V. (2004) Current strategies for cell delivery in cartilage and bone regeneration. *Curr. Opin. Biotechnol.* **15**, 411–418.
27. Raghunath, J., Salacinski, H. J., Sales, K. M., Butler, P. E., and Seifalian, A. M. (2005) Advancing cartilage tissue engineering: the application of stem cell technology. *Curr. Opin. Biotechnol.* **16**, 503–509.
28. Kuznetsov, S. and Robey, P. G. (1996) Species differences in growth requirements for bone marrow stromal fibroblast colony formation in vitro. *Calcif. Tissue Int.* **59**, 265–270.
29. Kraemer, P. M., Ray, F. A., Brothman, A. R., Bartholdi, M. F., and Cram, L. S. (1986) Spontaneous immortalization rate of cultured Chinese hamster cells. *J. Natl. Cancer Inst.* **76**, 703–709.
30. Johnstone, B., Hering, T. M., Caplan, A. I., Goldberg, V. M., and Yoo, J. U. (1998) In vitro chondrogenesis of bone marrow-derived mesenchymal progenitor cells. *Exp. Cell Res.* **238**, 265–272.
31. Muraglia, A., Corsi, A., Riminucci, M., Mastrogiacomo, M., Cancedda, R., Bianco, P., and Quarto, R. (2003) Formation of a chondro-osseous rudiment in micromass cultures of human bone-marrow stromal cells. *J. Cell Sci.* **116** (Pt. 14), 2949–2955.

6

Biomaterials/Scaffolds

Design of Bioactive, Multiphasic PCL/Collagen Type I and Type II–PCL-TCP/Collagen Composite Scaffolds for Functional Tissue Engineering of Osteochondral Repair Tissue by Using Electrospinning and FDM Techniques

Detlef Schumann, Andrew K. Ekaputra, Christopher X. F. Lam, and Dietmar W. Hutmacher

Summary

Current clinical therapies for traumatic or chronic injuries involving osteochondral tissue result in temporary pain reduction and filling of the defect but with biomechanically inferior repair tissue. Tissue engineering of osteochondral repair tissue using autologous cells and bioactive biomaterials has the potential to overcome the current limitations and results in native-like repair tissue with good integration capabilities. For this reason, we applied two modern biomaterial design techniques, namely, electrospinning and fused deposition modeling (FDM), to produce bioactive poly(ϵ-caprolactone)/collagen (PCL/Col) type I and type II–PCL-tri-calcium phosphate (TCP)/Col composites for precursor cell-based osteochondral repair. The application of these two design techniques (electrospinning and FDM) allowed us to specifically produce the a suitable three-dimensional (3D) environment for the cells to grow into a particular tissue (cartilage and bone) in vitro prior to in vivo implantation. We hypothesize that our new designed biomaterials, seeded with autologous bone marrow-derived precursor cells, in combination with bioreactor-stimulated cell-culture techniques can be used to produce clinically relevant osteochondral repair tissue.

Key Words: Electrospinning; FDM; Biomaterials; Composite scaffolds; PCL-TCP; Hyaluronic acid; Collagen type I and type II; Osteochondral repair; Functional tissue engineering; Bioactive; Precursor cells.

From: *Methods in Molecular Medicine, 2nd ed.: Tissue Engineering*
Edited by: H. Hauser and M. Fussenegger © Humana Press Inc., Totowa, NJ

1. Introduction

Osteochondral defects (i.e., those that affect both the articular cartilage and the underlying subchondral bone) represent a major challenge in orthopedic surgery. These defects, typically derived from traumatic injuries, osteochondritis dissecans, or chondromalacia, are often associated with the mechanical instability of the joint and therefore carry the risk of inducing osteoarthritic degenerative changes. Currently applied clinical methods to repair osteochondral defects such as shaving *(1)*, arthroscopic abrasion arthroplasty *(2)*, mosaicplastic (transplantation of healthy osteochondral plugs consisting of an articular layer and the underlying subchondral bone from a non-loaded area to the defect) *(3)*, or autologous chondrocyte transplantation *(4)* result in a filling of the defect, but with biomechanically inferior tissue. These techniques are also limited by the amount of material available, the donor site morbidity, the difficulty in matching the topology of the graft with the injured site, and its tissue integration. Tissue engineering of autologous osteochondral composites has the potential to overcome these limitations. Compared with the transplantation of cells alone, in vitro-grown tissue constructs offer the advantage of immediate functionality along with the capacity for integration with host tissues.

A biomimetic approach to tissue engineering of more complex structures such as osteochondral grafts with gradients of structural and functional properties is based on the biophysical regulation of precursor cells by exogenous signals generated by bioactive scaffolds and a bioreactor (*see* **Fig. 1**).

For this functional tissue engineering of osteochondral repair tissue, two-compartment (biphasic) composite biomaterials are needed to provide the precursor cells with the respectively ideal environment for the culture of a specific tissue. The cartilage compartment should consist of chondroconductive

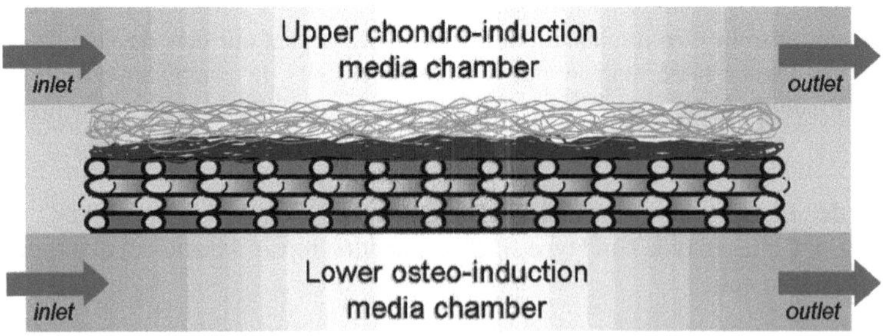

Fig. 1. Sketch of parallel-plate bioreactor with two independent flow inlets/outlets for the application of two different tissue differentiation induction media through different compartments of the used cell–scaffold construct.

materials such as PLGA, PGA, hyaluronan, or collagen (Col), whereas the bone compartment should consist of osteoconductive composite materials such as poly(ϵ-caprolactone)-tri-calcium phosphate (PCL-TCP)/Col or PLLA-HA copolymers.

A combination of two scaffold fabrication techniques, namely, electrospinning and fused deposition modeling (FDM), ushers a novel solution to the design of complex biomaterials for the challenging task of tissue-engineering osteochondral repair tissues.

Both techniques allow the design and fabrication of scaffolds for tissue-engineering applications ***(5–14)***.

It is well known that cells interact strongly with their environment (extracellular matrix). This structure, which is primarily composed of collagens, exhibits varying fiber diameters that is quite frequently one or more orders of magnitude smaller than the cell itself. Electrospinning allows the production of these small fiber diameters in sub-micron dimensions, down to about 0.05 microns (50 nm). In our special application, we take advantage of this biomaterial design technique to effectively produce a three-dimensional (3D) chondroconductive environment to enhance the chondrogenic differentiation capacity of bone marrow-derived precursor cells. We electrospun two major components of the extracellular matrix, namely, collagen type I and type II, together with the stabilizing polymer PCL in a one-step solution to manufacture the chondroconductive part of our biphasic scaffold for osteochondral tissue engineering on top of a rapid-prototyped bone-conductive PCL-TCP/Col compartment (*see* **Fig. 2**). To produce the "bone-induction-part," we used a previously published rapid-prototyping technique, namely, FDM. This method allowed the manufacture of highly reproducible scaffolds within precisely defined parameters ***(9, 12, 15–18)***. Subsequent collagen Col incorporation (by freeze drying) enhances the attachment and proliferation capacity of precursor cells.

Previous in vitro and in vivo studies undertaken by our group have shown the ability to use rapid-prototyped (using an FDM technique) PCL-TCP scaffolds

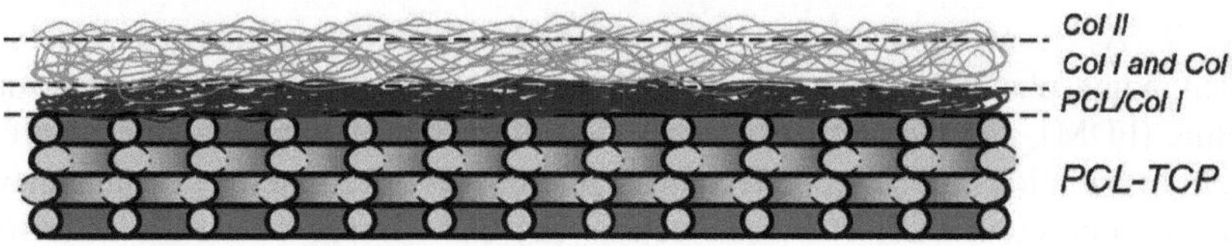

Fig. 2. Schematic illustration of multiphasic scaffold design for osteochondral tissue engineering. The cartilage compartment is built by using poly(ϵ-caprolactone)/collagen (PCL/Col) type I and type II and the bone part by PCL-tri-calcium phosphate (TCP)/Col.

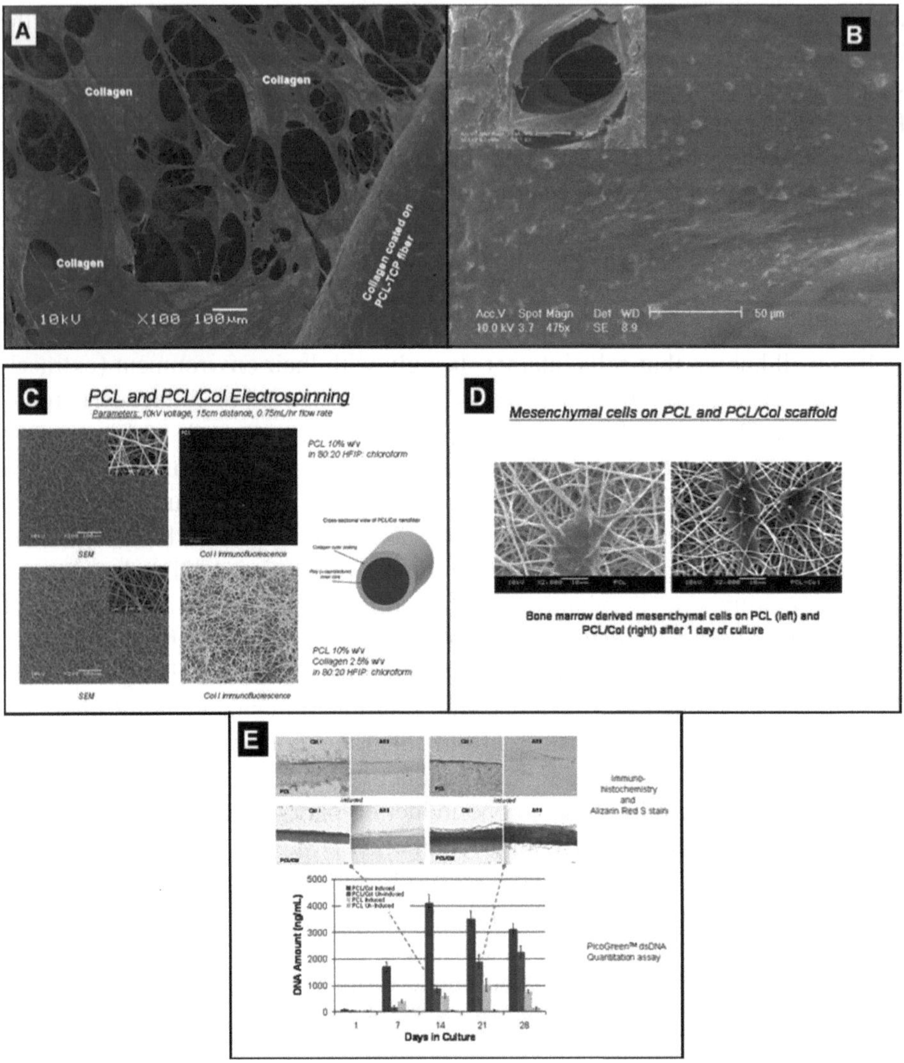

Fig. 3. (A) Scanning electron microscope (SEM) image of a poly(ϵ-caprolactone)-tri-calcium phosphate/collagen (PCL-TCP/Col) scaffold [fabricated by fused deposition modeling (FDM), and then collagen mesh was fabricated by lyophilization]. Uniform collagen distribution can be seen. Secondary micropores are formed by collagen layers within macropores of PCL-TCP architecture. (B) SEM image of a PCL-TCP/Bones marrow steomal cells (BMSC) construct cultured for 4 weeks under osteogenic conditions. Mineralized Extracellular Matrix (ECM) can be detected throughout the scaffold morphology and on surface of the scaffold bars and struts. (C) SEM micrographs

and also electrospun collagen for precursor cell-based tissue-engineering applications *(10)*. In **Fig. 3**, panel A shows a scanning electron microscopic image of a PCL-TCP/Col scaffold (fabricated by FDM and subsequent collagen incorporation). Thorough collagen distribution can be seen with secondary pore formation by incorporated collagen. Panel B shows a scanning electron microscopic image of a PCL-TCP scaffold seeded with bone marrow-derived precursor cells and cultured in vitro for 4 weeks under static osteogenic conditions. A thorough filling of the pores with mineralized extracellular matrix was found. Panel C shows the homogeneous distribution of coaxial fibers after electrospinning, panel D shows the mesenchymal precursor cell attachment behavior on the biomaterial, and panel E presents the tissue differentiation capacity of the electrospun matrix during osteogenic in vitro culture.

By combining the two scaffold fabrication techniques, such as electrospinning and FDM, we can take advantage of each method to manufacture a biphasic, multicomponent construct for tissue engineering of complex grafts such as osteochondral tissue by specifically designing the respective 3D environments for the cells to grow into the particular tissue.

2. Materials

2.1. Polymer Solution for Electrospinning

1. Electrospinning solvents: chloroform (Merck, Darmstadt, Germany) and 1,1,1,3,3,3-hexafluoro-2-propanol (HFIP; Fluka, Buchs, Switzerland).
2. PCL, MW: 80,000) from Absorbable Polymer International (Pelham, AL).
3. Poly(L-lactide-co-D,L-lactide; Boehringer Ingelheim GmbH, Germany).
4. Col type I particulates (purified, acid soluble) from skin (Nippon Meat Packers, Japan).
5. Col type II (purified, acid soluble) from Symatese biomateriaux, France.
6. Glass scintillation vials 20 ml and stirring bars 15 mm.

Fig. 3. of PCL/Col blend and PCL nanofibers showing gross surface topography at lower magnification (×200) and a close-up look of the nanofiber morphology at higher magnification (inset, ×5000). Also shown, confocal immunofluorescence against porcine collagen I of PCL/Col blend fibers where collagen is present on the surface of the PCL/Col fibers and a three-dimensional (3D) sketch of the hypothesized PCL/Col nanofiber structure. (D) SEM image showing bone marrow-derived mesenchymal precursor cell attachment on electrospinned PCL and PCL/Col scaffolds. (E) Histological micrographs (immunohistochemical collagen type I staining and alizarin red S staining) of in vitro osteogenic precursor cell tissue differentiation on electrospun PCL and PCL/Col scaffolds in relation to its proliferation in the course of culture time.

2.2. Electrospinning Apparatus

1. High-voltage power supply capable of generating voltage bias up to 30 kV (positive and negative) from Gamma High Voltage Research (Ormond Beach, FL).
2. Syringe pump from kd Scientific (Holliston, MA).
3. Teflon tubing 1/6 inch outer diameter from Spektra Teknik Pte., Singapore.
4. Syringes 5 ml and 21-G needles from BD (Franklin Lakes, NJ).
5. Laboratory lifts (15 cm × 15 cm), used as collection screen, from Troemner (Thorofare, NJ).
6. Aluminum foils.

2.3. Detection of Nanofibers by Scanning Electron Microscope

1. Samples of nanofiber sheets cut into about 1 cm × 1 cm dimensions.
2. Gold sputter coater (BAL-TEC AG, Liechtenstein).
3. Mounting stubs for sputtering and scanning electron microscope (SEM) viewing and double-sided adhesive tapes.
4. JEOL 5600 SEM system (JEOL, Tokyo, Japan) for viewing sputter-coated nanofiber samples.

2.4. Analysis of Col on the Electrospun Nanofiber via Immunofluorescence

1. Samples of nanofiber sheets cut into about 1 cm × 1 cm dimensions.
2. Mouse monoclonal primary antibody against porcine Col type I unconjugated from Sigma (St. Louis, MO).
3. Rabbit polyclonal secondary antibody against mouse immunoglobulins conjugated to fluorescein isothiocyanate (FITC) from DAKO Cytomation (Glostrup, Denmark).
4. Phosphate-buffered saline (PBS) and antibody diluent (DAKO Cytomation).
5. Laser scanning confocal microscope (LSCM) FV500 system by Olympus (Olympus Corp., Tokyo, Japan).

2.5. Fused Deposition Modeling

1. Medical grade PCL (mPCL) pellets (Aldrich Chemical Company) were used as received. The number average molecular weight (M_n) provided by the manufacturer was approximately 80,000, and density was 1.145 g/cm^3.
2. β-TCP in the form of micro-powder was obtained from Shanghai Rebone Biomaterials.
3. A Coulter Laser Diffraction LS100Q (working range: 0.4–1000 μm) particle size analyzer was used to determine the particle size distribution of the TCP powder. About 90% of the particles were lesser than 63 μm and 50% lesser than 16 μm.
4. FDM 3000 from Stratasys *(20)* (*see* **Fig. 6**).

2.6. Cell Culture and Differentiation Analysis

1. Ficoll gradient density separation (density 1.077 g/ml; Histopaque, Sigma Aldrich).
2. Proliferation medium: low-glucose Dulbecco's modified Eagle's medium (DMEM) (Gibco) supplemented with 10% fetal calf serum (PAN), 1% HEPES buffer (Gibco), and 1% Pen/Strep (Sigma Aldrich).
3. Chondrogenic differentiation medium: high-glucose DMEM (Gibco), +1% sodium pyruvate, 1% ITS+3 Liquid Media Supplements (Sigma Aldrich), +1% dexamethasone (stock: 0.00392 g dexamethasone in 10 ml 100% ethanol; usage: 200 μl stock solution in 19.8 ml low-glucose DMEM), 1% recombinant transforming growth factor ß1 (TGFß1; final concentration 10 ng/ml), 1% ascorbic acid (0.0794 g ascorbic acid in 20 ml Tyrodes solution).
4. Tyrodes solution: (1 l) 9.73 g Tyrodes salt, 0.2 g $NaHCO_3$, 10 ml Pen/Strep, 990 ml H_2O.
5. Osteogenic differentiation medium: high-glucose DMEM (Gibco), 50 μM ascorbate-2-phosphate, 10 mM β-glycerophosphate, 0.01 μM 1,25-dihydroxyvitamin D3.
6. Solution of trypsin (0.25%) and ethylenediaminetetraacetic acid (EDTA; 1 mM) (Gibco).
7. Alizarin red S solution (staining for calcium deposition): 2 g alizarin red S (Sigma Aldrich) in 100 ml distilled water. Adjust pH from 4.1 to 4.3 using 0.5% ammonium hydroxide.

2.8. Scaffold Preparation

2.8.1. NaOH Treatment

1. 5 M NaOH (Sigma Aldrich).
2. PBS (Gibco).

2.8.2. Col Incorporation by Lyophylization

1. Rat tail Col type I: 5 mg/ml in 0.5 M acetic acid (Sigma Pharmaceuticals).
2. 71.2 mg/ml sodium bicarbonate (volume ratio Col : sodium bicarbonate is 100:9).
3. Freeze dryer.

3. Methods

3.1. Electrospinning

Electrospinning is a simple and versatile technique for producing fibers with diameters in the sub-micrometer range. Vast array of materials such as polymers, ceramics, and their composites can be manufactured into ultrathin fibril structures using electrospinning method. Briefly, in solvent-based electrospinning, polymer solutions are fed through a charged capillary where they form a pendant drop at the tip. If the applied electric field is enough to overcome the surface tension of the pendant drop, liquid polymer jet will be formed from the nozzle. The jet is then subjected to continuous stretching and whipping process in a phenomenon called bending instability. During this process, the polymer is greatly elongated while solvent evaporates, creating fibers with diameters as small as tens of nanometers (***21***, ***22***). The applied voltage, solution flow rate, viscosity, polymer concentration, polymer molecular weight, and spinneret-to-collector distance are parameters crucial to the electrospinning process (***23***, ***24***). Thus, modifying the aforementioned processing variables will affect the resulting fiber morphology.

This section focuses on the basic electrospinning setup of polymers particularly in the field of biomaterials for tissue-engineering applications. Although materials and steps outlined here are specific for PCL and PCL/Col-based polymers, modifications can be made to the protocol for other polymer systems, for example, PLLA-PDLA, PLLA-PDLA/Col (*see* **Fig. 4**) with different solvent system, flow rates, bias voltage, etc.

3.1.1. Preparation of the Polymer Solution

1. For the electrospinning of PCL nanofibers, make a 5 ml 10% (w/v) solution by dissolving 500 mg PCL pellets in a 3:1 HFIP : chloroform solvent system. This is done in a 20-ml glass scintillation vial (*see* **note 1**).
2. For the electrospinning of a PCL/Col composite fiber, add 125 mg porcine Col particulates into the 10% PCL solution to give a 2.5% (w/v) concentration of Col.
3. Mix the solution with a magnetic stirring bar for about 1–2 h until components fully dissolve.

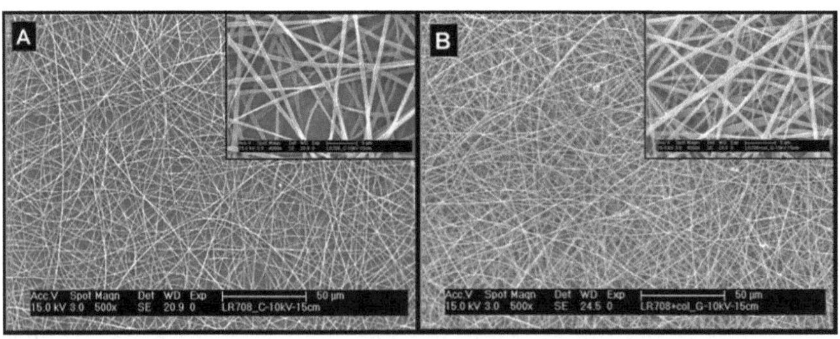

Fig. 4. (a) Scanning electron microscope (SEM) picture showing electrospun PLLA-PDLA (70/30) nanofiber mesh with lower (×500) and higher (×4000) magnification (upper right). (b) SEM picture showing electrospun PLLA-PDLA/Col nanofiber mesh with lower (×500) and higher (×4000) magnification (upper right).

3.1.2. Electrospinning of Polymers

1. A schematic setup of a basic electrospinning system is shown in **Fig. 5**. The polymer reservoir (in a syringe) is mounted on a syringe pump (B). A Teflon tubing of approximately 20 cm in length is used to connect the syringe containing the polymer reservoir to the charging capillary (C). An inverted stainless steel needle is used as the capillary. The capillary is connected, via a cable, to the positive terminal of the high-voltage power supply (A). A laboratory lift (D) is placed 15 cm below the capillary, which acts as the collection screen for the nanofiber. This laboratory lift is connected, again via a cable, to the ground terminal of the power supply. On the surface of the laboratory lift, a piece of aluminum foil is used to collect the spun fibers.
2. Load the polymer solution into a 5-ml syringe. Mount the syringe on the syringe pump in the electrospinning system.
3. Set the flow rate of the syringe pump to 0.75 ml per hour for a 5-ml syringe size.
4. Start the syringe pump and let the polymer fill the Teflon line until the tip of the capillary.
5. Once the polymer emerges from the capillary, turn on the high-voltage power supply and charge the capillary to +10 kV.
6. Nanofiber formation can be seen as whitish deposition on the aluminum foil collector (*see* **note 2**).

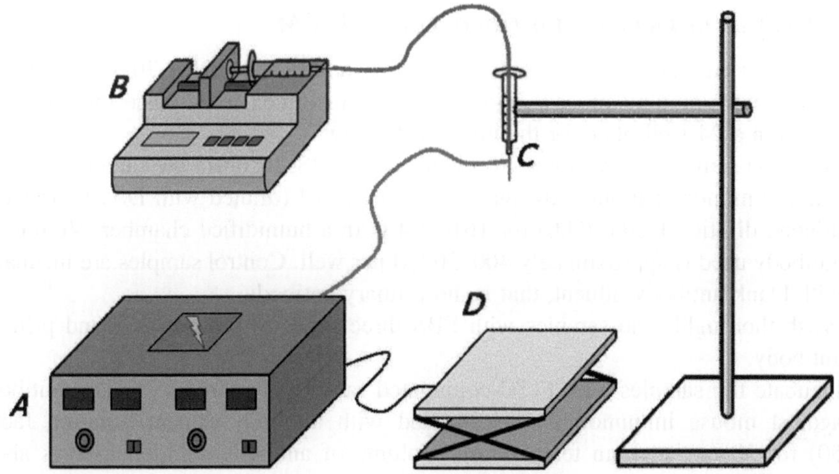

Fig. 5. A schematic illustration of a basic electrospinning system. Main components are a high-voltage power supply to generate voltage bias (a), a syringe pump to provide continuous flow of polymer to the spinneret (b), a metal capillary to charge the polymer (c), and a collection screen (d).

7. The electrospinning system is kept running until the desired thickness of the nanofiber sheet is achieved (prior optimization of correlation between thickness and electrospinning time is required, *see* **note 3**).
8. Once the desired thickness of the nanofiber sheet is achieved, flow of the polymer to the capillary and the application of high voltage are stopped by switching off the syringe pump and the power supply, respectively.
9. Store the collected nanofiber on the aluminum foil in a desiccator or vacuum chamber to remove any remaining solvent for at least 24 h. The nanofiber is ready for use (*see* **note 4**).

3.2. Electrospun Fibers Selection Under SEM

1. Cut the dried nanofiber sheet (PCL/Col, PCL, and PLLA-PDLA 70:30) obtained from electrospinning process into 1 cm × 1 cm dimensions.
2. Mount the pieces to be analyzed on SEM viewing stubs using a double-sided adhesive tape.
3. Sputter coat the samples before SEM viewing with gold using a BAL-TEC sputter coater for 80 s with 30 mA current.
4. Place the samples inside the SEM chamber and visualize under vacuum with 10 kV accelerating voltage.

3.3. Analysis of Electrospun Fibers Under LSCM

1. Cut the dried nanofiber sheet (PCL/Col, PCL, and PLLA-PDLA 70:30%) obtained from electrospinning process into 5 mm × 5 mm dimensions. Place the nanofiber pieces in a 24-well plate for the immunostaining.
2. Wash the samples a few times thoroughly with PBS. Incubate the samples with the primary monoclonal antibody against porcine Col I (diluted with DAKO antibody diluent, dilution factor 4000) for 16 h at 4°C in a humidified chamber. Volume of antibody used is approximately 400–500 μl per well. Control samples are incubated with blank antibody diluent, that is, no primary antibody.
3. Wash thoroughly the samples with PBS three times to remove unbound primary antibody.
4. Incubate the samples with FITC-conjugated rabbit polyclonal secondary antibody against mouse immunoglobulins (diluted with antibody diluent, dilution factor 50) for 30 min at room temperature. Volume of antibody solution used is about 400–500 μl per well (*see* **note 5**).
5. Wash thoroughly the samples with PBS three times to remove unbound fluorescently labeled secondary antibody.
6. View samples under laser scanning confocal microscope tuned to FITC emission wavelength.

3.4. Fused Deposition Modeling

We have adopted the FDM technique to fabricate tissue-engineering scaffolds by using the FDM 3000 from Stratasys (http://www.stratasys.com). The FDM utilizes thermoplastic-based raw materials to construct 3D objects and was invented by Crump SS of Stratasys ***(20)*** (*see* **Fig. 6**).

The working principle of the FDM is based on the deposition of the melted or molten state of a thermoplastic polymer. Classically, the polymer material is supplied to the FDM in the form of a filament, which is introduced into a heated chamber by a set of rollers that will force the filament through, and the pressure from the rollers would drive the molten polymer through a nozzle at the exit forming micro-struts, which would be laid in a pre-determined position to form the desired object (*see* **notes 6–8 and Figs 7 and 8**) (***9***, ***25***).

3.4.1. Preparation of PCL-TCP(20% wt) Blend

Method 1. Physical mixing:

1. The physical blend of PCL-TCP (20% wt) has to be optimized for the FDM fabrication method (*see* **Fig. 9**).
2. Physically mix the appropriate quantity of PCL polymer with TCP ceramic in a metal bowl at 120°C to achieve the blend.
3. Pass the blend through a screw-extruder twice to ensure uniformity in the particulate distribution.

Method 2. Solvent casting:

1. Prepare each small uniform batch by weighing 4 g TCP and 16 g PCL (20% wt) (*see* **note 9**).
2. Place the TCP powder into a 500-ml bottle and add 200 ml methylene chloride (CH_2Cl_2; J. T. Baker). Shake gently to allow the TCP to disperse evenly within the solvent, which acts as a dispersant, preventing the particles from agglomerating.
3. Add the PCL pellets slowly into the solution and place the mixture on an orbital shaker for 1 h at 150 rpm.
4. Pour the mixture out evenly into three aluminum plates (⌀16 cm) and leave it there to evaporate for 6 h, yielding three 6.5 g circular approximately 300 μm thick films.
5. Further dry the films in a vacuum oven at 35°C for 12 h to remove any residual CH_2Cl_2.

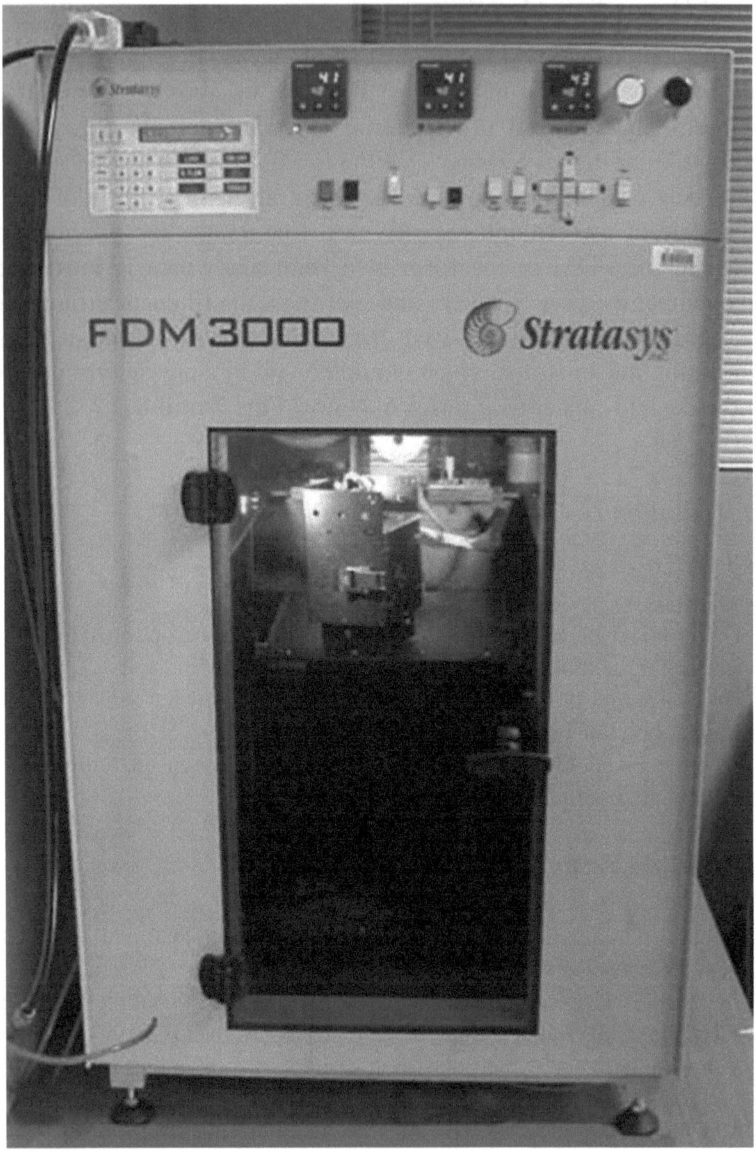

Fig. 6. Fused deposition modeling (FDM) 3000 from Stratasys (http://www.stratasys.com).

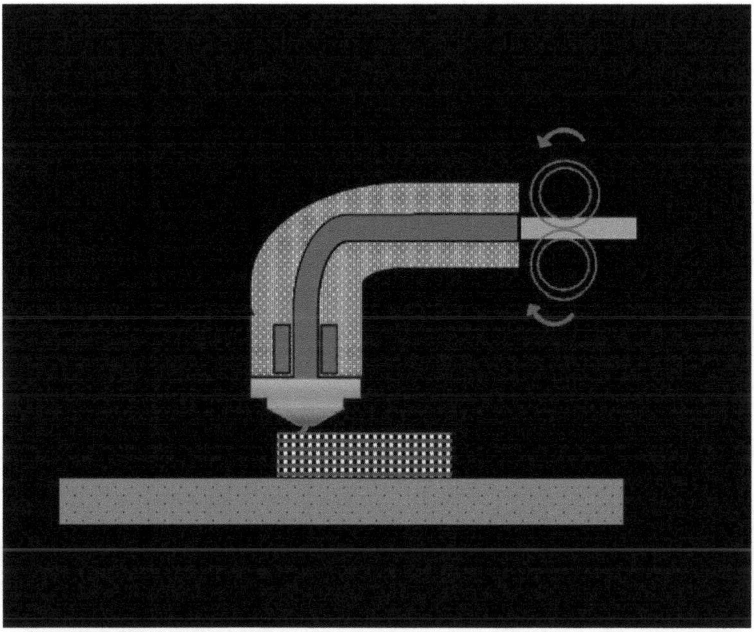

Fig. 7. Graphic illustration that demonstrates the basic principle of the fused deposition modeling (FDM) liquefier unit.

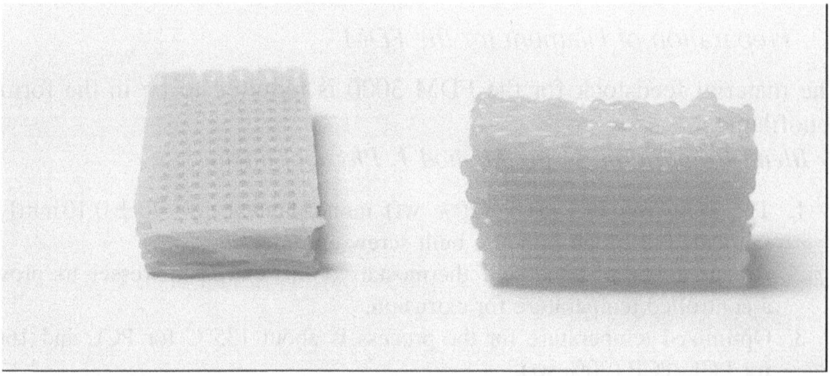

Fig. 8. Fused deposition modeling (FDM) fabricated poly(ε-caprolactone)-tri-calcium phosphate (PCL-TCP) scaffold used for bone tissue engineering.

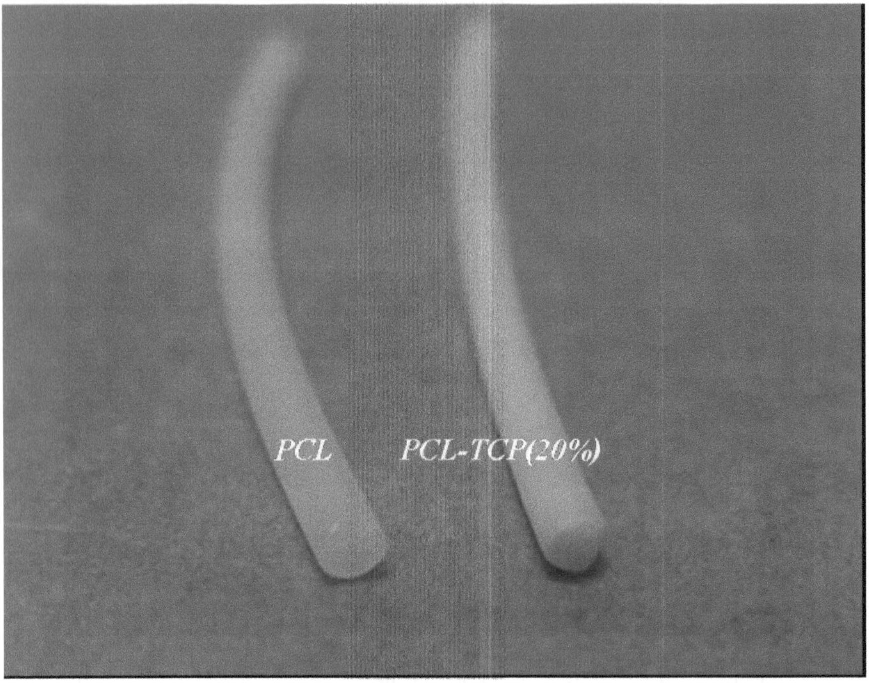

Fig. 9. Picture of a poly(ϵ-caprolactone) (PCL) and PCL-tri-calcium phosphate (PCL-TCP) (20% wt) filament showing the uniform cut cross-section and surface.

3.4.2. Preparation of Filament for the FDM

The material feedstock for the FDM 3000 is required to be in the form of a monofilament.

After Blend Preparation Using Method 1. Physical mixing:

1. The PCL and PCL-TCP (20% wt) monofilament (⌀1.70 ± 0.10 mm) are fabricated using an in-house built screw-extruder.
2. Attach a heater band and thermostat to the extrusion vessel to provide a controlled temperature for extrusion.
3. Optimized temperature for the process is about 135°C for PCL and 160°C for PCL-TCP (20% wt).
4. The screw-extruder has an exit nozzle of about ⌀1.9 mm.
5. Employ a water bath and a motorized drum to quench and collect the extruded filaments.

6. The height of the water level to the die exit has to be maintained between 40 and 80 mm.
7. The speed of the drum has to be maintained within a range of speeds to take in the slack and sustain the molten polymer filament in a straight alignment in the water bath, it should not draw the polymer melt out of the vessel.
8. It is essential for the filament diameter to be highly consistent and the filament itself void free to ensure regular material feed during the FDM extrusion process.
9. The combination of parameters such as temperature, screw-extrusion speed, size of nozzle, distance to water distance, water quenching, and speed of motorized-drum rotation need to be optimized to produce filaments within the usable range.
10. The filament feedstock has to be vacuum-dried and kept in a desiccator until usage.

After Blend Preparation Using Method 2. Solvent casting:

Extrusion of filaments for FDM by using in-house built screw-extruder.

1. The extrusion set comprises an extrusion vessel with an inner shaft of ∅30 mm, a die module with an inner tapered profile with a die exit of ∅1.9 mm and a piston rod with an epoxy base screwed on.
2. All parts are made of stainless steel, case hardened to withstand frictional wear, and finished with chrome for corrosion resistance.
3. Fasten the extrusion vessel to a jig and secure it to an Instron 4302 Material Testing System, which is used to provide the compressive force at 4 mm per minute to the piston rod secured to the crosshead.
4. Attach a heater band and thermostat to the extrusion vessel near the die module to provide a controlled temperature for extrusion.
5. Employ a water bath filled with ice and a motorized drum to quench the extrudate and collect the filaments.
6. The combination of optimized melt temperature, piston speed, die exit to water distance, ice-water quenching, and speed of motorized-drum rotation produces the $\varnothing 1.70 \pm 0.10$ mm filaments.
7. Chop PCL-TCP (20% wt) films into granulates using the Retsch SM2000 material processor before the extrusion.
8. Place chopped PCL granules in the heated chamber and pre-heat for 10 min.
9. This step serves to reduce the air bubbles within the molten polymer that would affect the continuity and surface texture of the extruded filament.
10. The height of the water level to the die exit has to be maintained between 40 and 80 mm.
11. The speed of the drum has to be maintained within a range of speeds to take in the slack and sustaining the molten polymer filament in a straight

alignment in the water bath. It should not draw the polymer melt out of the vessel.

12. It is essential for the filament diameter to be highly consistent and the filament itself void free to ensure regular material feed during the FDM extrusion process. The filament feedstock has to be vacuum dried and kept in a desiccator until usage.
13. Optimized extrusion temperatures for PCL and PCL-TCP (20% wt) materials are 135 and 160°C, respectively.

3.4.3. FDM Scaffold Fabrication

The FDM 3000 rapid-prototyping system from Stratasys (*see* **Figs 6 and 7**) was used to produce scaffolds using PCL-TCP.

1. Organize the prepared monofilaments of the biomaterials onto a spool and supply into the FDM 3000.
2. The material spool is placed at the rear of the FDM machine where the filament passes through a sensor guide to the front of the apparatus into the build chamber. These sensors are employed as a safety device to ensure conformity of the filament dimensions so as not to cause any feeding problems.
3. At the top of the build chamber is a carriage supported on a series of bars, which allows it to translate in the x–y direction, covering a maximum build volume of $1067\,(l) \times 660\,(b) \times 914\,(h)$ mm.
4. The "liquefier" unit of the FDM is attached on it (*see* **Fig. 7**), consisting of three main elements: a set of rollers, a "liquefier" heating channel, and the extrusion nozzle. A small D.C. motor powers the set of counter-rotating rollers that feed the filament material into the insulated thermostat-controlled resistance heater channel.
5. The T16 (0.016 inch or 400 μm) nozzle has to be used for both PCL and PCL-TCP. There are two other common nozzle sizes T10 and T12 (0.01 and 0.012 inch, respectively).
6. As the PCL filament melts at 60°C, the temperature at the inlet to the liquefier could reach a softening temperature. This would result in the softening of the filaments and buckling under the force of the rollers, before the filament could be pushed into the liquefier. An air jet has to be installed to reduce the inlet ambient temperature.
7. The FDM 3000 is used with the accompanying software, Insight™.
8. The scaffold or model can be designed on a CAD software or on Insight™. Insight™ has restricted design capabilities and is limited only to simple shapes.
9. If the scaffold was designed using a CAD software, it would be imported in the .stl (STereoLithography) format into Insight™ (*see* **Fig. 10**, step 1).
10. In Insight™ slice, the scaffold into horizontal layers and convert into .slc (SLiCe) format (*see* **Fig. 10**, step 2).

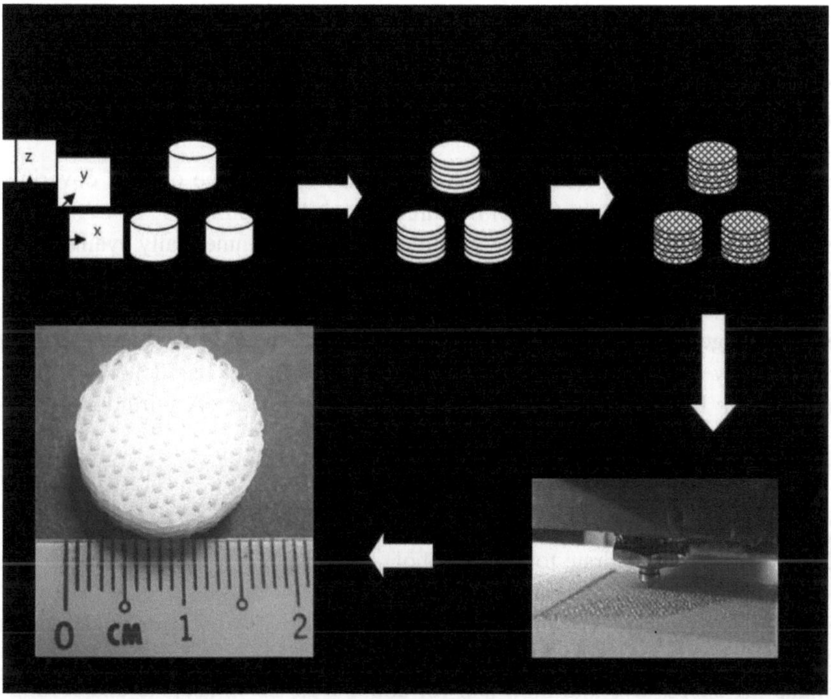

Fig. 10. Overview of the data processing stages in Insight™ before importing the data file of scaffold into the fused deposition modeling (FDM) machine.

11. Input the desired or optimized parameters and create the desired toolpaths for each layer and convert into .sml (Stratasys Machine Language) format for downloading to the FDM machine (*see* **Fig. 10**, step 3).
12. Once the data are downloaded into the FDM machine, the fabrication process starts (*see* **Fig. 10**, step 4).

3.5. Cell Harvest, Culture, and Differentiation

1. Puncture the iliac crest of 6-month-old Yorkshire pigs (60–80 kg) to harvest bone marrow. Aspirate bone marrow into a sterile 10-ml syringe containing 0.1 ml preservative-free sodium heparin. Store at 4°C.
2. At the latest 18 h after aspiration, the bone marrow precursor cells should be selected by using Ficoll® gradient density separation. For this, add the bone marrow to a 50-ml Falcon tube containing Ficoll® solution (density 1.077 g/ml). Cautiously collect the BMSC containing fraction (upper layer at the Ficoll–bone marrow interface) after centrifugation (500 × *g*, 20 min, no break) and mix with MSC proliferation

medium. Determine mononucleated cell counts with a hemacytometer following the staining of the cells with crystal violet in 0.1 M citric acid.

3. Seed 10^7 cells into 75-cm^2 cell-culture flasks, add 15 ml proliferation medium, and amplificate the cells in an incubator (37°C; 5% CO_2). Change the proliferation medium every 3 days.
4. Harvest the cells after reaching 80% confluency (percentage of cells covering the bottom of the cell-culture flasks) by careful trypsinization.
5. Seed the cells into the composite scaffolds by using commercially available Fibrin glue as a cell carrier (Fibrinogen Immuno, Baxter AG). The cell density is 3×10^6 cells per scaffold (size 40 mm × 15 mm × 2 mm).
6. Start osteogenic and chondrogenic differentiation after cell seeding by supply of respectively relevant induction media (*see* **sections** 2) in a parallel-flow bioreactor system (*see* **Fig. 1**). Renew the growth induction media every third day.
7. Harvest tissue samples after 14 and 28 days.

3.6. Alizarin Red S Staining for Calcium Deposition

1. Deparaffinize and hydrate to 70% alcohol.
2. Rinse rapidly in distilled water.
3. Staining of samples in alizarin red S solution for 30 s to 5 min under constant microscopical control for an orange-red color.
4. Shake off excess dye and blot. Dip for about ×15–20 in acetone and ×15–20 in acetone xylene.
5. Clear in xylene and mount in permount.

3.7. Scaffold Preparation

3.7.1. NaOH Treatment

1. Put one scaffold/well in a 24-well plate and treat with 1 ml/well of 5 M NaOH for 3 h at 37°C. The scaffolds are then washed with PBS (two times) and then dried under the laminar flow hood.

3.7.2. Col Incorporation by Lyophylization

1. Neutralize rat tail Col type I (5 mg/ml in 0.5 M acetic acid, Sigma Pharmaceuticals) by using 71.2 mg/ml sodium bicarbonate (volume ratio Col : sodium bicarbonate is 100:9) solution.
2. Add 70 μl Col/sodium bicarbonate mixture on to the NaOH-treated scaffolds.
3. After leaving the scaffold + Col construct for 30 min at room temperature, it will be frozen at −80°C for 2 days and lyophilized in a freeze dryer for at least 24 h.
4. The scaffolds are then stored in a dry cabinet and ultraviolet (UV) sterilized prior to use.

4. Notes

1. The concentration and the molecular weight of the polymer used for electrospinning affect the viscosity of the solution. At low viscosities (<1 poise), surface tension predominates and formation of droplets instead of fibers is more likely. At high viscosities (>20 poise), viscoelastic forces predominate and may impair the controlled flow of material to the capillary ***(23)***. This may provide a rough guide on the processing window on the polymer solution as far as viscosity is concerned
2. Aluminum collector provides a basic and simple collection method of the nanofiber. Other collectors may be used, and in some system of collectors, nanofiber deposition can be somewhat controlled. For example, Theron et al. ***(26)*** described a grounded and tapered wheel-like bobbin as a method to collect aligned nanofibers by electrospinning. Other examples include rotating mandrel systems (***27***, ***28***) and dual collection rings ***(?)***.
3. Optimization can be done for a particular polymer system by running the electrospinning process for certain timepoints and measuring the thickness of fibers obtained at each timepoint. Measurement of sheet thickness can be done by electron microscopy, sectioning the sheet followed by optical measurements under microscope.
4. As the nanofiber is deposited on aluminum foils it may be desirable to remove the foils for further analysis or usage of the nanofiber mesh. Aluminum foil can be removed by soaking the nanofiber sheets in 70% EtOH and peeling the white nanofiber sheet gently from one corner of the foil.
5. Incubation of the samples with the fluorescently conjugated antibody should be done, whenever possible, in the dark to minimize photobleaching of the dye. Subsequent steps and transport of samples should also be done in the dark.
6. The basic parameters for PCL and PCL-TCP scaffold fabrication are broadly classified into four categories in **Table 1**. Most of the parameters were interdependent and required optimizations for different materials used in the FDM process.
7. There are many parameters that control the FDM process, and their relationship should be clearly understood to develop the desired accurate porous model with minimal problems. Some variable parameters, which determine the architectural structure of the fabricated scaffold, are listed in **Table 2**.
8. A balance needs to be attained with this set of parameters such that the laid molten PCL would be consistent, uniform, and stable. Currently based on our parameters, we are able to attain filament width of 250–965 μ m and layer thickness of 50–762μm. The PCL material shows a complete fusion between the laid microfilaments, even after 3 days of NaOH (5 M) treatment (*see* **Fig. 11**).

Table 1
Fused Deposition Modeling (FDM) Fabrication Process Parameters

Build parameters	PCL	PCL-TCP (20% wt)
FDM nozzle	T16	T16
Liquefier temperature (°C)	135	220
Flow rate (%)	75	75
Air gap (mm)	0.635	0.635

PCL, poly(ϵ-caprolactone); TCP, tri-calcium phosphate.

Table 2
Parameters Defining the Architectural Structure of Fused Deposition Modeling (FDM) Designed Scaffolds

Parameter	Effects
Liquefier temperature	Even in its molten state, poly(ϵ-caprolactone) (PCL) is extremely viscous. Raising the temperature beyond its melting temperature was the only way to reduce the viscosity.
Nozzle size	Determines the range of diameters of the microfilaments extruded from the nozzle.
Flow rate	Speed at which molten material is extruded from the nozzle. Basically, the flow rate of the molten polymer would be determined by the speed of the rollers that drive the new filament into the liquefier and the molten material out of the nozzle. However, the pressure built up inside the liquefier needs to be regulated, as a too high pressure may result in a backflow of molten material.
Air gap	The "air gap" determines the "spacing" distance between the laid microfilaments. It therefore determines the porosity of the final scaffold and the stability of the laid filaments. As this "air gap" is also the space that the following layer of laid filaments has to bridge, the filament may sag if it was set too large.
Translation speed of the nozzle	This parameter works in cooperation with the flow rate to enable consistent and uniform laying down of the microfilaments as they are extruded. It can also be regulated to allow some stretching of the microfilaments to further reduce their diameters.

9. Solvent casting: the blending and casting were maintained in small uniform lots to ensure uniformity and consistency as opposed to bulk mixing that would

compromise uniform distribution because of TCP settling in the PCL solution. Disadvantages are

- possibility of residual solvent in the polymer,
- time-consuming process,
- costly as it consumes large quantity of solvent, and
- cast sheet need to be cut to size for feed into the extrusion machine.

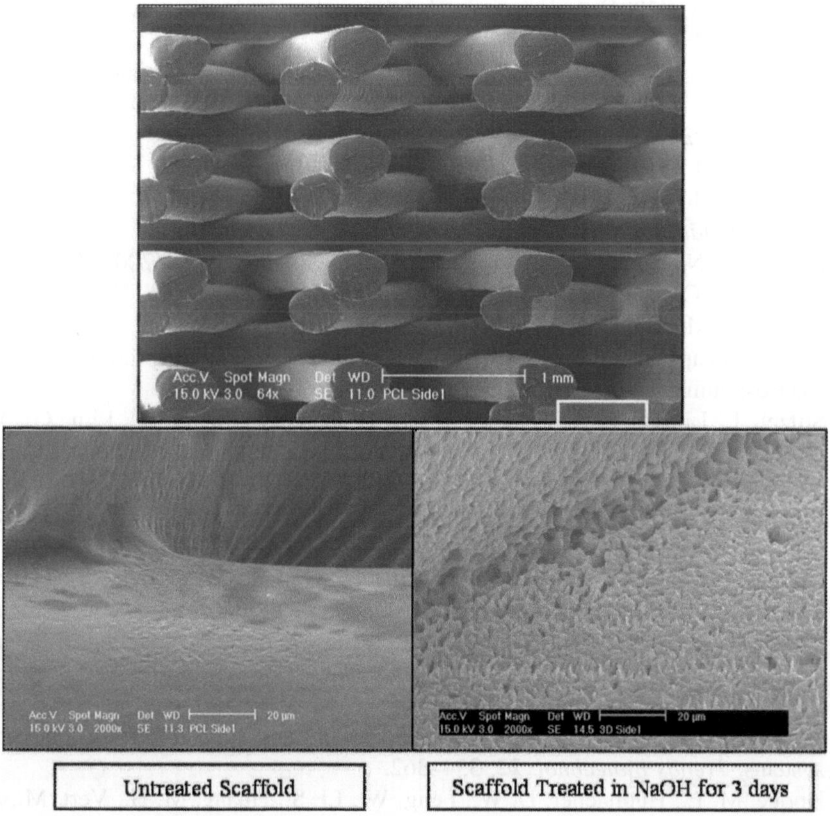

Fig. 11. Scanning electron microscope (SEM) micrographs showing the fusion of microfilaments between each scaffold layer. Complete fusion of the poly(ϵ-caprolactone) (PCL) material was observed between the microfilaments within the scaffold architecture. After 3 days of NaOH (5 M) treatment, although the surface roughness was increased through the degradation, the fusion joints remain well integrated.

Acknowledgments

The authors thank the Electron Microscopy Unit, National University of Singapore, for assistance with the microscopy work. Mr. Tan Kim Cheng (Temaesk Polytechnic, Singapore) for support in the FDM scaffold fabrication.

References

1. Messner, K. (1996) The longterm prognosis for severe damage to weight-bearing cartilage in the knee. A 14-year clinical and radiographic follow-up in 28 young athletes. *Acta Orthop. Scand.* **67**, 165–168.
2. Johnson, L. L. (2001) Arthroscopic abrasion arthroplasty: a review. *Clin. Orthop. Relat. Res.* **391**, 306–317.
3. Hangody, L. (1997) Arthroscopic autogenous osteochondral mosaicplasty for the treatment of femoral condylar articular defects. *Knee Surg. Sports Traumatol. Arthrosc.* **5**, 262–267.
4. Brittberg, M., Tallheden, T., Sjogren-Jansson, B., Lindahl, A., and Peterson, L. (2001) Autologous chondrocytes used for articular cartilage repair: an update. *Clin. Orthop. Relat. Res.* **391**, 337–348.
5. Huang, L., Nagapudi, K., Apkarian, R. P., and Chaikof, E. L. (2001) Engineered collagen-PEO nanofibers and fabrics. *J. Biomater. Sci. Polym. Ed.* **12**(9), 979–993.
6. Huang, Z. M., He, C. L., Yang, A., Zhang, Y., Han, X. J., Yin, J., and Wu, Q. (2006) Encapsulating drugs in biodegradable ultrafine fibers through co-axial electrospinning. *J. Biomed. Mater. Res. A* **77**, 169–179.
7. Stitzel, J., Liu, J., Lee, S. J., Komura, M., Berry, J., Soker, S., Lim, G., Van Dyke, M., Czerw, R., Yoo, J. J., and Atala, A. (2006) Controlled fabrication of a biological vascular substitute. *Biomaterials* **27**, 1088–1094.
8. Matthews, J. A., Wnek, G. E., Simpson, D. G., and Bowlin, G. L. (2002) Electrospinning of collagen nanofibers. *Biomacromolecules* **3**, 232–238.
9. Hutmacher, D. W., Schantz, T., Zein, I., Ng, K. W., Teoh, S. H., and Tan, K. C. (2001) Mechanical properties and cell cultural response of polycaprolactone scaffolds designed and fabricated via fused deposition modeling. *J. Biomed. Mater. Res.* **55**, 203–216.
10. Hutmacher, D. W., Sittinger, M., and Risbud, M. V. (2004) Scaffold-based tissue engineering: rationale for computer-aided design and solid free-form fabrication systems. *Trends Biotechnol.* **22**, 354–362.
11. Hoque, M. E., Hutmacher, D. W., Feng, W., Li, S., Huang, M. H., Vert, M., and Wong, Y. S. (2005) Fabrication using a rapid prototyping system and in vitro characterization of PEG-PCL-PLA scaffolds for tissue engineering. *J. Biomater. Sci. Polym. Ed.* **16**, 1595–1610.
12. Schantz, J. T., Brandwood, A., Hutmacher, D. W., Khor, H. L., and Bittner, K. (2005) Osteogenic differentiation of mesenchymal progenitor cells in computer designed fibrin-polymer-ceramic scaffolds manufactured by fused deposition modeling. *J. Mater. Sci. Mater. Med.* **16**, 807–819.

13. Shao, X. X., Hutmacher, D. W., Ho, S. T., Goh, J. C., and Lee, E. H. (2006) Evaluation of a hybrid scaffold/cell construct in repair of high-load-bearing osteochondral defects in rabbits. *Biomaterials* **27**, 1071–1080.
14. van Lieshout, M. I., Vaz, C. M., Rutten, M. C., Peters, G. W., and Baaijens, F. P. (2006) Electrospinning versus knitting: two scaffolds for tissue engineering of the aortic valve. *J. Biomater. Sci. Polym. Ed.* **17**, 77–89.
15. Rai, B., Teoh, S. H., Hutmacher, D. W., Cao, T., and Ho, K. H. (2005) Novel PCL-based honeycomb scaffolds as drug delivery systems for rhBMP-2. *Biomaterials* **26**, 3739–3748.
16. Rai, B., Teoh, S. H., Ho, K. H., Hutmacher, D. W., Cao, T., Chen, F., and Yacob, K. (2004) The effect of rhBMP-2 on canine osteoblasts seeded onto 3D bioactive polycaprolactone scaffolds. *Biomaterials* **25**, 5499–5506.
17. Rohner, D., Hutmacher, D. W., Cheng, T. K., Oberholzer, M., and Hammer, B. (2003) In vivo efficacy of bone-marrow-coated polycaprolactone scaffolds for the reconstruction of orbital defects in the pig. *J. Biomed. Mater. Res. B Appl. Biomater.* **66**, 574–580.
18. Endres, M., Hutmacher, D. W., Salgado, A. J., Kaps, C., Ringe, J., Reis, R. L., Sittinger, M., Brandwood, A., and Schantz, J. T. (2003) Osteogenic induction of human bone marrow-derived mesenchymal progenitor cells in novel synthetic polymer-hydrogel matrices. *Tissue Eng.* **9**, 689–702.
19. Crump, S. (1992) Apparatus and method for creating three-dimensional objects. US Patent 5121329.
20. Reneker, D. H., Yarin, A. L., Fong, H., and Koombhongse, S. (2000) Bending instability of electrically liquid jets of polymer solutions in electrospinning. *J. Appl. Physiol.* **8**, 4531–4547.
21. Li, D. and Xia, Y. (2004) Electrospinning of nanofibers: re-inventing the wheel? *Adv. Mater.* **16**, 1151–1170.
22. Deitzel, J. M., Kleinmeyer, J., Harris, D., and Beck Tan N. C. (2001) The effect of processing variables on the morphology of electrospun nanofiber and textiles. *Polymer* **42**, 261–272.
23. Theron, S. A., Zussman, E., and Yarin, A. L. (2004) Experimental investigation of the governing parameters in the electrospinning of polymer solutions. *Polymer* **45,** 2017–2030.
24. Zein, I., Hutmacher, D. W., Tan, K. C., and Teoh, S. H. (2002) Fused deposition modeling of novel scaffold architectures for tissue engineering applications. *Biomaterials* **23**, 1169–1185.
25. Theron, S. A., Zussman, E., and Yarin, A. L. (2001) Electrostatic field-assisted alignment of electrospun nanofibers. *Nanotechnology* **12,** 384–390.
26. Mo, X. and Weber, H.-J. (2004) Electrospinning P(LLA-CL) nanofiber: a tubular scaffold fabrication with circumferential alignment. *Macromol. Symp.* **217**, 413–416.

27. Fennessey, S. F. and Farris, J. (2004) Fabrication of aligned and molecularly oriented electrospun polyacrylonitrile nanofibers and the mechanical behavior of their twisted yarns. *Polymer* **45**, 4217–4225.
28. Dalton, P., Klee, D., and Möller, M. (2005) Electrospinning with dual collection rings. *Polymer* **46**, 611–614.

7

Synthetic Hydrogel Matrices for Guided Bladder Tissue Regeneration

Catharina A. M. Adelöw and Peter Frey

Summary

Tissue engineering aims to provide a temporary scaffold for repair at the site of injury or disease that is able to support cell attachment and growth while synthesis of matrix proteins and reorganization take place. Although relatively successful, bladder tissue engineering suffers from the formation of scar tissue at the scaffold implant site partly due to the phenotypic switch of smooth muscle cells (SMCs) from a quiescent contractile phenotype to a synthetic proliferative phenotype, known as myofibroblast. We hypothesize that culturing human SMCs in enzymatically degradable poly(ethylene) glycol (PEG) hydrogels modified with integrin-binding peptides, and in co-culture with human urothelial cells (UCs), will offer some insight as to the required environment for their subsequent differentiation into quiescent SMCs. We have established protocols for isolation, culture, and characterization of human bladder UCs, SMCs, and fibroblasts and investigated co-culture conditions for SMCs and UCs. The optimal PEG hydrogel properties, promoting growth of these cells, have been investigated by varying the amounts of cell adhesion peptide, PEG, and crosslinker and examined using light and fluorescence microscopy. Furthermore, the cell organization within and on top of gels 14 days post seeding has been examined by histology and immunohistochemistry. We have investigated a co-culture model for UCs and SMCs integrated into PEG hydrogels, mimicking a section of the bladder wall for reconstructive purposes that also could contribute to the understanding of the underlying basic mechanisms of SMC differentiation.

Key Words: Tissue engineering; Human; Bladder; Smooth muscle cell; Urothelium; Poly(ethylene) glycol; Hydrogel; Integrin-binding peptide.

From: *Methods in Molecular Medicine, 2nd ed.: Tissue Engineering*
Edited by: H. Hauser and M. Fussenegger © Humana Press Inc., Totowa, NJ

1. Introduction

Bladder diseases and congenital malformations affect approximately 400 million people worldwide *(1)*. The clinical demand for bladder repair in the United States only has been estimated to be 57,200/year *(2)*. Whereas bladder cancer is among the 10 most common cancers, bladder disorders such as bladder exstrophy, congenital and acquired neurogenic bladder, and interstitial cystitis amount to the critical need for bladder tissue substitutes. Intestinal or gastric tissue is still the most commonly used replacement, although frequently met with metabolic complications or cancer due to inadequate wall properties. Other approaches to bladder augmentation include the use of synthetic- *(1)* [polylactic acid (PLLA) and polyglycolic acid (PGA)] or animal-derived materials *(3,4)* (collagen, fibrin) applied as cell-delivery vehicles or as acellular matrices. These materials have to fulfill a range of requirements concerning biocompatibility, porosity, degradability, and mechanical properties. Knowing that tissue development is regulated through the interplay of various signals, including soluble signaling molecules, insoluble ligands, mechanical cues, and cell–cell interactions, the material should also be inductive, that is, contain the necessary biological information to promote the desired cellular processes. Some of these processes include interaction of appropriate cell populations, cell migration, ingrowth of blood vessels, and site-specific differentiation of in situ bladder cells. It is evident that bladder tissue, constituted by urothelial-, extracellular matrix-, fibroblast-, and smooth muscle-layers, is not trivial to replace as characteristics such as its impermeability for toxic substances in the urine, extensive compliance, and ability to change the luminal surface area rapidly are not easily achievable. The common outcome of bladder tissue substitutes seems to fail at reestablishing normal smooth muscle function. Numerous studies have shown the regeneration of normal bladder urothelium *(5,6)*, whereas the success in restoring functional bladder SMCs remains scarce *(7–10)* as these cells have a tendency to turn into a synthetic proliferative phenotype instead of the required quiescent contractile phenotype. The synthetic proliferative cells produce an excessive amount of extracellular matrix proteins that eventually result in non-compliant fibrous scar tissue with decreased compliance and therefore reduced functional bladder volume and higher urinary storage pressures.

In addition to protocols for isolation, culture, and characterization of human bladder UCs and SMCs, we have investigated a co-culture model for these cells integrated into bioactive poly(ethylene) glycol (PEG) hydrogels functionalized with cell adhesion peptides and matrix metalloproteinase (MMP)-sensitive degradation sites, mimicking a section of the bladder wall for reconstructive purposes that also could contribute to the understanding of the underlying basic mechanism of SMC differentiation.

2. Materials

2.1. Transport and Digestion of Human Bladder Tissue

1. Transport medium: Hank's balanced salt solution (HBSS) without Ca^{2+} and Mg^{2+} (Gibco, Invitrogen, Basel, CH) is supplemented with 5% penicillin streptomycin (Gibco, Invitrogen).
2. Liberase Blendzyme I (Roche, Basel, CH) is dissolved in HBSS at 0.5 mg/ml and stored in 1.6-ml aliquots at −20°C. One aliquot is thawed and diluted in 13.6 ml HBSS for use.
3. Alpha minimal essential medium (AMEM) with L-glutamine and without deoxyribonucleic and ribonucleic acids (Gibco, Invitrogen) is supplemented with 10% bovine growth serum (BGS; Hyclone, Perbio, Lausanne, CH), 5% penicillin streptomycin (Gibco, Invitrogen), and 0.025% amphotericin B (Gibco, Invitrogen).
4. Cholera toxin (CT; Fluka, Buchs SG, CH) is reconstituted in 1 ml sterile distilled water. Working solutions, 30 μg/ml, are made by diluting 150 μl CT to 5 ml in keratinocyte serum-free medium (KSFM). Use at 1:1000 to give a final concentration of 30 ng/ml CT. The stock solution is stored at 4°C for up to 1 year after reconstitution.
5. KSFM (Gibco, Invitrogen) is supplemented with 50 ng/ml bovine pituitary extract (BPE), 5 ng/ml endothelial growth factor (EGF; Gibco, Invitrogen), and 30 ng/ml CT (*see* **section 2.1.**, item 4).
6. Petri dishes, non-tissue culture treated, 15 mm × 60 mm (Fisher, Allschwil, CH).
7. Sterile scalpels and anatomical forceps.
8. Cell strainer (Falcon, Milian, Geneva, CH).
9. Syringe, 5 ml (Omnifix®, Braun, Schlieren, CH).
10. Falcon® tubes, 15 $ and 50 ml (Falcon).

2.2. Magnetic Bead Preparation, Cell Isolation, and Primary Culture

1. Phosphate-buffered saline (PBS; Gibco, Invitrogen).
2. Sodium azide (NaN_3; Fluka, Buchs SG) is dissolved in Millipore water (MPW) at 1 M.
3. Sterile syringe filters, 0.22 μM (TPP, Fisher), and syringes, 5, 10, and 20 ml (Omnifix®, Braun).
4. Washing buffer PBS/0.1% bovine serum albumin (BSA; Fluka, Buchs SG) is dissolved at 5 mg/ml in a mixture of 50 ml × 10 PBS and 450 ml MPW, that is, 2.5 g BSA is weighed out and layered on top of the PBS/water solution. BSA is

allowed to dissolve without stirring at room temperature for 1 h. The solution is filtered (*see* **section 2.2.**, item 3) and stored at 4°C.

5. Bead storage buffer—PBS/0.1% BSA/0.02% NaN_3: PBS/BSA is prepared as the washing buffer (*see* **section 2.2.**, item 4) and 0.5 ml 1 M NaN_3 is added. The solution is filtered (*see* **section 2.2.**, item 3) and stored at 4°C.
6. CELLection Dynabeads® pan mouse immunoglobulin G (IgG; Dynal Biotech, Invitrogen AG, Basel, CH).
7. Primary monoclonal mouse anti-human epithelial antibody EpBer4 (Dakocytomation, Baar, CH).
8. Primary monoclonal mouse anti-human CD90/THY-1 antibody (Dianova, Hamburg, Germany).
9. Dynal small magnetic separator (Dynal Biotech, Invitrogen).
10. Eppendorf tubes, 1.5 ml (Eppendorf, Basel, CH).
11. Trypsin (Gibco, Invitrogen).
12. Trypsin inhibitor (Fluka, Buchs SG) is reconstituted at 10 mg/ml in PBS and sterile filtered (*see* **section 2.2.**, item 3). The stock solution is diluted 1:10 in KSFM and used at equal volume as the volume trypsin used for dissociation. The solution is stored at 4°C, and remains stable at 4°C for up to 3 years after reconstitution.
13. AMEM (*see* **section 2.1.**, item 3).
14. CT (*see* **section 2.1.**, item 4).
15. KSFM (*see* **section 2.1.**, item 5).
16. Cell flasks, 75 and 150 ml (TPP, Omnilab, Chavannes-de-Bogis, CH).
17. Cell flasks Primaria®, 25 and 75 ml (Falcon) for UC culture.
18. Neubauer hemacytometer (Optik Labor, Milian, Geneva, CH).

2.3. Cell Characterization by Fluorescence-Activated Cell Sorting Analysis

1. Formaldehyde 37% (Brunschwig, Basel, CH): A 1.5% solution is prepared in PBS fresh for each experiment.
2. Permeabilization buffer: A 10% saponin solution is prepared by dissolving 5 g saponin (Sigma, Epalinges, CH) in 50 ml PBS at 37°C until the saponin has dissolved completely. The mixture is sterile filtered (*see* **section 2.1.**, item 3). To prepare the permeabilization buffer, 5 ml 10% saponin in PBS is mixed with 95 ml PBS/BSA/azide buffer (*see* **section 2.2.**, item 5) and stored at 4°C.
3. Blocking serum: goat serum (Fluka, Buchs SG) is diluted at 10% in PBS.
4. Primary monoclonal mouse anti-human epithelial antibody EpBer4 (*see* **section 2.2.**, item 7) is used at 1:300 dilutions in PBS/0.1% BSA (*see* **section 2.2.**, item 4).
5. Primary monoclonal mouse anti-human pancytokeratin antibody (Abcam, Cambridge, UK) is used at 1:200 dilutions in PBS/0.1% BSA (*see* **section 2.2.**, item 4).

6. Primary monoclonal mouse anti-human cytokeratin 8 antibody (Dakocytomation) is used at 1:50 dilutions in PBS/0.1% BSA (*see* **section 2.2.**, item 4).
7. Primary monoclonal mouse anti-human smooth muscle alpha actin antibody (Abcam) is used at 1:100 dilutions in PBS/0.1% BSA (*see* **section 2.2.**, item 4).
8. Primary monoclonal mouse anti-human myosin antibody (Abcam) is used at 1:100 dilutions in PBS/0.1% BSA (*see* **section 2.2.**, item 4).
9. Primary monoclonal mouse anti-human calponin antibody (Abcam) is used at 1:100 dilutions in PBS/0.1% BSA (*see* **section 2.2.**, item 4).
10. Primary monoclonal mouse anti-human CD90/THY-1 antibody (*see* **section 2.2.**, item 8) is used at 1:100 dilutions in PBS/0.1% BSA (*see* **section 2.2.**, item 4).
11. Primary monoclonal mouse anti-human CD105 antibody (Endoglin; Abcam) is used at 1:200 dilutions in PBS/0.1% BSA (*see* **section 2.2.**, item 4).
12. Primary monoclonal mouse anti-human CD29 antibody (β1; Abcam) is used at 1:500 dilutions in PBS/0.1% BSA (*see* **section 2.2.**, item 4).
13. Primary monoclonal mouse anti-human CD61 antibody (β3; BD Bioscience, Basel, CH) is used at 1:200 dilutions in PBS/0.1% BSA (*see* **section 2.2.**, item 4).
14. Primary monoclonal mouse anti-human CD49a antibody (α1; BD Bioscience) is used at 1:200 dilutions in PBS/0.1% BSA (*see* **section 2.2.**, item 4).
15. Primary monoclonal mouse anti-human CD49c antibody (α3; BD Bioscience) is used at 1:200 dilutions in PBS/0.1% BSA (*see* **section 2.2.**, item 4).
16. Primary monoclonal mouse anti-human CD49e antibody (α5; BD Bioscience) is used at 1:200 dilutions in PBS/0.1% BSA (*see* **section 2.2.**, item 4).
17. Primary monoclonal mouse anti-human CD51 antibody (αv; Abcam) is used at 1:200 dilutions in PBS/0.1% BSA (*see* **section 2.2.**, item 4).
18. Secondary goat anti-mouse IgG AF488 antibody (Molecular Probes, Invitrogen) is used at 1:400 dilutions in PBS/0.1% BSA (*see* **section 2.2.**, item 4).
19. Nuclear stain: Hoechst 33342 (MP, Invitrogen, Basel, CH) is diluted to 2 μg/ml in HBSS (Gibco, Invitrogen, Basel, CH).

2.4. PEG Hydrogel Formation and Cell Co-Cultures in Hydrogels

1. Triethanolamine (TEOA) buffer (Fluka, Buchs SG: A 0.3 M solution is prepared by pipetting 0.8 ml TEOA with a positive displacement pipette into 19.2 ml MPW in a Falcon® tube. The pH is set to 8 by adding 37% (fuming) hydrochloride (HCl; Fluka, Buchs SG) dropwise (*see* **note 1**) and examining pH with a pH meter (Hanna Instruments, Germany). The ready solution is filtered (*see* **section 2.2.**, item 3) and stored under argon at room temperature.
2. PEG-RGDSP aliquots are prepared by reaction of 11.3 mg PEG-VS (Fluka, Buchs SG) with 0.25 mg RGD peptide (Ac-Gly-Cys-Gly-Trp-Gly-Arg-Gly-Asp-Ser-Pro-Gly-NH_2, NEOMPS, Strasbourg, France) as previously described by Lutolf et al. ***(11)***. A schematic of the PEG hydrogel components and functionalization is shown in **Fig. 1**.

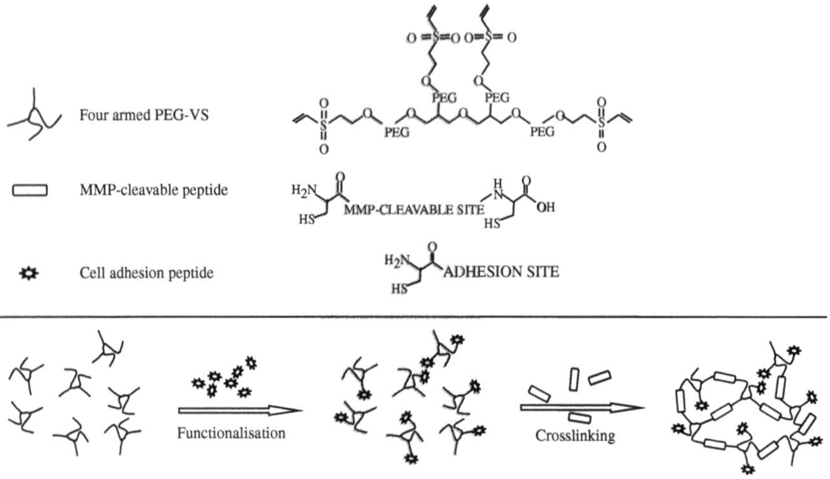

Fig. 1. Poly(ethylene) glycol (PEG) hydrogel functionalization and formation. The vinyl sulfone functionalized PEG is reacted with cysteine-rich adhesion peptides, and subsequently the PEG hydrogel is formed by crosslinking with matrix metalloproteinase (MMP) cleavable peptides.

3. MMP-sensitive crosslinker (Ac-Gly-Cys-Arg-Glu-Gly-Pro-Gln-Gly-Ile-Trp-Gly-Gln-Glu-Arg-Cys-Gly-NH_2; NEOMPS): 1.8 mg aliquots are prepared as previously described by Lutolf et al. ***(11)***.
4. Sigmacote®-coated microscope slides are prepared by dip coating untreated microscope slides (Menzel GmbH, Braunschweig) in Sigmacote® silicon solution (Fluka, Buchs SG, CH) (*see* **note 2**) for 3–5 min. The slides are air-dried, spray cleaned with 70% ethanol and dried with Kim wipe®.
5. Cover glass spacers are made by joining six cover glass slides, 22 mm × 22 mm (Menzel GmbH) with tape.
6. Paper clamps.
7. Spatula.
8. Biopsy punch, Ø6 mm (Provet, Ag, Lyssach, CH).
9. Ultra low binding 96-well plates (Costar, Corning, Fisher).
10. Co-culture medium: AMEM (*see* **section 2.1.**, item 3) supplemented with CT (*see* **section 2.1.**, item 4).

2.5. Histology and Immunohistochemistry

1. OCT Tissue TEK® (Digitana SA, Yverdon-les-Bains, CH).
2. Tissue Tek Cryomold, 10 mm × 10 mm × 5 mm (Medite technique médicale SA, Nunningen, CH).
3. Plastic tray.

4. Isopentane (Fluka, Buchs SG).
5. Liquid nitrogen.
6. Microscope slides (Menzel GmbH).
7. Hematoxylin and eosin (H&E; Fluka, Basel, CH).

3. Methods

Different techniques for the isolation of cells from bladder tissue have previously been described ***(12–14)*** including micro-dissection, tissue digestion, and cell selection by culture in cell type-specific medium. However, these methods are regularly met with contamination of the primary SMC culture by fibroblasts. Owing to their proliferative and synthetic character, fibroblasts easily overgrow smooth muscle cells (SMCs) in culture. Cell isolation by magnetic bead cell separation is believed to yield a higher purity and viability of target cells ***(15)***. As the characterization of cell phenotypes, and hence the purity of the culture, is critical for this study, we have used the latter technique for isolation of primary cells from human bladder tissue.

Urothelial cells (UCs) were grown in standardized serum-free cultures to compare the difference in phenotype of UCs cultured in serum-containing to serum-free medium by fluorescence-activated cell sorting (FACS).

The advantage of synthetic materials for tissue regeneration is the ability of controlling the exact content of the implant and being able to determine the effect of certain molecules on wound repair and tissue regeneration. A biodegradable PEG hydrogel is such a synthetic material that provides the opportunity of tailoring a specific cellular niche in a three-dimensional environment. As PEG has limited cellular interactions, it provides an excellent scaffold for the controlled incorporation of biological signals such as integrin-binding peptides and growth factors for guiding tissue formation and repair ***(16–19)***. Incorporation of degradation sites sensitive to specific MMPs further completes a system of cell-demanded degradation of the gel as synthesis of new matrix proteins takes place. Different sets of integrins have been reported to be associated with the differentiated quiescent SMC phenotype as well as with the de-differentiated synthetic SMC phenotype ***(20,21)***. Hence, in developing a scaffold for delivery of undifferentiated or differentiated cells to the affected site, the determination of changes in cell phenotypes due to exposure to various stimuli such as co-culture is critical.

3.1. Harvesting, Transport, and Digestion of Bladder Tissue

1. At open surgery in children undergoing surgical correction of congenital malformations such as vesicoureteral reflux, open the bladder longitudinally and separate the detrusor muscle from the urothelium by sharp dissection with Steven's scissors

and by blunt dissection with anatomical forceps. Opening the bladder, excise an approximately 2 cm × 4 cm piece of the bulging urothelium and place it immediately into transport medium. Thereafter, excise 1 cm × 3 cm pieces of the previously separated detrusor muscle and transfer them into transport medium.
2. Bring the collected tissue specimens (50-ml Falcon® tubes with transport medium) on ice to the laboratory (*see* **note 3**).
3. Transfer the specimens into Ø60-mm Petri dishes with 3 ml fresh transport medium. Rinse once in transport medium.
4. Cut the tissue into small uniform pieces approximately 1–2 mm^2 with sterile scalpels and forceps for even tissue dissociation.
5. Rinse one to two times in 3 ml transport medium.
6. Aspirate transport medium and add 3 ml Liberase Blendzyme® I to the Petri dish. Incubate at 37°C for 1–2 h until the tissue is fully digested (*see* **note 4**).
7. Use a 5-ml syringe to filter cell suspension through a 70-μm cell strainer into a 50-ml Falcon® tube (on ice) by forcing suspension through strainer (*see* **note 5**).
8. Centrifuge at 174 × *g* for 5 min.
9. Resuspend cells in 2 ml medium and transfer into a centrifuge tube. Continue with cell isolation immediately.

3.2. Magnetic Bead Preparation

1. Resuspend IgG-coated beads well by vortexing.
2. Aliquot desired amount (100 μl = 4×10^7 beads). Make one aliquot for each antibody.
3. Wash in 1 ml washing buffer, apply magnet, and aspirate the washing buffer. Repeat two times and resuspend in 100 μl washing buffer.
4. Add antibody to bead solution (4×10^7 beads):

   i. Add 12.5 μl primary monoclonal mouse anti-human epithelial antibody EpBer4.
   ii. Add 4 μl primary monoclonal mouse anti-human CD90/THY-1 antibody.

5. Seal top of tubes with parafilm and incubate at 4°C under end-over-end rotation overnight.
6. Wash four times in 1 ml washing buffer at 4°C under end-over-end rotation for 30 min. Apply magnet to discard waste buffer.
7. Resuspend in 100 μl washing buffer (*see* **note 6**).

3.3. Cell Isolation and Primary Culture of Urothelium, SMCs, and Fibroblasts

1. To isolate urothelium: add 15–20 μl epithelial (EpBer4 conjugated) beads per 2 ml tube of whole tissue cell suspension (*see* **section 3.1.**, item 9) and rotate end-over-end at 4°C for 15 min.
2. Separate with magnetic separator 1–1.5 min and collect supernatant containing fibroblasts and SMCs in a separate tube.

3. Wash magnetic beads in washing buffer two times. Apply magnet, collect first wash, and add to the previously collected supernatant (*see* **section 3.3.**, item 2).
4. Resuspend bead-conjugated cells in complete medium (AMEM or KSFM) and plate into Primaria® flasks at a seeding density of 4×10^4 cells/cm^2.
5. Isolate fibroblasts from SMCs: add 15–20 µl fibroblast (Thy-1 conjugated) beads per 2 ml tube with the supernatant and first wash (*see* **section 3.3.**, items 2 and 3) and rotate end-over-end at 4°C for 15 min. Separate by applying the magnet 1–1.5 min.
6. Collect the supernatant and first wash, containing the SMCs, into a separate tube. Plate cells into cell flasks at a seeding density of 1×10^4 cells/cm^2.
7. Resuspend bead-conjugated cells in complete medium (AMEM) and plate into cell flasks at a seeding density of 1×10^4 cells/cm^2.
8. Maintain cultures in a humidified atmosphere at 37°C and 5% CO_2.
9. After 24 h, wash two times in 10 ml HBSS to remove non-adherent cells and beads. Add fresh medium. Change medium every 3 days.
10. First and second purification: at 70–80% cell confluence, add 25–30 µl of antibody-conjugated beads (*see* **note 7**) to each of the primary cultures and incubate at 37°C for 15–20 min. Shake the flask gently and assess by inverted phase-contrast microscopy if beads have attached.
11. Wash the cells two to three times gently with washing buffer. Add 3 ml trypsin and incubate at 37°C until cells detach and form a single cell suspension.
12. Add 3 ml complete medium and mix well. Transfer the suspension into two Eppendorf tubes. Separate cells by applying the magnet 1–1.5 min.
13. Collect the supernatant (containing non-bead-conjugated cells) into a separate tube.
14. Wash the beads two times in PBS. Collect the first wash and add to the supernatant.
15. Resuspend bead-conjugated cells in complete medium and plate in flasks for expansion. After plating for 3 h, wash cells in PBS two to three times to remove beads.
16. Transfer the collected supernatant and first wash into Eppendorf tubes and apply the magnet for 1–1.5 min.
17. Again, collect the supernatant into a new tube and plate the cells into a flask for expansion.
18. Perform a second purification step with cells at passage 2.

3.4. Subculture of Human UCs

1. Cells are ready for subculturing when cell monolayers reach 70–80% confluence. Examples of the morphology of UCs cultured in serum-free as well as serum-containing medium is shown in **Fig. 2**.
2. Aspirate medium and wash cells in 5 ml PBS once.
3. Add 3 ml trypsin and incubate at 37°C until cells detach and form a single cell suspension. Tap the flask on the side to quicken the detachment.

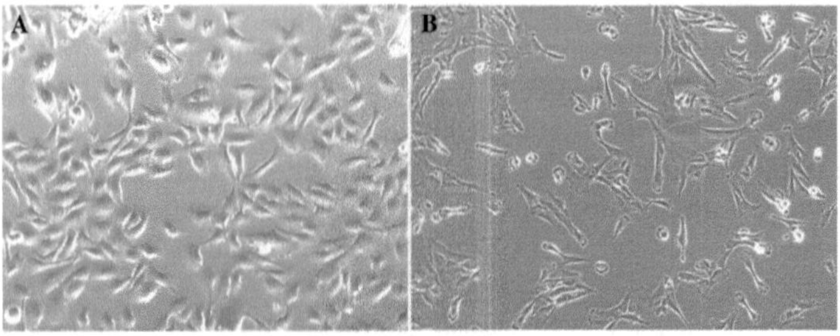

Fig. 2. Inverted phase-contrast microscopy. (a) Monolayer of human urothelial cells (UCs) isolated and cultured in keratinocyte serum-free medium (KSFM) under serum-free conditions (×20 magnification) shows a pronounced cobble stone morphology. (b) Monolayer of human UCs isolated and cultured under serum-containing conditions [alpha minimal essential medium (AMEM)] (×10 magnification) shows a more elongated morphology and is more proliferative than UCs cultured in serum-free KSFM.

4. For urothelium cultured in serum-containing medium, add 3 ml supplemented AMEM, and in case of serum-free culture, add 3 ml serum-free KSFM with trypsin inhibitor and transfer to a Falcon® tube. Count cells using the Neubauer hemacytometer and an inverted phase-contrast microscope. Centrifuge at 174 × *g* for 5 min.
5. Aspirate medium and resuspend in fresh medium (supplemented AMEM or KSFM). Plate cells at a seeding density of 4×10^4 cells/cm^2 in Primaria® flasks. Maintain cultures in a humidified atmosphere at 37°C and 5% CO_2.

3.5. Subculture of Human SMCs and Fibroblasts

1. Cells are ready for subculturing when cell monolayers reach 70–80% confluence.
2. Aspirate medium and wash cells in 5 ml PBS once.
3. Add 3 ml trypsin and incubate at 37°C until cells detach and form a single cell suspension. Tap the flask on the side to quicken the detachment.
4. Add 3 ml supplemented AMEM and transfer to a Falcon® tube. Count cells using the Neubauer hemacytometer and an inverted phase-contrast microscope. Centrifuge at 174 × *g* for 5 min.
5. Aspirate medium and resuspend in fresh complete medium, AMEM. Plate cells in cell flasks at a seeding density of 1×10^4 cells/cm^2. Maintain cultures in a humidified atmosphere at 37°C and 5% CO_2.

3.6. Cell Characterization by FACS Analysis

1. Detach and count cells (*see* **sections 3.4. and 3.5.**).
2. Centrifuge cells at 174 × *g* for 5 min, aspirate medium, and wash in PBS once. Centrifuge again and aspirate supernatant.
3. Fixation: resuspend cells in 500 µl 1.5% formaldehyde and mix well by pipetting. Incubate the cells in the dark at 4°C for 20 min.

4. Centrifuge cells at 695 × *g* for 5 min and remove supernatant (*see* **note 8**). Wash once with 500 μl ice-cold PBS/BSA/azide buffer (see **section 2.2.**, item 5).
5. Permeabilization: centrifuge at 695 × *g* for 5 min and resuspend in 500 μl permeabilization buffer. Mix gently with pipette. Incubate cells in the dark at room temperature for 30 min.
6. Centrifuge cells at 695 × *g* for 5 min and remove supernatant. Wash once in PBS/BSA/azide buffer.
7. Immunofluorescent staining: centrifuge cells at 695 × *g* for 5 min and remove supernatant. Add 100 μl blocking serum and incubate for 30 min to block unspecific binding.
8. Add primary antibody (*see* **section 2.3.**, items 5–19, and **note 9**) directly to the tube and incubate at 37°C for 40 min. Leave controls without addition of primary antibody.
9. Wash cells two times in 500 μl PBS. Centrifuge at 695 × *g* for 5 min. Aspirate supernatant.
10. Add secondary antibody (final dilution 1:400) to all samples including controls and incubate at room temperature for 30 min.
11. Wash cells two times in 500 μl PBS and resuspend in 200 μl PBS/BSA/azide buffer (*see* **section 2.2.**, item 5).
12. Analyze samples by flow cytometry (488 FITC laser, wavelength 515–545 nm).

3.7. Cell Seeding and Culture in PEG Hydrogels

1. Trypsinize and count SMCs to obtain 300,000 cells within each PEG hydrogel precursor (200 μl). Centrifuge cell suspension at 174 × *g* for 5 min, aspirate supernatant, and resuspend in 20 μl supplemented medium.
2. Set pipettes and place to use in the following order: 20, 49.6, 132, and 200 μl. Place the pipette tips easily accessible.
3. Take out the crosslinker and PEG-RGD aliquots from the freezer. Place the crosslinker on ice. Hold the PEG in the hand until thawed.
4. Add 20 μl PBS to the PEG-RGD aliquot. Vortex and spin quickly in a bench-top centrifuge to bring aliquot to the bottom of the tube. Put back on ice.
5. Add 49.6 μl TEOA to the crosslinker. Vortex thoroughly without stopping until completely dissolved. Add the 20 μl SMC suspension. Vortex and spin down.
6. Add the PEG-RGD to the crosslinker. Mix well with the pipette tip. Spin down to eliminate bubbles.
7. Pipette the hydrogel precursor solution onto a Sigmacote®-coated microscope slide. Place cover glass spacers on each side of the slide and add a second microscope slide on top.
8. Seal with two paper clamps on the edges and place in the incubator for 15 min for the gel to polymerize. A schematic of the gel formation procedure is shown in **Fig. 3**.
9. Take out and pipette 200 μl AMEM medium in between the slides. Tilt the slides and introduce the pipette tip as far as possible.

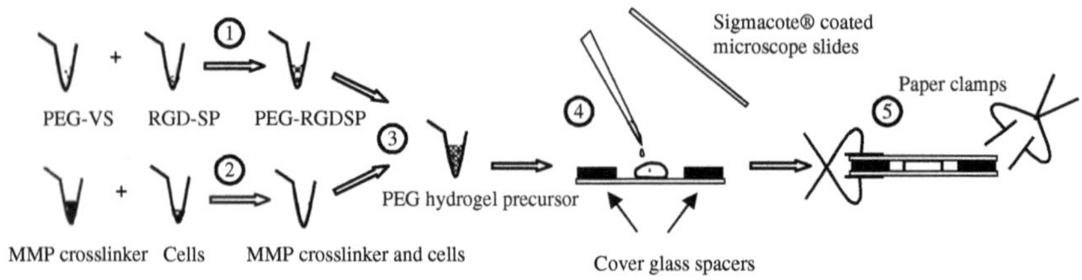

Fig. 3. Scheme for gel preparation: (1) PEG-VS is dissolved in triethanolamine (TEOA) buffer and reacted with cysteine-rich RGDSP at room temperature during 60 min. (2) Matrix metalloproteinase (MMP) cleavable crosslinker is dissolved in buffer and mixed with cells. (3) PEG-RGDSP is quickly mixed with the crosslinker/cell suspension. (4) The cell/hydrogel precursor solution is pipetted onto Sigmacote®-coated microscope slides. (5) Polymerization takes place between slides separated by cover glass spacers at 37°C during 15 min.

10. Release the clips and carefully detach the gel with a spatula. Transfer the gel into a 6-well plate for swelling in co-culture medium for 3 h.
11. Cut out smaller gels with a 6-mm biopsy punch and transfer with a spatula into wells of an ultra low binding 96-well plate. Add 150 μl AMEM into each well.
12. Trypsinize and count UCs to obtain 200,000 cells for seeding on top of each small hydrogel (cut out to fit a 96-well plate). Centrifuge cell suspension at 174 × *g* for 5 min, aspirate supernatant, and resuspend in 20 μl co-culture medium.
13. Remove medium carefully from each hydrogel. Seed the UCs (200,000/gel in 20 μl) on top of each gel. Incubate at 37°C and 5% CO_2 for 2 h for cells to adhere. Add 130 μl co-culture medium gently to each well and continue incubation.
14. Change medium every 3 days. Examples of SMCs growing inside PEG hydrogels and UCs growing on top of PEG hydrogels are shown in **Fig. 4**.

3.8. Histology and Immunohistochemistry of Cell-PEG Matrices

1. Medium is carefully removed from the hydrogel/cell samples, and the cell-matrix constructs are embedded in OCT Tissue TEK® in transparent moulds.
2. Isopentane is poured into a plastic tray placed in liquid nitrogen. Once the isopentane has frozen (i.e., solidified and taken a white color), the moulds are gently immersed into the isopentane until samples are frozen. Samples are stored at −20°C.
3. Freeze sections of 6 μm are generated on a cryostat and mounted on microscope slides.

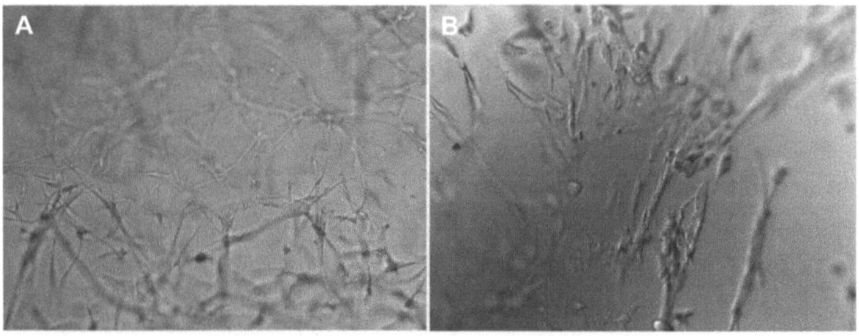

Fig. 4. Inverted phase-contrast microscopy. (a) Human smooth muscle cells (SMCs) spreading within the poly(ethylene) glycol (PEG) hydrogel 9 days post seeding (×10 magnification). (b) Human urothelial cells (UCs) growing on top of the PEG hydrogel 9 days post seeding (×20 magnification).

4. Sections are stained with H&E routine staining to visualize cell nucleus and cytoplasm, respectively, and by immunohistochemistry (antibodies, *see* **section 2.3.**, items 4–19) to determine cell phenotypes. Examples of H&E-stained sections of SMCs cultured inside and UCs cultured on top of PEG hydrogels are shown in **Fig. 5.**

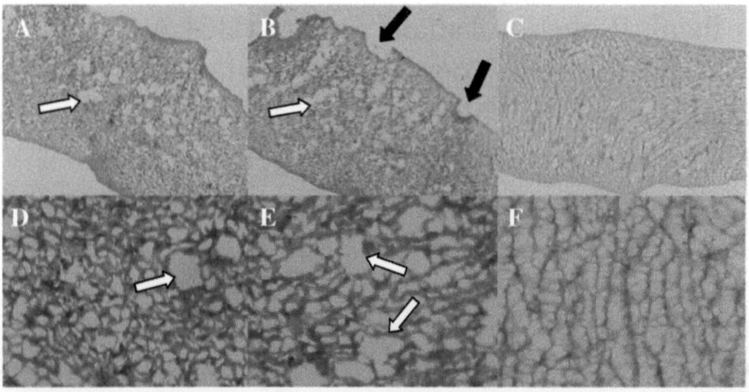

Fig. 5. Histology. To determine cellular distribution within poly(ethylene) glycol (PEG) gels, 6 μM thin cryosections were prepared and stained with hematoxylin and eosin (H&E). Human smooth muscle cells (SMCs) grown alone within PEG gels after 2 weeks in culture: (a) (×4 magnification) and (d) (×20 magnification). SMCs grown within PEG gels after 2 weeks in co-culture with urothelial cells (UCs) seeded on top of the gels (b) (×4 magnification) and (e) (×20 magnification). As a control, PEG gels without cells were processed and stained: (c) (×4 magnification) and (f) (×20 magnification). Cell-grown gels show degradation by SMCs within (white arrows) (a, b, d, and e) and UCs on top (black arrows) (b) of PEG gels due to cellular matrix metalloproteinase (MMP) excretion. Control gels do not demonstrate signs of degradation.

Notes

1. The pH of TEOA is initially around 9–10 and requires 20–40 droplets HCl to lower to pH 8.
2. The Sigmacote® solution can be poured back into the flask after use and be reused a couple of times.
3. The tissue should be processed, that is, digestion and cell isolation should be carried out as soon as possible after surgery, preferentially within 3 h.
4. Scrape the tissue pieces gently with a scalpel to accelerate tissue digestion.
5. Force the suspension through the strainer by "rubbing" the syringe tip on the filter.
6. Resuspend beads in original volume (100 μl) storage buffer with NaN_3 if storing more than 1 month, to avoid contamination.
7. Add EpBer4-conjugated beads to the urothelial primary cultures to select out non-epithelial cells, add CD90/THY-1-conjugated beads to the primary fibroblast cultures to select out non-fibroblastic cells, and add EpBer4- and CD90/THY-1-conjugated beads to primary cultures of SMCs to select out UCs and fibroblasts.
8. Leave 50–100 μl solution in the tubes to make sure not to lose any cells during washes, etc.
9. All antibodies are light sensitive.

Acknowledgments

The authors thank Professor Tatiana Segura (UCLA, CA, USA), Professor Jeffrey Hubbell, and Carolyn Yong (EPFL, Switzerland) for their scientific advice and technical support. This work was supported by the Swiss National Science Foundation NCCR 404640-101113.

References

1. Oberpenning, F., Meng, J., Yoo, J. J., and Atala, A. (1999) De novo reconstitution of a functional mammalian urinary bladder by tissue engineering. *Nat. Biotechnol.* **17**, 149–155.
2. Kim, W. J. H. (2000) Cellular signaling in tissue regeneration. *Yonsei Med. J.* **41**, 692–703.
3. Badylak, S. F. (2004) Xenogeneic extracellular matrix as a scaffold for tissue reconstruction. *Transpl. Immunol.* **12**, 367–377.
4. Kanematsu, A., Yamamoto, S., Ozeki, M., Noguchi, T., Kanatani, I., Ogawa, O., and Tabata, Y. (2004) Collagenous matrices as release carriers of exogenous growth factors. *Biomaterials* **25**, 4513–4520.
5. Cross, W. R., Eardley, I., Leese, H. J., and Southgate, J. (2005) A biomimetic tissue from cultured normal human urothelial cells: analysis of physiological function. *Am. J. Physiol. Renal Physiol.* **289**, F459–F468.

6. Sutherland, R. S., Baskin, L. S., Hayward, S. W., and Cunha, G. R. (1996) Regeneration of bladder urothelium, smooth muscle, blood vessels and nerves into an acellular tissue matrix. *J. Urol.* **156**, 571–577.
7. Kropp, B. P., Badylak, S., and Thor, K. B. (1995) Regenerative bladder augmentation: a review of the initial preclinical studies with porcine small intestinal submucosa. *Adv. Exp. Med. Biol.* **385**, 229–235.
8. Kropp, B. P., Rippy, M. K., Badylak, S. F., Adams, M. C., Keating, M. A., Rink, R. C., and Thor, K. B. (1996) Regenerative urinary bladder augmentation using small intestinal submucosa: urodynamic and histopathologic assessment in long-term canine bladder augmentations. *J. Urol.* **155**, 2098–2104.
9. Probst, M., Dahiya, R., Carrier, S., and Tanagho, E. A. (1997) Reproduction of functional smooth muscle tissue and partial bladder replacement. *Br. J. Urol.* **79**, 505–515.
10. Yoo, J. J., Meng, J., Oberpenning, F., and Atala, A. (1998) Bladder augmentation using allogenic bladder submucosa seeded with cells. *Urology* **51**, 221–225.
11. Lutolf, M. P. and Hubbell, J. A. (2003) Synthesis and physicochemical characterization of end-linked poly(ethylene glycol)-co-peptide hydrogels formed by Michael-type addition. *Biomacromolecules* **4**, 713–722.
12. Hubschmid, U., Leong-Morgenthaler, P. M., Basset-Dardare, A., Ruault, S., and Frey, P. (2005) In vitro growth of human urinary tract smooth muscle cells on laminin and collagen type I-coated membranes under static and dynamic conditions. *Tissue Eng.* **11**, 161–171.
13. Southgate, J., Masters, J. R. W., and Trejdosiewicz, L. K. (2002) Culture of human urothelium, in *Culture of Epithelial Cells* (Freshney, R. I. F. and Freshney, M. G., eds), Wiley-Liss, New York, NY, pp. 381–399.
14. Sugasi, S., Lesbros, Y., Bisson, I., Zhang, Y. Y., Kucera, P., and Frey, P. (2000) In vitro engineering of human stratified urothelium: analysis of its morphology and function. *J. Urol.* **164**, 951–957.
15. Gomm, J. J., Browne, P. J., Coope, R. C., Liu, Q. Y., Buluwela, L., and Coombes, R. C. (1995) Isolation of pure populations of epithelial and myoepithelial cells from the normal human mammary gland using immunomagnetic separation with Dynabeads. *Anal. Biochem.* **226**, 91–99.
16. Lutolf, M. P., Raeber, G. P., Zisch, A. H., Tirelli, N., and Hubbell, J. A. (2003) Cell-responsive synthetic hydrogels. *Adv. Mater.* **15**, 888–892.
17. Lutolf, M. P., Weber, F. E., Schmoekel, H. G., Schense, J. C., Kohler, T., Muller, R., and Hubbell, J. A. (2003) Repair of bone defects using synthetic mimetics of collagenous extracellular matrices. *Nat. Biotechnol.* **21**, 513–518.
18. Zisch, A. H., Lutolf, M. P., Ehrbar, M., Raeber, G. P., Rizzi, S. C., Davies, N., Schmokel, H., Bezuidenhout, D., Djonov, V., Zilla, P., and Hubbell, J. A. (2003) Cell-demanded release of VEGF from synthetic, biointeractive cell ingrowth matrices for vascularized tissue growth. *FASEB J.* **17**, 2260–2262.

19. Zisch, A. H., Lutolf, M. P., and Hubbell, J. A. (2003) Biopolymeric delivery matrices for angiogenic growth factors. *Cardiovasc. Pathol.* **12**, 295–310.
20. Blindt, R., Krott, N., Hanrath, P., vom Dahl, J., van Eys, G., and Bosserhoff, A. K. (2002) Expression patterns of integrins on quiescent and invasive smooth muscle cells and impact on cell locomotion. *J. Mol. Cell. Cardiol.* **34**, 1633–1644.
21. Moiseeva, E. P. (2001) Adhesion receptors of vascular smooth muscle cells and their functions. *Cardiovasc. Res.* **52**, 372–386.

8

Generation of Multicellular Tumor Spheroids by the Hanging-Drop Method

Nicholas E. Timmins and Lars K. Nielsen

Summary

Owing to their in vivo-like characteristics, three-dimensional (3D) multicellular tumor spheroid (MCTS) cultures are gaining increasing popularity as an in vitro model of tumors. A straightforward and simple approach to the cultivation of these MCTS is the hanging-drop method. Cells are suspended in droplets of medium, where they develop into coherent 3D aggregates and are readily accessed for analysis. In addition to being simple, the method eliminates surface interactions with an underlying substratum (e.g., polystyrene plastic or agarose), requires only a low number of starting cells, and is highly reproducible. This method has also been applied to the co-cultivation of mixed cell populations, including the co-cultivation of endothelial cells and tumor cells as a model of early tumor angiogenesis.

Key Words: MCTS; Multicellular tumor spheroid; Tumor; Angiogenesis; Cancer; 3D cell culture.

1. Introduction

First utilized for the study of tumors by Sutherland and colleagues in the 1970s *(1–4)*, the multicellular tumor spheroid (MCTS) has established itself as an excellent in vitro model of tumors. These small three-dimensional (3D) spherical cellular aggregates recapture many features of real tumor, and respond to anti-tumor treatments in a manner that corresponds well to clinical findings *(5)*.

Transport limitations within MCTS result in the formation of environmental gradients, with consequent changes in cellular phenotype and status, mimicking that observed for tumors in vivo. Concurrently, the enhanced cell–cell and cell–extracellular matrix contacts that 3D cultivation permits, maintain the in

From: *Methods in Molecular Medicine, 2nd ed.: Tissue Engineering*
Edited by: H. Hauser and M. Fussenegger © Humana Press Inc., Totowa, NJ

vivo-like phenotype of cells. By contrast, widely used monolayer cell cultures fail to generate an appropriate environment, cell phenotype and status differ significantly from those found in tumors, and the response to anti-tumor therapies bears little resemblance to that for tumors in vivo.

Although the importance of MCTS in anti-tumor research has been firmly established, the increased complexity of these cultures compared with monolayer cultures has hindered widespread use. A recent development in the field has been the adaptation of "hanging-drop" cultures used by embryologists for the cultivation of MCTS in an efficient and reproducible manner. This method is applicable to a wide variety of cell lines, avoids interference from potential material artifacts or uncontrolled mechanical forces (*6*, *7*), and opens up new avenues for the co-cultivation of mixed cell population MCTS *(8)*.

The method relies on the generation of small drops of cell suspension which are then inverted. Gravity-enforced cell settling, cell–cell adhesion, and subsequent growth result in the formation of homogeneously sized MCTS, the kinetics of which can be readily followed in situ. MCTS are easily harvested for histological evaluation or direct visualization using advanced light microscopy methods.

2. Materials

2.1. Hanging-Drop Culture

1. Cell-culture medium (*see* **note 1**).
2. Cell dissociation solution, for example, 0.25% trypsin/ethylenediaminetetraacetic acid (EDTA)-4Na.
3. Phosphate-buffered saline (PBS) without Ca^{2+} or Mg^{2+}.
4. MicroWell MiniTrays (60 wells; Nunc).

2.2. Cryosectioning

1. 4% paraformaldehyde (PFA) in PBS.
2. Tissue-Tek OCT compound (Sakura Finetek).
3. 30% sucrose solution in PBS.
4. Superfrost Plus glass microscope slides (Menzel-Gläser).
5. Cryomolds.
6. Dry ice.

2.3. Whole Mount for Confocal Microscopy

1. 4% PFA.
2. PBS.
3. Blocking buffer: 6% bovine serum albumin (BSA) + 0.1% Triton-X100 in PBS.
4. Labeling buffer: 1% BSA + 0.1% Triton-X100 in PBS.
5. Antibodies.
6. 80% glycerol in PBS.

3. Methods

3.1. Hanging-Drop Culture

A simple technique requiring little more than a steady pipetting hand, the hanging-drop method is illustrated in **Fig. 1**.

1. Starting from a monolayer cell culture, dissociate the cells by rinsing with PBS followed by incubation with dissociation solution (e.g., trypsin/EDTA).
2. Pellet the cells by centrifugation, discard the supernatant, resuspend the cells in a small volume of cell-culture medium to obtain a single-cell suspension, and enumerate by the preferred method (e.g., hemocytometer).
3. Dilute the cell suspension with culture medium to obtain the desired cell density (*see* **note 2**).
4. Dispense 20 μl cell suspension into the wells of a 60-well minitray. The use of a multistep or multichannel pipette compatible with this plate configuration is recommended.
5. Place a lid on the minitray and invert the entire minitray.
6. Place the minitray inside a larger container along with a small dish of water (e.g., eight minitrays fit in a bioassay tray, *see* **note 3**). Replace the lid on the larger container (*see* **note 4**).

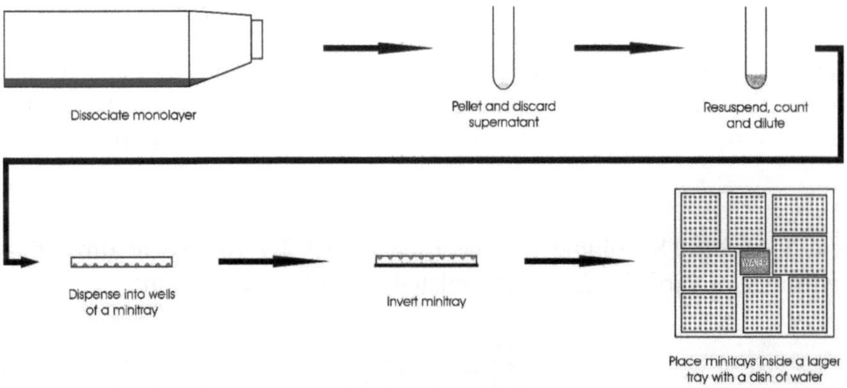

Fig. 1. Preparation of multicellular tumor spheroids (MCTS) by the hanging-drop method. Monolayers are dissociated and pelleted by centrifugation. Cells are then resuspended in growth medium at the desired density, and 20 μl suspension dispensed into each well of a 60-well minitray. With the lid in place, the minitray is inverted and placed inside a large bioassay tray (or other container) with a small reservoir of water. These are then placed in an incubator for cultivation of MCTS.

3.2. Growth Kinetics

Owing to gravity, cells within a hanging-drop settle at the nadir of the medium/air interface. Here they aggregate and, with subsequent proliferation and deposition of ECM, develop into MCTS (*see* **Fig. 2**). This process is readily observed with a phase-contrast microscope and kinetics determined using a digital camera and image processing software.

1. At the desired time points, retrieve cultures from the incubator, and maintaining them in the inverted position, image directly in the well (*see* **note 5**), ensuring good contrast between the MCTS border and the background.
2. Using the preferred analysis software, determine the average diameter (or alternatively determine the projected area of the MCTS and back calculate to an equivalent diameter, *see* **Eq. 1**) and roundness (MCTS perimeter divided by the perimeter of a perfect circle with the same radius, *see* **Eq. 2**).

$$D = \sqrt{4\frac{A}{\pi}} \quad (1)$$

$$\Phi = \frac{P}{\pi D}, \quad (2)$$

where D = diameter; A = projected area; P = measured MCTS perimeter; and Φ = roundness.

A plot of MCTS diameter versus time describes the growth pattern (*see* **Fig. 3**) and typically fits a Gompertzian kinetic model (*see* **Eq. 3**), as is common for MCTS and small avascular tumors in vivo (***9***, ***10***). A plot of roundness versus time (*see* **Fig. 3**) describes the aggregation tendency of the cells and is also indicative of health. In combination, these kinetics data can be used as a measure of therapeutic impact as demonstrated in **Fig. 3** (*see* **note 6**).

$$V_t = V_0 \cdot \mathrm{e}^{\frac{\alpha}{\beta}(1-e^{-\beta t})}, \quad (3)$$

where V_t = MCTS volume at time t; V_0 = MCTS volume at time 0; α = growth-related variable; β = death-related variable; and t = time.

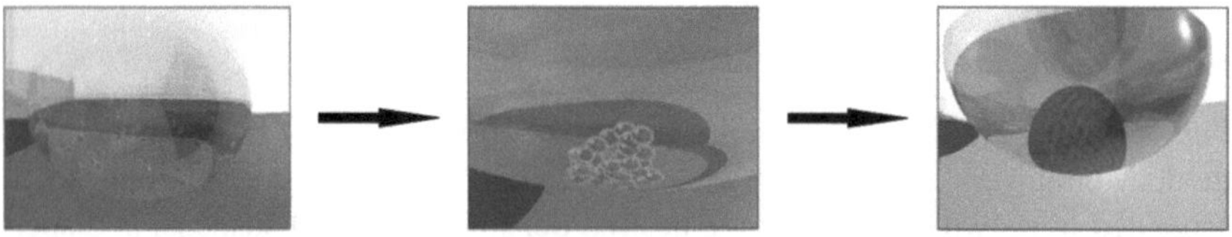

Fig. 2. Cells in a hanging-drop gravity settle at the medium/air interface. Here they aggregate, and with subsequent proliferation and deposition of ECM, develop into a multicellular tumor spheroid (MCTS).

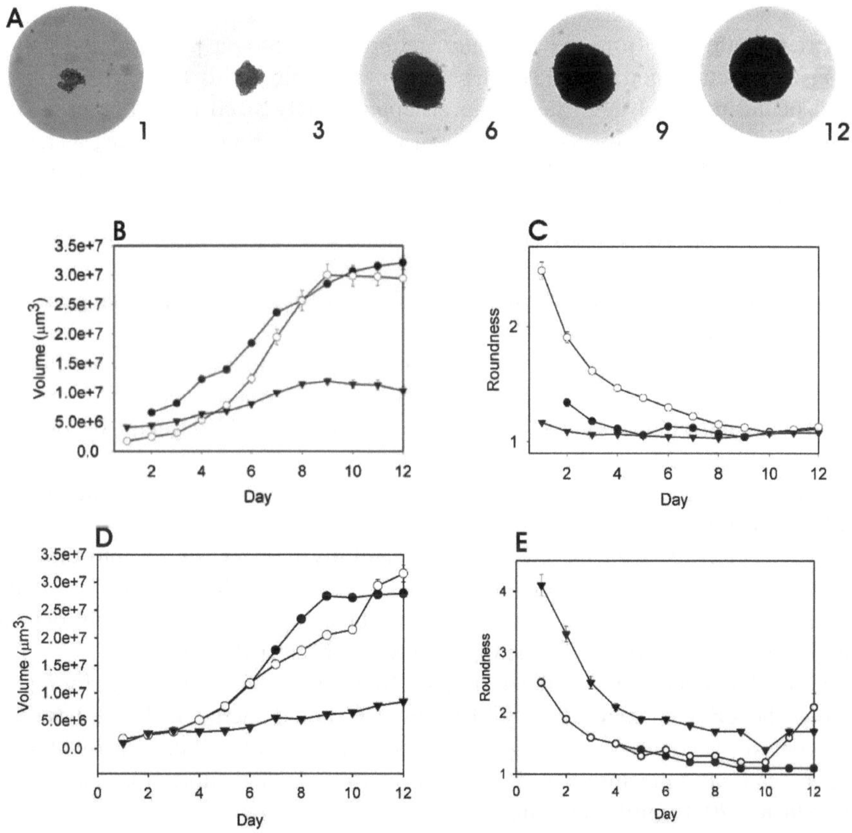

Fig. 3. (a) Phase-contrast images of HepG2 multicellular tumor spheroid (MCTS) developing in hanging drops. Numbers indicate days in hanging-drop culture. (b) Growth of hanging-drop MCTS from HepG2 (○), 293 (▼), and HCT116 (•) cell suspensions over a 12-day period. (c) Aggregation/roundness of the same cultures. (d and e) Effect of experimental anti-tumor agent (compound X) on growth and roundness of HepG2 hanging-drop MCTS respectively. Two milligrams per milliliter compound X was either included in the initial cell suspension (▼) or introduced at day 5 (○). Control cultures without compound X were maintained in parallel (•). Inclusion of compound X in the initial cell suspension severely inhibits MCTS growth and reduces the extent of aggregation achieved. In cultures where compound X was introduced at day 5, a decrease in growth rate is observed followed by a sudden increase in MCTS size after day 10. This corresponds to a sudden increase in roundness at the same time, indicating deterioration of the MCTS. Visual observation confirmed massive deterioration of MCTS in these cultures, with extensive cell debris present in the drops.

3.3. Harvesting and Preparation for Cryosectioning

1. Retrieve minitrays from the incubator, return to an upright position, and using a transfer pipette flood with PBS. Use the transfer pipette to flush MCTS to one end of the tray and then transfer to an appropriately sized tube. For very small MCTS, it may be necessary to pick them individually from wells using a standard single-channel pipette (simply draw in the entire contents of each well and transfer to a tube).
2. Allow MCTS to settle under gravity (5–10 min) and remove PBS. If MCTS do not settle, spin at low G.
3. Immerse the harvested MCTS in 4% PFA for 10 min. Remove PFA and rinse three times for 5 min in fresh PBS.
4. Add a volume of 30% sucrose solution sufficient such that all MCTS are floating, and allow to stand at 4° C until the MCTS have settled at the bottom of the tube. Add an equal volume of OCT to the sucrose solution. Mix gently by pipetting (avoid air bubbles) and allow to stand at 4° C overnight.
5. Prepare cryomolds using a small amount of OCT to cover the bottom, and place on a bed of dry ice to harden.
6. Using pipette tips with the ends cut off, carefully transfer MCTS into the mold. Draw up the MCTS in sucrose/OCT solution, wait briefly for the MCTS to collect at the bottom of the tip (take care to avoid blocking). Dispense slowly onto the mold such that as little OCT/sucrose as possible is carried over.
7. The MCTS should form a frozen droplet. Remove the mold from the dry ice and allow the OCT to melt slightly. The droplet should merge with the base (this prevents air becoming trapped around the edges). Carefully fill the mold with OCT, taking care not to disturb the MCTS, and return to dry ice. Store frozen samples at less than −70° C until sectioning.
8. Using a cryomicrotome, section the frozen blocks. The region where MCTS are present, usually becomes visible to the eye as a slightly yellowish area among the white OCT. Large or numerous MCTS are directly visible. In general, 10 μm sections are easily obtainable.
9. Transfer sections to Superfrost Plus glass microscope slides and proceed with the usual method of staining for visualization.

3.4. Whole Mount for Confocal Microscopy

1. Harvest MCTS with PBS and fix as described in **section 3.3**. (steps 1–3).
2. Block and permeabilize MCTS by immersing in blocking buffer for 30 min.
3. Remove blocking buffer and immerse in primary antibody diluted in labeling buffer.
4. Remove primary antibody solution and wash thoroughly with PBS three times for 5 min.
5. Immerse MCTS in secondary antibody diluted in labeling buffer for 1 h.

6. Remove secondary antibody solution and wash thoroughly with PBS three times for 5 min.
7. Using pieces of coverslips and an adhesive (e.g., nail varnish), build a chamber on a glass microscope slide, of a depth just great enough to accommodate the MCTS. Transfer MCTS to the chamber in PBS and then draw off the PBS.
8. Fill the chamber with a clearing agent of the choice (e.g., 80% glycerol in PBS), place a coverslip over the chamber, and seal with nail varnish.
9. After allowing sufficient time for clearing, the samples are ready for imaging.

3.5. Co-Culture with Human Umbilical Vein Endothelial Cell

The angiogenic potential of many MCTS in vivo is widely reported on, and aspects of this process are observable in vitro when MCTS are co-cultivated with human umbilical vein endothelial cells (HUVEC, *see* **Fig. 4**). HUVEC migrate into the MCTS, and capillary-like networks exhibiting luminal features can subsequently be observed.

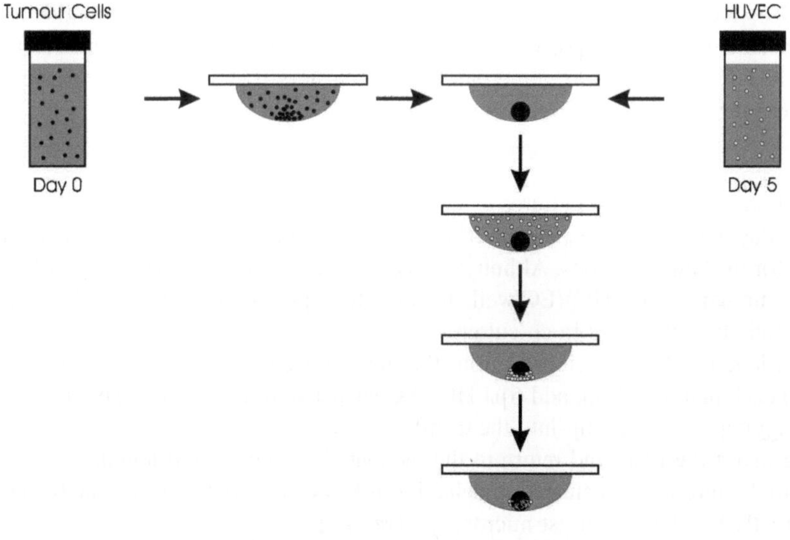

Fig. 4. Schematic for co-culture of human umbilical vein endothelial cells (HUVEC) with multicellular tumor spheroid (MCTS). MCTS hanging-drop culture is initiated as normal, and after 5 days in culture, a suspension of HUVEC is introduced. Typically these HUVEC collect adjacent to the MCTS and merge with it.

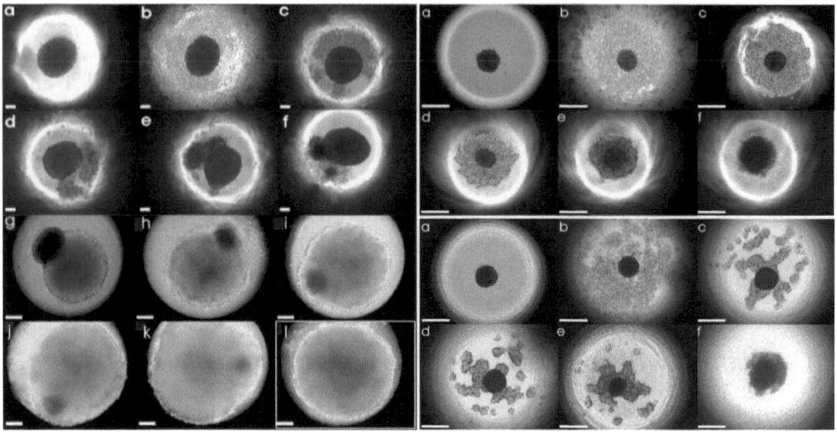

Fig. 5. Clustering and migration behavior of human umbilical vein endothelial cells (HUVEC) added to day 5 hanging-drop multicellular tumor spheroid (MCTS) observed by phase-contrast microscopy. Left, HCT116 MCTS (bar = 100 μm) *(8)*; top right, HepG2 MCTS (bar = 250 μm); lower right, 293 MCTS (bar = 150 μm). (a) Prior to HUVEC addition; (b) 15 min after HUVEC addition; (c–k) 2, 4, 8, 24, 48, 72, 96, and 120 h after addition of HUVEC, respectively; and (l) 120-h control without HUVEC. Distinctive differences are evident between the different MCTS types. Whereas for HCT116, HUVEC form large individual secondary clusters that penetrate into the MCTS as a whole, and HUVEC enter HepG2 MCTS en masse. In the presence of 293 MCTS, HUVEC form numerous smaller clusters that subsequently coalesce and surround the MCTS before entering into it.

1. Initiate hanging-drop cultures as described in **section 3.1**. and cultivate for 5 days.
2. On day 5, prepare a suspension of HUVEC (*see* **notes 7 and 8**) in the same medium as for the hanging drops. Although density is dependent on experimental objectives, in our hands 10^4 HUVEC/well allowed for easy identification of HUVEC and endothelial structures in co-cultures.
3. Retrieve the MCTS minitray from the incubator and return to an upright position. To each hanging drop, add 6 μl HUVEC suspension (care should be taken to avoid dragging the pipette tip into the drop).
4. Re-invert the plate and return to the incubator for the desired length of time.
5. Initial migration and clustering behavior of HUVEC into the MCTS can be observed directly by phase-contrast microscopy (*see* **Fig. 5**).
6. At the desired time point following introduction of HUVEC, MCTS can be harvested as has been described in **section 3.3**. (steps 1 and 2) and processed for histological evaluation, confocal microscopy, or other forms of evaluation (*see* **Fig. 6**).

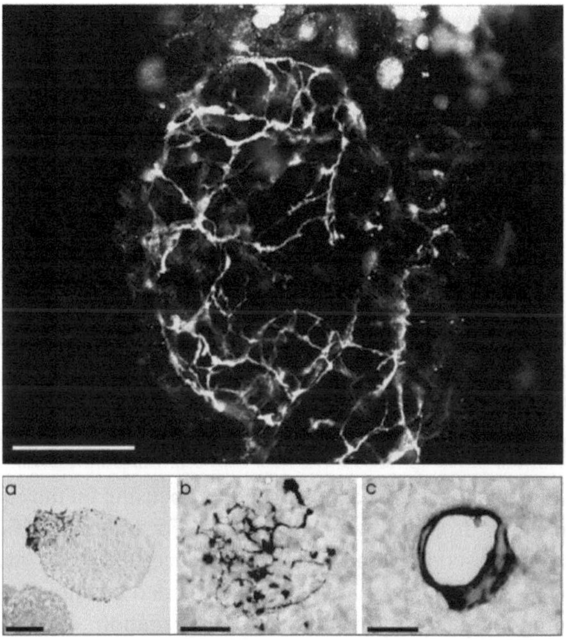

Fig. 6. Endothelial structures in HCT116 multicellular tumor spheroid (MCTS) *(8)*. (Top) Optical section of anti-von Willebrand factor (anti-vWF) labeled whole mount MCTS showing endothelial network, bar = 50 μm. (Bottom) anti-vWF (black) labeled cryosections confirm the presence of these networks (a and b) and reveal lumen-like structures (c). Scale bars, a = 100 μm, b and c = 25 μm.

7. As for monocultures, test agents (e.g., pro- or anti-angiogenic agents) can be introduced to the individual hanging drops (*see* **note 6**).

Notes

1. Although cell-culture medium may be specific for a given cell line, excellent results are typically achieved with high-glucose Dulbecco's modified Eagle's medium (DMEM; Gibco Invitrogen) with 10% fetal bovine serum (FBS). If other media fail to result in MCTS formation, simply switching to DMEM is often highly beneficial. Similarly, increasing FBS concentrations typically enhances aggregation, although the physiological implications of high serum concentrations should be borne in mind. In some circumstances, cultivation in serum-free medium is also possible with the addition of 1% BSA.
2. An initial cell density of 2.5×10^4 cells/ml (500 cells/hanging drop, 20 μl/drop) generally results in the formation of well-formed spheroids. To increase or reduce time required to achieve a specific size, the initial cell density can be altered

accordingly. In the case of very rapidly proliferating cells, very low cell densities may be required to obtain well-formed MCTS. Large numbers of cells resulting from expansion tend to form large irregular masses. Similarly, cultures of poor viability or otherwise containing debris will aggregate poorly, as the debris prevent contact between cells and small clusters that would otherwise coalesce.

3. Placed directly in an incubator, evaporation from the minitrays is rapid. Placement inside a larger container with a reservoir of water significantly reduces the rate of evaporation. Evaporation rate is also influenced by the effectiveness of incubator humidification and incubator usage (e.g., high use or low use). The rate at which evaporation takes place is easily determined by recording the mass of a loaded minitray over 12 days and plotting the change in mass with time (*see* **Fig. 7**). Feeding of fresh medium, when appropriate, is easily achieved by careful addition of a small volume to each drop.
4. If necessary, perforate the lid to permit gas exchange (e.g., air-tight plastic storage containers).
5. Many upright microscopes are easily capable of focusing on the equatorial plane of MCTS in a hanging drop at low magnification. For high magnification and when using inverted microscopes, long working distance objectives may be necessary. When observing the cultures, maintain in an inverted position. Disturbance of the drops by turning the plates over will disrupt aggregation.

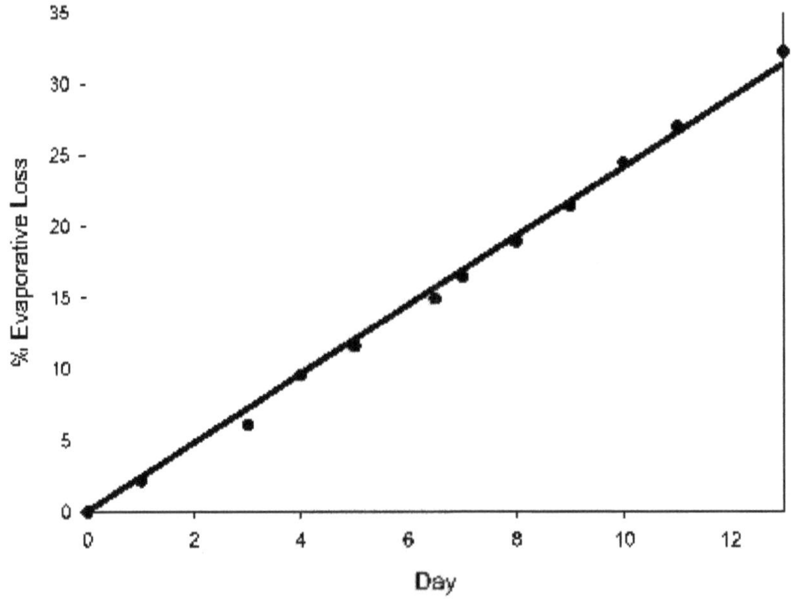

Fig. 7. Evaporative loss from a 60-well minitray over a 12-day period (line = linear fit).

6. Test agents (e.g., chemotherapeutics) can be introduced into this initial cell suspension or added later in a small volume of fresh medium. Choose a concentration of the test agent that will give the appropriate final concentration on dilution into the drop culture, making allowances for evaporation (*see* **note 3**).
7. The endothelial phenotype of HUVEC is progressively lost in monolayer cultures, and the freshest possible cells should be used. As a rule of thumb, cells of passage number greater than 6 should not be used. The medium used for monolayer expansion of HUVEC can also impact upon behavior in co-culture, DMEM/F12 supplemented with 1 ng/ml rhVEGF, 1 ng/ml rhbFGF, 1 ng/ml rhEGF, 10 μg/ml hydrocortisone, 100 μg/ml heparin, 100 μg/ml endothelial cell growth supplement (Sigma, E0760), and 10% FBS performed well in our hands.
8. HUVEC can also be introduced into the initial MCTS cell suspension. However, in those cell lines tested by us, this did not result in the formation of endothelial networks.

References

1. Inch, W. R., McCredie, J. A., and Sutherland, R. M. (1970) Growth of nodular carcinomas in rodents compared with multi-cell spheroids in tissue culture. *Growth* **34**, 271–282.
2. Sutherland, R. M. and Durand, R. E. (1976) Radiation response of multicell spheroids–an in vitro tumour model. *Curr. Top. Radiat. Res. Q.* **11**, 87–139.
3. Sutherland, R. M., Inch, W. R., McCredie, J. A., and Kruuv, J. (1970) A multi-component radiation survival curve using an in vitro tumour model. *Int. J. Radiat. Biol. Relat. Stud. Phys. Chem. Med.* **18,** 491–495.
4. Sutherland, R. M., McCredie, J. A., and Inch, W. R. (1971) Growth of multicell spheroids in tissue culture as a model of nodular carcinomas. *J. Natl. Cancer Inst.* **46,** 113–120.
5. Desoize, B. and Jardillier, J. (2000) Multicellular resistance: a paradigm for clinical resistance? *Crit. Rev. Oncol. Hematol.* **36,** 193–207.
6. Kelm, J. M., Timmins, N. E., Brown, C. J., Fussenegger, M., and Nielsen, L. K. (2003) Method for generation of homogeneous multicellular tumor spheroids applicable to a wide variety of cell types. *Biotechnol. Bioeng.* **83,** 173–180.
7. Timmins, N. E., Maguire, T. L., Grimmond, S. M., and Nielsen, L. K. (2004) Identification of three gene candidates for multicellular resistance in colon carcinoma. *Cytotechnology* **46,** 9–18.
8. Timmins, N. E., Dietmair, S., and Nielsen, L. K. (2004) Hanging-drop multicellular spheroids as a model of tumour angiogenesis. *Angiogenesis* **7,** 97–103.
9. Kunz-Schughart, L. A., Kreutz, M., and Knuechel, R. (1998) Multicellular spheroids: a three-dimensional in vitro culture system to study tumour biology. *Int. J. Exp. Pathol.* **79,** 1–23.
10. Sutherland, R. M. and Durand, R. E. (1984) Growth and cellular characteristics of multicell spheroids. *Recent Results Cancer Res.* **95,** 24–49.

9

In Vitro Vascularization of Human Connective Microtissues

Jens M. Kelm, Wolfgang Moritz, Doerthe Schmidt, Simon P. Hoerstrup, and Martin Fussenegger

Summary

Vascularization is one of the most central processes enabling multicellular life. Owing to the complexity of vascularization regulatory networks, minor control imbalances often have severe pathologic consequences ranging from ischemic diseases to cancer. Tissue engineers are immediately confronted with vascularization as artificial tissues of a clinically relevant size require a vascular system to ensure vital physiologic logistics throughout the entire tissue and to enable rapid connection to the host vasculature following implantation. Using human umbilical vein endothelial cells (HUVECs) coated onto a human aortic fibroblast (HAF) core microtissue generated by gravity-enforced self-assembly in hanging drops, we created a tissue-culture system for studying capillary network formation. We provide comprehensive technical insight into the design and analysis of prototype vascularization in multicell-type-based microtissues. Detailed understanding of generic processes managing capillary formation in human tissue culture may foster advances in the development of clinical tissue implants.

Key Words: Self-assembly; Hanging drop technology; Angiogenesis; Microscale tissue engineering; Vascular endothelial growth factor.

1. Introduction

Growing mammalian cells in a way to form artificial tissues in a reliable manner remains a major challenge for the tissue engineering community, which is determined to deliver clinically licensed tissue replacements in the not-too-distant future *(1)*. It is generally accepted that tissues shape matters, and on these grounds, an entire science has been developed to design scaffold structures, which enable inoculated cell populations to form tissues with desired shape.

From: *Methods in Molecular Medicine, 2nd ed.: Tissue Engineering*
Edited by: H. Hauser and M. Fussenegger © Humana Press Inc., Totowa, NJ

Although pioneering scaffolds simply provided a three-dimensional substrate *(2)* for cell implants, their latest-generation prototypes are functionalized to enable physiologically triggered tissue-mimetic functions at the implantation site *(3)* and to disintegrate over time or on demand *(4,5)*. As the scaffold technology is getting increasingly mature, novel challenges appear at the horizon: how could synthetic tissues be vascularized to enable production of tissue implants with increased size and to link them to the host vasculature following implantation *(6)*.

Gravity-enforced self-assembly of monodispersed cells to microtissues in hanging drops is gathering momentum as an alternative to using scaffolds for prototype tissue engineering initiatives. This technology capitalizes on the fundamental self-assembly processes evolved to organize tissue structures during embryogenesis *(7,8)*. Similar to the cell mass in early blastular stage, different cell types within a microtissue have been observed to self-organize according to the differential adhesion hypothesis, which suggests that intramicrotissue structures form as a result of different surface proteins expressed on different cell types *(9–11)* . Coating of human umbilical vein endothelial cells (HUVECs) onto a core spheroid composed of human aortic fibroblasts (HAFs) in hanging drops triggers a unique self-organization event resulting in a HUVEC-composed capillary network penetrating the entire HAF-based microtissue. We provide detailed protocols associated with in vitro vascularization of human microtissue to foster advances in prototype tissue engineering *(12)*.

2. Materials

2.1. Cell Culture and Isolation of Primary Cells

1. HUVEC (PromoCell, Heidelberg, Germany; cat. no. C-12200).
2. Biopsy of the human aorta (provided according to Swiss legislation).
3. Endothelial Cell Growth Medium Kit (PromoCell; cat. no. C-22110) and 10% fetal calf serum (FCS, PAA Laboratories, Linz, Austria; cat. no. A-15-022, lot no. A01129-242).
4. Dulbecco's modified Eagle's medium (DMEM, Invitrogen, Carlsbad, CA; cat. no. 41966-029) supplemented with 10% FCS (PAA Laboratories; cat. no. A-15-022, lot no. A01129-242) and 1% penicillin/streptomycin.
5. Collagenase A solution (Roche, Basel, Switzerland; cat. no. 1088785). Dissolve 2 mg collagenase A in 1 ml serum-free DMEM. Always prepare a fresh solution and keep on ice.
6. Sterile phosphate-buffered saline (PBS): 150 mM NaCl, 6.5 mM $Na_2HPO_4 \cdot 2H_2O$, 2.7 mM KCl, and 1.5 mM KH_2PO_4, pH 7.4 (Sigma-Aldrich, Buchs, Switzerland; cat. no. P3813).
7. Trypsin/ethylenediaminetetraacetic acid solution (EDTA) (0.05% trypsin, 0.02% EDTA in PBS; PAN Biotech GmbH, Aidenbach, Germany; cat. no. P10-023100).

8. Penicillin/streptomycin solution ×100 (Invitrogen; cat. no. 15140-122).
9. Gentamycine ×1000 (PAN Biotech GmbH; cat. no. P06-13050).
10. Fungizone® ×100 (Amphotericin B; PAN Biotech GmbH; cat. no. P06-01050).
11. Dimethylsulfoxide (DMSO) (Sigma-Aldrich; cat. no. D2438).

2.2. Microtissue Production in Hanging Drops

1. 60-well Terasaki plates (Greiner Bio-One, St. Gallen, Switzerland; cat. no. 653161) (*see* **note 1**.)
2. Casy® cell counter (Schaerfe System GmbH, Reutlingen, Germany; such as CASY® Model TT).
3. Improved Neubauer chamber (Brand, Wertheim, Germany; cat. no. 717806).
4. Variable 8-channel pipettor, 2–125 μl (Matrix Impact2, Matrix, Cheshire, UK; cat. no. 2021).
5. Sterile tip tubes for multichannel pipettor.
6. Whatman filter paper (3 Chr), 4 cm × 7.5 cm in size, sterile (Whatman®, Brentford, UK; cat. no. 3001-917).

2.3. Vascular Endothelial Growth Factor Profiling

1. 96-well high-binding plate (Greiner Bio-One; cat. no. 655-061) coated with the vascular endothelial growth factor (VEGF)-specific capture antibody (ELISA Development System human VEGF, R&D Systems, Minneapolis, MN; cat. no. DY293B). Store capture antibody at −20°C. In brief, dilute capture antibody at 1:180 in PBS (adjust final volume according to the sample number) and pipet 100 μl/well of the 96-well plate. Seal the plate and incubate overnight at room temperature. Wash each well three times with 400 μl wash buffer (*see* **note 2**). Block each well with 300 μl reagent diluent and incubate for 1 h at room temperature. Wash three times with 400 μl wash buffer. 96-well plates are now ready for use (*see* **note 3**).
2. VEGF detection antibody (ELISA Development System human VEGF, DY293B; R&D Systems). Store aliquots at −20°C.
3. Streptavidin-conjugated horseradish peroxidase (ELISA Development System human VEGF, DY293B; R&D Systems). Store at 2–8°C.
4. VEGF standard (ELISA Development System human VEGF, DY293B; R&D Systems). Store aliquots at −20°C.
5. PBS: 150 mM NaCl, 6.5 mM $Na_2HPO_4 \cdot 2H_2O$, 2.7 mM KCl, and 1.5 mM KH_2PO_4, pH 7.4 (Sigma Chemicals, St. Louis, MO).
6. Wash buffer: 0.05% Tween® 20 in PBS, pH 7.2–7.4.
7. Reagent diluent: 1% BSA in PBS, pH 7.2–7.4. Store at 2–8°C.
8. Substrate solution (1:1 mixture of H_2O_2 and tetramethylbenzidine; R&D Systems; cat. no. DY994).
9. Microplate reader (e.g., TECAN GENios Pro™, Tecan AG, Maennedorf, Switzerland, or any equipment recording at 450 nm).

2.4. Confocal Microscopy of Microtissues

1. 2% paraformaldehyde (PFA) solution in PBS: Dissolve 2 g reagent-grade PFA powder (Sigma-Aldrich; cat. no. P6148) into an Erlenmeyer flask and adjust to 100 ml using deionized water. Heat the solution to 70°C and mix it using a magnetic stir bar. Perform this operation in a safety cabinet. Add one drop of 10 M NaOH (Sigma-Aldrich; cat. no. 72068); the solution should immediately clarify. Cool down the solution to room temperature and adjust it to pH 7.4.
2. Permeabilization stock solution: 10% Triton-X 100 (Sigma-Aldrich; cat. no. T8787) in PBS. For permeabilization, always prepare a fresh solution by diluting the stock solution to 0.05% in PBS.
3. Tris-buffered saline (TBS): 20 mM Tris base, 155 mM NaCl, 2 mM ethylene glycol tetraacetic acid (EGTA), and 2 mM $MgCl_2$.
4. Bovine serum albumin (Sigma-Aldrich; cat. no. B4287).
5. Lisbeth's medium: Tris-buffered glycerol [a 3:7 mixture of 0.1 M Tris–HCl (pH 9.5) and glycerol supplemented with 50 mg/ml *n*-propyl-gallat (Sigma-Aldrich; cat. no. P3130)].
6. A633-Phalloidin (Molecular Probes Inc., Eugene, OR; cat. no. A22284).
7. Primary antibodies: (i) Mouse monoclonal anti-human CD31 antibody (Sigma Chemicals; cat. no. P8590, Clone WM-59) and (ii) rabbit polyclonal anti-human von Willebrand factor (vWF) antibody (Sigma Chemicals; cat. no. F3520).
8. Secondary antibodies: (i) Cy3™-coupled anti-mouse antibody (Jackson Immunochemicals, West Grove, PA; cat. no. 115-165-146) and (ii) fluoreszein isothiocyanat (FITC)-coupled anti-rabbit (ICN Pharmaceuticals, Hyland, CA; cat. no. 55646).
9. Confocal microscope (e.g., Leica SP2 with excitation wavelength of 433, 520, and 633 nm; Leica Microsystems, Glattbrugg, Switzerland). The imaging system consists of an inverted fluorescence microscope equipped with a ×20 oil immersion objective, a confocal scanner featuring an argon and helium–neon laser, and a Silicon Graphics Workstation with Imaris three-dimensional multichannel image processing software ***(13)***, which has been used to analyze the confocal data.

2.5. Histologic Analysis of Microtissues

1. 2% (w/v) PFA solution (see Chapter 2.4, section 1).
2. 15% sucrose solution (w/v) in H_2O.
3. High-speed microcentrifuge equipped with swing-out or horizontal fixed-angle rotor (e.g., Beckmann Microfuge B).
4. 1.5 ml Eppendorf tubes.
5. PBS.
6. 1% agarose (w/v) in PBS (Sigma-Aldrich; cat. no. A9539).
7. Dye solution: 0.05% (w/v) bromophenol blue (Sigma-Aldrich; cat. no. B0126) in 10 mM Tris–HCl (pH 7.4) and 30% glycerol.
8. Disposable plastic pipette (Copan, Corona, CA; cat. no. 219C).

9. SuperFrost® Plus slide (Menzel-Glaeser, Braunschweig, Germany).
10. Microtome ultracut device (such as Leica RM 2155, Leica Microsystems).

3. Methods

Initially established for the production of embryoid bodies to study the differentiation potential of stem cells ***(14)***, assembly of microtissues in hanging drops was found to be compatible with a variety of cell types and is becoming increasingly popular among tissue engineers for de novo design of artificial tissues ***(15–19)*** (*see* **Fig. 1**). Key advantages of hanging drop-based microtissue design include (i) precise size control by titration of monodispersed cells, (ii) the dispensability of scaffolds, and (iii) the mobility of individual cells during assembly, which facilitates correct relative cell positioning and development of polarity in microtissues consisting of several cell types.

To get insight into endothelial cell migration and capillary development, we established a microtissue-based coculture system in which core spheroids assembled from HAFs are coated with HUVECs. HUVEC-coated HAF microtissues developed capillary-like networks in the absence of supplementary growth factors (*see* **Fig. 2**).

3.1. Primary Cell Culture

3.1.1. Isolation of HAFs

1. Transfer the tissue of a human aortic biopsy immediately into gentamycine-containing (200 μg/ml) PBS and amphotericin-containing (2.5 μg/ml) PBS and incubate for 15 min at room temperature. Eliminate adventitia and fat from the vessel segments of the human aorta (*see* **note 4**).
2. Endothelial cells have to be removed prior to HAF isolation. Rinse the vessel lumen with PBS, ligate the aorta with either a silk thread or a vessel clamp, depending on

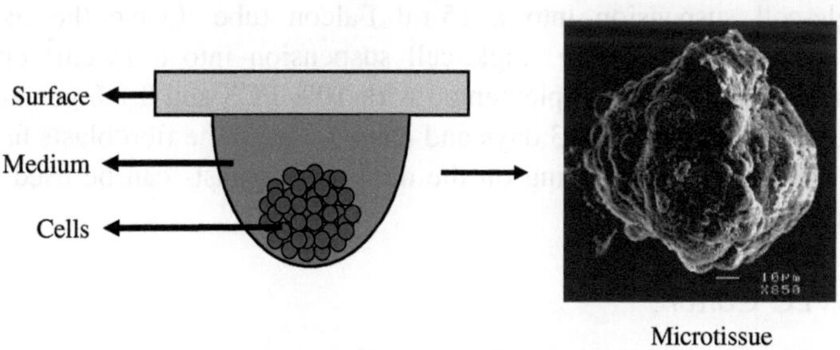

Fig. 1. Schematic representation of gravity-enforced self-assembly of monodispersed mammalian cells to microtissues in hanging drops.

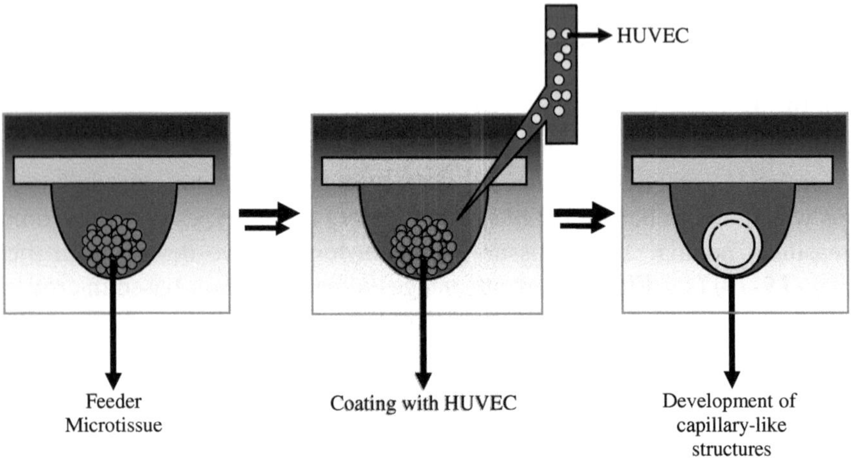

Fig. 2. Diagram showing human umbilical vein endothelial cell (HUVEC)-mediated coating of core human aortic fibroblast (HAF) spheroids and resulting development of a capillary-like network.

the size of the biopsy, and fill the lumen with collagenase A solution. Incubate for 20 min at 37°C.

3. After collagenase A treatment, open the vessel and release the cell suspension. Flush the inner lumen of the aorta with 10 ml DMEM containing 20% FCS (*see* **note 5**).
4. Mince the tissue and transfer 5–7 3 mm × 3 mm pieces into a cell-culture dish (3.5 cm in diameter) and cultivate them in DMEM supplemented with 10% FCS, 1% penicillin/streptomycin solution, and 1% Fungizone®.
5. Detach HAFs, which had grown out of the tissue pieces, at 70–80% confluence (10–14 days of culture) for 3–5 min with 1 ml trypsin.
6. Inactivate trypsin by the addition of 4 ml 10% FCS-containing DMEM and transfer the single-cell suspension into a 15-ml Falcon tube. Leave the tissue pieces sedimenting and transfer the single-cell suspension into a 75-cm^2 culture flask containing 10 ml DMEM supplemented with 10% FCS and 0.1% gentamycine.
7. Exchange the medium every 3 days and passage when the fibroblasts have reached 70–80% confluence. Depending on the donor, fibroblasts can be used for experiments up to passage 4–6.

3.1.2. HUVEC Culture

Prepare a frozen stock for off-the-shelf use to ensure constant HUVEC quality throughout a particular set of experiments. HUVECs purchased from PromoCell are cultivated up to passage 4 and frozen in aliquots of 2.5×10^5

cells/vial in freezing medium (45 ml FCS supplemented with 5 ml DMSO) according to standard protocols.

1. 2–3 days before coating of HAF core spheroids, thaw one vial of HUVECs in a water bath set to 37°C until only a few ice crystals remain.
2. Transfer thawed cells into a 25-cm^2 cell-culture dish containing 5 ml prewarmed endothelial cell growth medium and incubate for 2–3 days at 37°C in a humidified 5% CO_2-containing atmosphere.
3. Change medium once after 24 h.

3.2. Production of HAF Core Spheroids

The size of HAF-composed microtissues depends on the number of cells inoculated for gravity-enforced self-assembly in hanging drops. Studies on in vitro vascularization typically require HAF cores spheroids assembled from 10,000 cells. Because the cell number–size correlation may differ in a donor-dependent manner, it has to be established for every donor batch.

1. Wash HAFs expanded in monolayer cultures twice with PBS and trypsinize for 3–5 min (*see* **note 6**). Inactivate trypsin with 10% FCS-containing DMEM and collect the HAF single-cell suspension in a Falcon® tube.
2. Determine the HAF cell concentration by using either a CASY® TT cell counter or an improved Neubauer chamber.
3. Dilute the cell suspension to reach cell densities required for the desired microtissue size in a 20-μl hanging drop: 500 cells/microtissue (2.5×10^4 cells/ml), 2500 cells/microtissue (1.25×10^5 cells/ml), 5000 cells/microtissue (2.5×10^5 cells/ml), and 10,000 cells/microtissue (5×10^5 cells/ml).
4. Pipette 20 μl single-cell suspension per well of a 60-well plate using an 8-channel pipettor (or single pipette) (*see* **note 7**). Place a Whatman filter paper in the lid and moisten with 1 ml PBS. Cultivate the plates upside down for 48 h at 37°C in a humidified 5% CO_2-containing atmosphere to allow HAF-based microtissue formation.
5. Cultivate HAFs in hanging drops for 2 days if the microtissues are coated with HUVECs or for 4 days for quantification of VEGF production in differently sized microtissues.

3.3. VEGF Profiling

VEGF is a key protein managing neovascularization under physiologic and pathologic conditions *(20)*. The biological impact of VEGF depends on its production level: loss of one *vegf* allele results in fatal vascular defects during embryogenesis *(21)*, suboptimal VEGF levels lead to postnatal angiogenesis and ischemic heart disease *(22)*, and excessive VEGF production results in tumor formation *(23)*. To provide insight into tissue-based vascularization, we profiled VEGF production in HAF microtissues of different size (*see* **Fig. 3**).

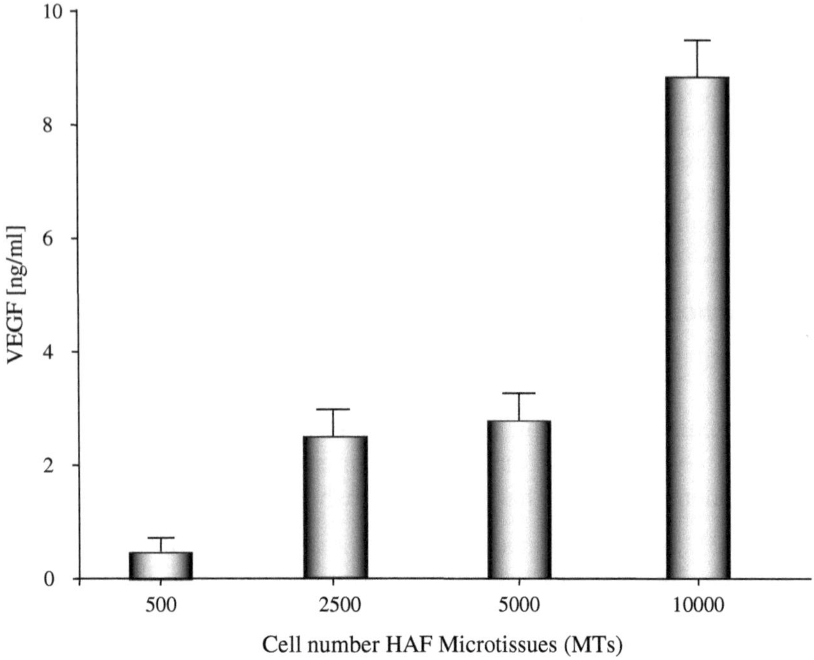

Fig. 3. Quantification of vascular endothelial growth factor (VEGF) production in the supernatant of differently sized human aortic fibroblast (HAF)-composed microtissues assembled from 500, 2500, and 5000 cells and 10,000 monodispersed cells.

1. Collect the supernatant of microtissue culture and centrifuge for 5 min at 4°C and 4000 rpm (1780 g) to remove cells and cell debris.
2. In the meantime, dilute the VEGF standard in reagent diluent (1:200, 1:400, 1:1000, and 1:4000) and pipette 100 μl of each dilution in triplicate into a 96-well high-binding plate.
3. Pipette 100 μl desired cell-culture supernatants in triplicate into a 96-well high-binding plate. Include 10% FCS-supplemented DMEM and pure reagent diluent as control. Incubate for 2 h at room temperature.
4. Wash each well three times with wash buffer.
5. Add 100 μl diluted detection antibody (1:180 in reagent diluent) into each well and incubate for 2 h at room temperature.
6. Wash three times with wash buffer.
7. Add 100 μl working dilution of the streptavidin-conjugated horseradish peroxidase to each well. Incubate for 20 min at room temperature in the dark.
8. Wash three times with wash buffer.
9. Add 100 μl substrate solution (mix reagent A and B in the ratio 1:1 immediately before use) and incubate for 20 min at room temperature in the dark.

10. Add 50 μl 2N H_2SO_4 to stop the enzymatic reaction.
11. Immediately determine the optical density of each well using microplate reader set to 450 nm. VEGF concentrations are calculated by correlation with the standard curve.

3.4. Coating of the HAF Core Spheroids With HUVECs

Two days after cultivation of monodispersed HAFs in hanging drops, they have self-assembled to spheroids and are ready for HUVEC-based coating. This is achieved by adding monodispersed HUVECs to the HAF-composed core spheroids.

1. HUVECs are washed once with PBS (*see* **note 8**).
2. Add 750 μl trypsin into a 25-cm^2 culture flask.
3. Incubate HUVECs for 2 min at 37°C.
4. Inactivate trypsin by adding 4.25 ml 10% FCS-containing DMEM and transfer the cell suspension into a 15-ml Falcon tube.
5. Determine the HUVEC cell concentration by using either a CASY® TTC cell counter or an improved Neubauer chamber.
6. Centrifuge the cell suspension at 1200 rpm (160 g) for 3 min, discard the supernatant by aspiration, and resuspend the cell pellet with the appropriate volume of 10% FCS-containing DMEM to adjust HUVECs to 2×10^5 cells/ml.
7. Add 5 μl HUVEC cell suspension to each HAF-composed core spheroid (1000 HUVECs/HAF spheroid) cultured in individual wells of a 60-well plate. Moisten the Whatman filter paper in the lid with 1 ml PBS and continue to incubate the 60-well plate upside down for 6 days at 37°C in humidified 5% CO_2-containing atmosphere to allow HUVEC coating of HAF core spheroids (*see* **note 9**).

3.5. Confocal Microscopy

Confocal microscopy-based immunofluorescence analysis enables three-dimensional projections of structures 80–100 μm deep inside microtissues (*see* **Fig. 4**).

1. Harvest microtissues into a 10-ml dish and transfer into a 15-ml Falcon® tube. Wash twice with PBS.
2. Fix the microtissues for 1 h in PBS containing 2% PFA.
3. Wash fixed microtissues three times for 5 min in phosphate-buffered Triton-X 100 (PBT, PBS containing 0.002% Triton-X 100).
4. Permeabilize the microtissues for 60 min in PBS containing 0.05% Triton-X 100.
5. Dilute primary antibodies specific for vWF and CD31 in 1% BSA-containing TBS (vWF, 1:100; CD31 1:100). Incubate microtissues with desired antibody solution overnight at 4°C.
6. Wash three times with PBS.

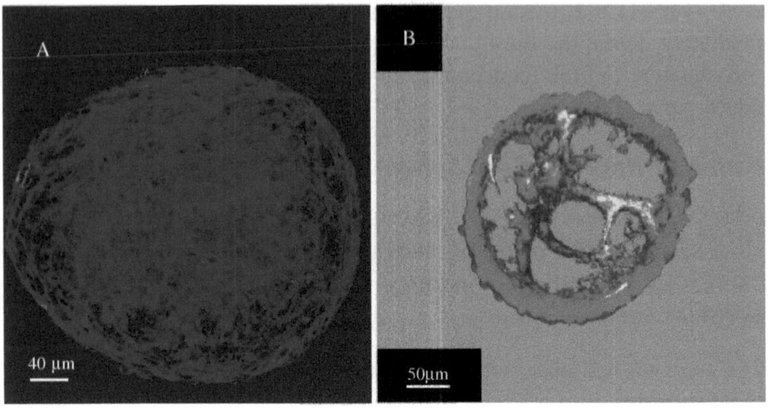

Fig. 4. Immunofluorescence-based confocal micrographs of human aortic fibroblast (HAF)-composed microtissues. (a) Section of a HAF microtissue, which was stained for F-actin with A633-coupled phalloidin. (b) Three-dimensional confocal projection of the capillary-like structures stained for CD31 (shown in red) and von Willebrand factor (vWF) (shown in green).

7. Primary antibodies are visualized using Cy3™-coupled anti-mouse (1:250) and FITC-coupled anti-rabbit (1:100) secondary antibodies. A633-coupled phalloidin (1:100) is used to visualize the entire microtissue by staining F-actin.
8. Wash three times with PBS.
9. Use 0.5 mm spacers (made of silicon) to prevent crunching of the microtissues between slide and cover slip. A drop of Lisbeths's medium is placed between the spacers and the microtissues placed into the drop (*see* **note 10**). The cover slip is sealed with nail varnish. The samples can be stored for several weeks at 4°C in the dark prior to analysis.
10. Analyze samples using a confocal microscope (e.g., Leica SP2).

3.6. Histologic Analysis of Microtissues

Detailed assessment of tissue morphology combined with precise expression profiling of specific biomarkers is essential for characterization of microtissues. In many situations, the amount or size of microtissues is not sufficient to apply regular embedding procedures. We have therefore developed a straightforward protocol to prepare minute amounts of microtissues for histological analysis (*see* **Fig. 5**).

1. Fix microtissues in 2% PFA for 1 h on ice, wash them once with 15% sucrose for 1 h on ice, and resuspend them in PBS.
2. Prepare 1% agarose solution in PBS, boil, and let cool down to approximately 40–45°C with constant stirring.

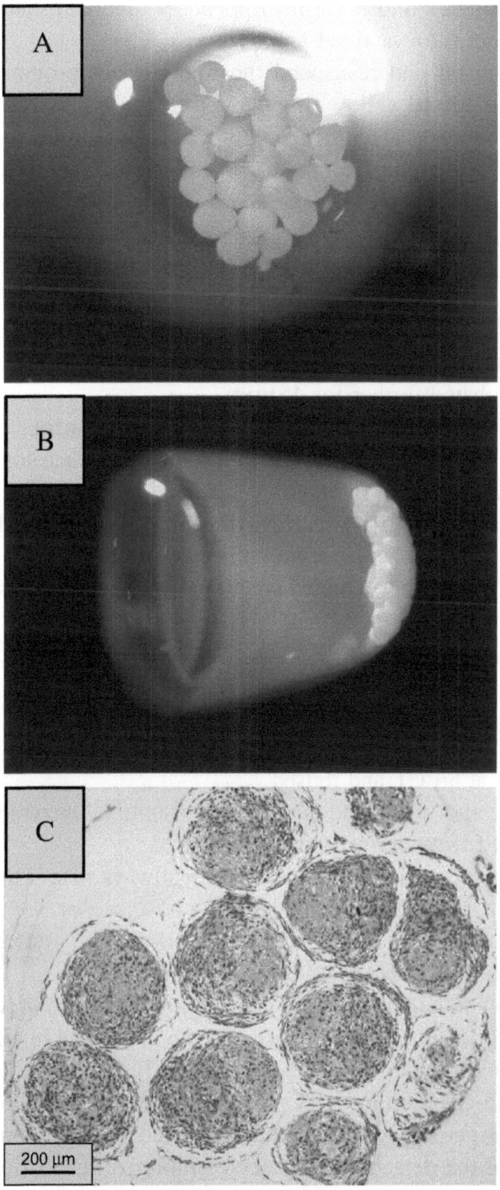

Fig. 5. Human aortic fibroblast (HAF)-composed microtissues cast into agarose prior to paraffin embedding (a and b). Hematoxylin and Eosin (Sigma-Aldrich, Buchs, Switzerland; cat. no. 03972) stained sections generated from paraffin-embedded microtissue-containing agarose plugs (c).

3. Transfer microtissues into a 1.5-ml Eppendorf tube and quick spin them for 3 s using a swing-out rotor or fixed-angle rotor.
4. Remove supernatant and resuspend the microtissues promptly in 30 μl agarose solution using a 200-μl yellow tip enlarged by cutting its tail off. Quick spin the sample for 2 s by using a centrifuge with a horizontal rotor.
5. Remove the agarose plug from the Eppendorf tube by adding 250 μl PBS to prevent drying out and remove the agarose plug with disposable plastic pipette (*see* **note 11**).
6. Transfer the agarose plug into a conventional paraffin-embedding cassette and dehydrate the tissue sample by serial incubation in ×2 70% Ethanol (EtOH), ×2 90% EtOH, ×2 100% EtOH, and ×2 toluidine (each step 10 min). After dehydration, samples are placed into an embedding mold filled with liquid paraffin (60°C) and incubated further for 45 min.
7. Assemble cassette lid and replenish the mold with paraffin.
8. Cut 3–5 μm sections of the paraffin block using a microtome (e.g., Leica RM 2155) and have them float on distilled water adjusted to 40°C.
9. Transfer the sections onto a SuperFrost® slide. Allow the slides to dry overnight and store them at room temperature until use.
10. Rehydrate slides and process them using standard (immuno)-histochemical staining protocols.

Notes

1. Plates can be washed with deionized water after use, sterilized with cold gas (ethylene oxide at 56°C), and reused.
2. Complete removal of the liquid is required for optimal enzyme-linked immunoadsorbent assay (ELISA) performance.
3. Plates can be stored at 4°C, but sensitivity is always best if they are prepared for immediate use.
4. Using a stereomicroscope (e.g., Leica MS5, Leica Microsystems) for removing fat and adventitia results in HAF cultures of higher purity.
5. In case aortic endothelial cells could be used for other purposes, centrifuge the flush solution for 5 min at 1200, resuspend the cell pellet in 10 ml endothelial cell growth medium (EBM-2; Cambrex, Verviers, Belgium; cat. no. CC-3156), and incubate in a 25-cm^2 culture flask. Reseed the aortic endothelial cells at 6×10^4 endothelial cells per cm^2 once they have reached 70–80% confluence.
6. Visually control the detachment process—minimize incubation with trypsin as it compromises microtissue assembly performance.
7. The distance of the single wells in the 60-well plate is too narrow to use all 8-channels of the multichannel pipette. Therefore, it is required to load only five channels with pipette tips. The distance of the pipettes is adjusted, so that the tips fit into every second well.

8. HUVECs adhere not too strong to the culture dish surface. A single washing step is therefore sufficient.
9. Remoisten the Whatman filter paper with a few drops of H_2O every 2 days to prevent the dry up of the hanging drops.
10. Because microtissues adhere strongly to plastics, rinse the pipette tips with 1% BSA-containing PBS before pipetting the microtissues. Such treatment prevents adhesion of the microtissues to the inner tip surface.
11. Dip the tip of the agarose plug into a dye solution (e.g., bromophenol blue) to facilitate location of the microtissues during production of sections (optional).

References

1. Griffith, L. G. and Naughton, G. (2002) Tissue engineering – current challenges and expanding opportunities. *Science* **295**, 1009–1014.
2. Hollister, S. J. (2005) Porous scaffold design for tissue engineering. *Nat. Mater.* **4**, 518–524.
3. Lutolf, M. P. and Hubbell, J. A. (2005) Synthetic biomaterials as instructive extracellular microenvironments for morphogenesis in tissue engineering. *Nat. Biotechnol.* **23,** 47–55.
4. Neuenschwander, S. and Hoerstrup, S. P. (2004) Heart valve tissue engineering. *Transpl. Immunol.* **12,** 359–365.
5. O'Reilly, M. S., Boehm, T., Shing, Y., Fukai, N., Vasios, G., Lane, W. S., Flynn, E., Birkhead, J. R., Olsen, B. R., and Folkman, J. (1997) Endostatin: an endogenous inhibitor of angiogenesis and tumor growth. *Cell* **88,** 277–285.
6. Vogel, V. and Baneyx, G. (2003) The tissue engineering puzzle: a molecular perspective. *Annu. Rev. Biomed. Eng.* **5,** 441–463.
7. Jakab, K., Neagu, A., Mironov, V., Markwald, R. R., and Forgacs, G. (2004) Engineering biological structures of prescribed shape using self-assembling multicellular systems. *Proc. Natl. Acad. Sci. U. S. A.* **101**, 2864–2869.
8. Wang, F., Dumstrei, K., Haag, T., and Hartenstein, V. (2004) The role of DE-cadherin during cellularization, germ layer formation and early neurogenesis in the Drosophila embryo. *Dev. Biol.* **270**, 350–363.
9. Duguay, D., Foty, R. A., and Steinberg, M. S. (2003) Cadherin-mediated cell adhesion and tissue segregation: qualitative and quantitative determinants. *Dev. Biol.* **253**, 309–323.
10. Foty, R. A. and Steinberg, M. S. (2005) The differential adhesion hypothesis: a direct evaluation. *Dev. Biol.* **278**, 255–263.
11. Steinberg, M. S. (1963) Reconstruction of tissues by dissociated cells. Some morphogenetic tissue movements and the sorting out of embryonic cells may have a common explanation. *Science* **141**, 401–408.
12. Kelm, J. M., Diaz Sanchez-Bustamante, C., Ehler, E., Hoerstrup, S. P., Djonov, V., Ittner, L., and Fussenegger, M. (2005) VEGF profiling and angiogenesis in human microtissues. *J. Biotechnol.* **118**, 213–229.

13. Messerli, J. M., van der Voort, H. T., Rungger-Brandle, E., and Perriard, J. C. (1993) Three-dimensional visualization of multi-channel volume data: the amSFP algorithm. *Cytometry* **14,** 725–735.
14. Keller, G. M. (1995) In vitro differentiation of embryonic stem cells. *Curr. Opin. Cell Biol.* **7**, 862–869.
15. Kelm, J. M., Ehler, E., Nielsen, L. K., Schlatter, S., Perriard, J. C., and Fussenegger, M. (2004) Design of artificial myocardial microtissues. *Tissue Eng.* **10**, 201–214.
16. Kelm, J. M., and Fussenegger, M. (2004) Microscale tissue engineering using gravity-enforced cell assembly. *Trends Biotechnol.* **22,** 195–202.
17. Kelm, J. M., Ittner, L. M., Born, W., Djonov, V., and Fussenegger, M. (2006) Self-assembly of sensory neurons into ganglia-like microtissues. *J. Biotechnol.* **121,** 86–101.
18. Layer, P. G., Robitzki, A., Rothermel, A., and Willbold, E. (2002) Of layers and spheres: the reaggregate approach in tissue engineering. *Trends Neurosci.* **25,** 131–134.
19. Timmins, N. E., Dietmair, S., and Nielsen, L. K. (2004) Hanging-drop multicellular spheroids as a model of tumour angiogenesis. *Angiogenesis* **7,** 97–103.
20. Gerber, H. P., Hillan, K. J., Ryan, A. M., Kowalski, J., Keller, G. A., Rangell, L., Wright, B. D., Radtke, F., Aguet, M., and Ferrara, N. (1999) VEGF is required for growth and survival in neonatal mice. *Development* **126,** 1149–1159.
21. Ferrara, N. and Alitalo, K. (1999) Clinical applications of angiogenic growth factors and their inhibitors. *Nat. Med.* **5,** 1359–1364.
22. Carmeliet, P., Ng, Y. S., Nuyens, D., Theilmeier, G., Brusselmans, K., Cornelissen, I., Ehler, E., Kakkar, V. V., Stalmans, I., Mattot, V., Perriard, J. C., Dewerchin, M., Flameng, W., Nagy, A., Lupu, F., Moons, L., Collen, D., D'Amore, P. A., and Shima, D. T. (1999) Impaired myocardial angiogenesis and ischemic cardiomyopathy in mice lacking the vascular endothelial growth factor isoforms VEGF164 and VEGF188. *Nat. Med.* **5,** 495–502.
23. Wong, A. K., Alfert, M., Castrillon, D. H., Shen, Q., Holash, J., Yancopoulos, G. D., and Chin, L. (2001) Excessive tumor-elaborated VEGF and its neutralization define a lethal paraneoplastic syndrome. *Proc. Natl. Acad. Sci. U. S. A.* **98,** 7481–7486.

10

Artificial Skin

M. Föhn and H. Bannasch

Summary

Replacement of skin has been one of the most challenging aims for surgeons ever since the introduction of skin grafts in 1871. It took more than one century until the breakthrough of Rheinwald and Green in 1975 that opened new possibilities of skin replacement. The combination of cell culture and polymer chemistry finally led to the field of tissue engineering. Many researchers all over the world have been fascinated by the chance of creating a skin-like substitute ex vivo without any further harm to the patients, especially those with massive burns. Many different approaches to create new substitutes and further improvements in genetical and stem cell research led to today's skin equivalents. But still, the "gold standard" for wound coverage is the autologous split-thickness skin graft. Future research will aim at originating biologically and physiologically equal skin substitutes for the treatment of severe burns and chronic ulcers.

Key Words: Tissue engineering; Skin substitutes; Burns; Chronic ulcers; Keratinocyte culture; Fibrin sealant.

1. Introduction

One of the ultimate tasks of surgeons has been the replacement of skin due to thermal burns and chronic ulcers. In recent years, skin replacement has evolved from the original autografts and allografts to sophisticated biosynthetic and tissue-engineered living skin equivalents.

Nevertheless, the "gold standard" for resurfacing damaged skin remains the autologous transplantation of split-thickness or full-thickness skin grafts. For more complex restoration, surgical procedures such as pedicle and free flaps are necessary. The problems with these procedures are the limited resource of autologous skin especially in patients with extensive burn injuries. Therefore,

From: *Methods in Molecular Medicine, 2nd ed.: Tissue Engineering*
Edited by: H. Hauser and M. Fussenegger © Humana Press Inc., Totowa, NJ

allogeneic skin has been provided by skin and tissue banks as a temporary cover. Only with this additional skin support an effective and ensured treatment of burn patients may be guaranteed.

More than 30 years ago, Rheinwald and Green were the first to describe a method for the cultivation of epidermal cells in vitro *(1)*, which represented a landmark in the development of tissue engineering.

Tissue engineering is a fairly new interdisciplinary field that originated in the disciplines of cell culture and polymer chemistry. Its principles are the in vitro cultivation of autologous cells, the combination of these cells and a delivery vehicle and/or matrix, and the differentiation of this new-formed "tissue" in vitro or in vivo. These matrices may consist of biodegradable structures such as collagen, polylactid acid, polyglycol acid, glycosaminoglycans, acellular dermis, hyaluronic acid, fibrin, and others. All of these combinations have one purpose in common: the construction of a skin equivalent to normal anatomy and physiology.

Until today, tissue-engineered skin substitutes only partially restore anatomic structures and physiological functions. Consequently, the full potential for skin tissue engineering has not yet been explored *(2)*. But patients need the complete restoration of skin structures to regain an appropriate interface between the organism and its environment and to get suitable protection against all external influences.

Full-thickness skin regeneration by means of tissue engineering requires a material capable of restoring both epidermal barrier function and dermal properties of mechanical stability and elasticity *(3)*. Ideally, it should also contain skin adnexes.

2. Cultivated Epithelial Autografts—First Steps Toward Skin

The historical development of skin cell cultures followed different ways. In 1889, Ljunggren discovered that skin fragments could be kept viable in ascites and could be retransplanted to the donors after a prolonged period of time (*4*, *5*). Others showed that fragments of skin on substrates, kept in suitable solutions, produced cell outgrowth *(6–8)*. Many others renewed and improved the harvesting and cultivation techniques of epidermal cells (*9*, *10*).

In 1975, Rheinwald and Green revolutionized the human keratinocyte culture techniques *(1)*. They cultivated dissociated keratinocytes on a lethally irradiated mouse fibroblast feeder layer. Thus, they obtained a multilayered epithelial sheet graft, which represents an established clinical procedure to achieve permanent wound closure (*see* **Fig. 1**). Prior to grafting, confluent and differentiated sheets are transferred to a backing material. A biopsy of 1,5 cm^2

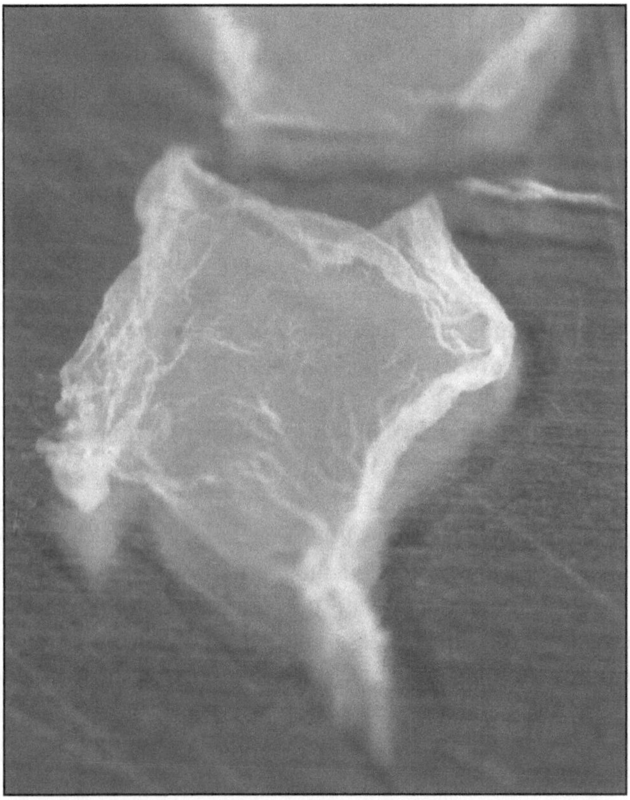

Fig. 1. Conventional cultivated epithelial autograft (CEA) floating in buffer after enzymatic detachment from the culture flask bottom prior to grafting.

approximately takes 3 weeks to cultivate a sufficient number of autologous epithelial sheets to cover an adult with a 70% total burn surface area (TBSA).

For more than 20 years, burn patients have been treated worldwide with cultured autologous keratinocytes in a multilayered thin sheet—so-called cultivated epithelial autografts (CEAs). These skin grafts have proved to be able to achieve a permanent wound closure ***(11–13)***.

Despite promising results in the early 1980s, CEAs are not to be considered as a definitive solution to durable skin restoration. One of the main reasons is the time period required (3–4 weeks) to obtain sufficient graftable multilayered CEAs in appropriate quantity in burn patients. Meanwhile, a dressing with synthetic materials or allogenic cadaver skin has to be used as a temporary wound coverage ***(14)***. The surgical handling of these fragile CEAs is difficult, and there is a high risk of infection combined with high manufacturing costs *(15)*. Numerous reports on the clinical use of cultured epidermal sheets have been published with variable results. Early take rates of cultured epithelium are reported from 49 to 80% depending on the individual technique and the grafted parts of the body. Moreover, this "neoepidermis" is very fragile and vulnerable to shear stress, so that even years later spontaneous blistering will occur (***13***, ***16***).

Long-term studies with histological evaluation have shown that a delayed reconstitution of the basement membrane combined with a delayed or missing regeneration of rete ridges are supposed to be the main factors responsible for this phenomenon (***17***, ***18***, ***19***). Wound infection is the main cause of early graft failure. Especially within the first few days after transplantation, the fragile noncornified epithelium and the missing dermoepidermal junctions are very susceptible to damaging effects of bacteria. Other reasons are poor healing potency of the transplants *(19)* and the condition of the wound bed (***20***, ***21***).

Another disadvantage of CEAs is the process of harvesting the sheet grafts. To gain CEAs, it is essential to detach them from the bottom of the culture flask. This includes an enzymatic step that leads to a temporary loss of α6β4-integrins. These integrins are necessary for the adhesion of the cellular sheet *(19)*. Furthermore, there is evidence that the proliferating potency of these multilayered structures is decreased *(21)*.

To partially compensate the decreased wound-healing potency of the CEAs, preconditioning of the wound bed is necessary to increase the take rate of the cultivated cells. A possible method is to cover the wound areas after necrectomy with allogenic cadaver skin or Integra® (Integra Life Science Corp., Plainsboro, NJ) as a temporary dermal substitute. These materials are partially integrated into the wound bed, and they enhance the take rate of CEAs. This led to improved clinical results within the last 20 years (***22***, ***23***).

In conclusion, some of the frequent problems of this method, such as the difficult handling, fragility, and poor take rates, are partly attributed to the necessity of enzymatic sheet detachment from the bottom of the culture vessel, the lack of a dermal component, and the differentiation of the cells during the cultivation period in the incubator *(19)*.

3. Further Improvements and Alternative Techniques

The use of cultured nonconfluent cells instead of CEAs has recently gained much attention (***24***, ***25***). But the actual foundation of future developments for cultured cells was laid in the late 19th century. In 1895, von Mangoldt was the first to describe a technique of "epithelial cell seeding." He harvested epithelial cells or cell clusters by scraping skin from a patient's forearm until it started to bleed. These cell compounds were suspended in blood or serum and successfully applied to chronic wounds and cavities *(26)*. Despite further improvements and varieties during the next decades, none led to a wide clinical application.

The hypothesis has been forwarded that subconfluent noncontact-inhibited cells have a higher proliferative and wound-healing potency than a confluent cell layer, where cellular differentiation has been completed. This hypothesis is supported by the findings of a higher proliferation potency and a lower expression of differentiation markers in subconfluent cultures (***10***, ***27***, ***28***).

The enzymatic detachment of the cells leads to a temporary loss of α6β4-integrins and thus a retarded formation of the dermoepidermal junctions. Another applicability is the use of trypsinized dissociated subconfluent cells using either culture medium (***29***, ***30***) or fibrin as a delivery vehicle (***24***, ***25***). By this applicative form, the proliferative potential of the trypsinized cells is preserved as no final differentiation has yet occurred. Thus, despite the enzymatic dissociation that has taken place, the reepithelialization potential of the cells is still intact. Fibrin sealant as a biological two-component glue system has been used successfully in the treatment of burns for the last 20 years *(31)*. Moreover, fibrin as a natural matrix in the wound-healing process represents an ideal transport vehicle and guidance structure for the reorganization of cells at the same time *(32–34)*.

The cultivation and transplantation of cultured autologous keratinocytes as a single-cell suspension in a fibrin glue matrix, combined with allogenic skin grafting, were developed by our group (***24***, ***25***) and extensively investigated in the athymic nude mouse model *(35)* and applied clinically (*see* **Fig. 2**). Wounds can be reliably reepithelialized after a cultivation period of only 2 weeks. The development of a basement membrane has been demonstrated to take place after about 28 days.

Such keratinocyte cell suspension grafting simplifies the handling of the cells very much. In combination with allogenic cadaver skin, a stable neoepithelium can be achieved. The allogenic epithelial cells as the most important part of graft-versus-host reaction in skin grafting are rejected. The structural dermal part of the allogenic skin is partly integrated and serves as a scaffold for dermal reconstitution *(25)*. Clinical experience applying this method to both burn patients and patients with chronic wounds has been promising but restricted in number *(36)*. However, allogenic cadaver skin supply is limited and carries the problem of disease transmission. There is still a lack of clinical data showing a superiority of subconfluent cultivated single-cell suspensions when compared with conventional CEAs (***21***, ***32***).

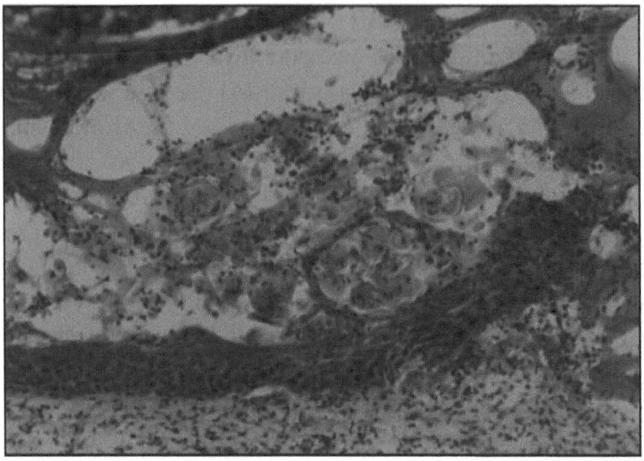

Fig. 2. H&E stain of biopsy taken from the wound center of the animal shown in Fig. 3 1 week after transplantation. Proliferating keratinocytes form a not yet fully stratified human neoepithelium. Above remnants of fibrin sealant as delivery vehicle can be seen.

Clinical results after CEA application have led to a general agreement among most clinicians that a dermal substitute is needed to enhance the quality of epithelial grafts. It seems apparent that applying a functional skin replacement rather than a pure epidermal replacement underlines the actual need for sufficient dermal support *(21)*. Efforts therefore focus on the combination of dermal substitutes and cultivated keratinocytes in vitro prior to grafting *(37–39)*. Handling properties of composite grafts have been improved, and in addition, the use of enzymes to remove the sheets from the culture vessel surface is no longer required. Various materials as "dermal scaffolds" have been used, including dermis-derived lattices, collagen-derived matrices, and cultured substrates (*2*, *3*).

These promising results of combining epidermal and dermal components lead to the concept of combined autologous keratinocytes and a dermal substrate—so-called composite grafts.

4. Composite Grafts—First and Simple Full-Thickness Skin Substitutes

As pointed out before, the enzymatic detachment of conventional CEAs leads to a temporary loss of $\alpha 6\beta 4$-integrins, which are responsible for the cellular adhesion. To avoid the enzymatic step during the cultivation procedure and to simplify the handling, keratinocyte cultivation for wound coverage was combined with various natural or synthetic carrier materials such as polyurethane membranes ***(21)***, silicon collagen membranes ***(40)***, hyaluronic acid-based membranes, collagen sponges ***(2)***, and fibrin glue (***25***, ***41***, ***32***). On these biomaterials, keratinocytes can be cultured subconfluently and—by avoiding the enzymatic detachment from the culture flask—maintain important biological characteristics. The ability of human keratinocytes to migrate represents an elementary part of the reepithelialization process of wounds during healing ***(33)***.

Despite the reepithelialization process, different biomaterials seeded with keratinocytes and/or fibroblasts (*see* **Table 1**) have shown experimental advantages regarding the regeneration of dermis and basement membrane complexes (***42***, ***43***, ***37***). These results led to the development of full-thickness skin substi-

Table 1
Exemplary overview on some commercially available skin substitutes

Product name	Composition	Distributor
Biobrane™	thin semipermeable silicone membrane bonded to nylon fabric, collagen covered	Bertek Pharmaceuticals
AlloDerm™	Acellular Dermis	Lifecell Corporation
LaserSkin™	Hyaluronic Acid	Fidia Biopolymers
Integra™	Collagen-GAG + Silicon	Integra Life Sciences
TissuFoil E™	Equine Collagen	Baxter Biosurgery
MatriDerm™	Collagen + Elastic Fibres	Dr. Suwelack Skin & Health Care AG
Epidex™	auto HK-sheets on silicone membrane	Modex Therapeutics
ApliGraf™	Collagen-Gel + allo HF + cult. allo HK	Organogenesis, Inc.
Comp Cult Skin™	Collagen + allo HF + allo HK	Ortec International,Inc.
DermaGraft™	PGA/PLA + allo HF	Advanced Tissue Sciences
EpiCell™	Allogenic Dermis + cult. auto HK	Genzyme Biosurgery
TransCyte™	allo HF seeded on nylon mesh	Smith and Nephew

allo = allogenic , auto = autologous , cult. = cultivated
HK = human keratinocytes , HF = human fibroblasts
GAG = Glycosaminoglycans; PGA/PLA = Polyglycol acid/Polylactid acid

tutes consisting of cultivated keratinocytes, extracellular dermal matrix, and eventually cultivated fibroblasts (***26***, ***45***).

The experimental and clinical data on the use of composite grafts are promising; however, there is still a gap between worldwide experimental efforts and the number of clinical applications. This is likely due to the unreliable take of these complex constructs with their enhanced need for nutrition per diffusion—a central problem the whole field of tissue engineering is faced with. One possible solution could be the addition of vascular endothelial cells to gain sufficient nutritional supply for the epidermal cells at an early stage. The combination of different cell types results in a complex combination of culture media in vitro to secure the best nutritional support for the cultivated cells. Additionally, the economic efforts to generate these composite grafts are higher than using split-thickness skin grafts or allogenic dermis as an underlying substrate ***(38)***.

In vitro experiments demonstrated the importance of an intact basement membrane, dermal structure, and viable keratinocytes for optimal cell growth. These findings were included in the development of a dermal lattice composed of collagen and glycosaminoglycans (collagen–GAG). Inoculation of the matrix with fibroblasts enabled an acceptable matrix for keratinocytes ***(37)***, a reliable result that has so far proved successful in experienced clinical centers ***(46)***.

The only commercially available matrix with an intact basement membrane is AlloDerm™ (LifeCell Corporation, The Woodlands, TX). This acellular processed freeze-dried human dermis preserves positive reactions to laminin and collagen type IV and VII in native immunohistological stainings (*see* **Figs 3–5**). The preparatory process also preserves the three-dimensional structure as well as the collagen and elastic bundles of the dermis ***(47)***. It has been used in soft tissue augmentation procedures for several years. So far, no immunologic reaction has been reported. AlloDerm™ has successfully been grafted in combination with a split-thickness skin graft overlay in a pig animal model and in burn patients (***44***, ***48***). Moreover, AlloDerm™ promotes engraftment of human keratinocytes, which were cultured on a synthetic hydrophilic dressing ***(22)***.

Further and still ongoing investigations of our group also showed good experimental results in combining MatriDerm™ (a collagen–elastin matrix, Dr. Suwelack Skin and Health Care, Billerbeck, Germany) and human cells. These experiments are still in working process (*see* **Figs 6–8**).

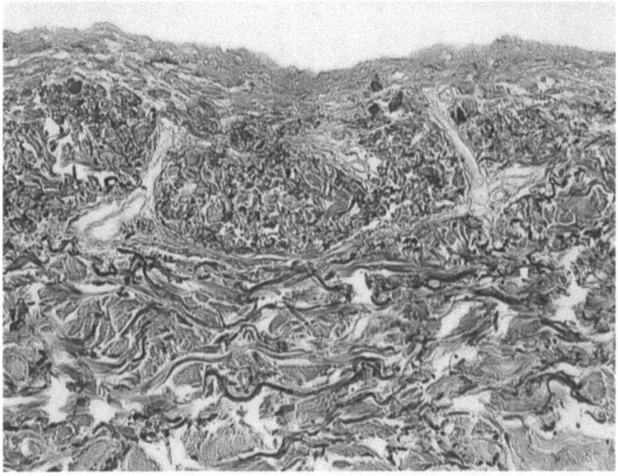

Fig. 3. Elastica-van-Gieson stain of native Alloderm™. Note the well preserved rete ridges and dermal structure with collagen bundles and different distribution of elastic fibers between papillary and reticular stratum.

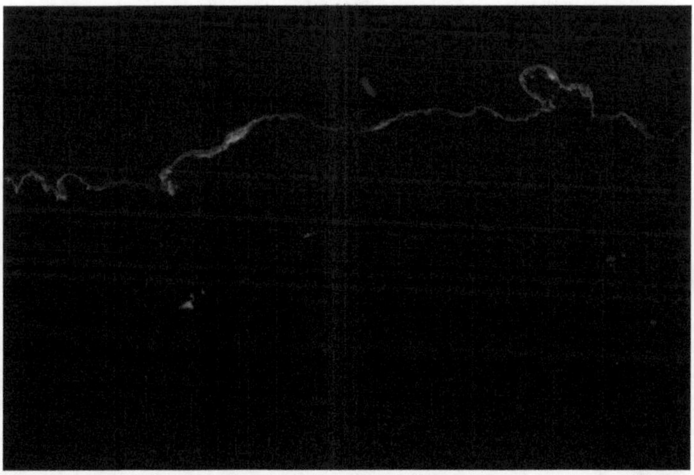

Fig. 4. Collagen-VII immunohistochemistry of the basement membrane in native AlloDerm™, ×200.

Fig. 5. Composite graft consisting of AlloDerm™ and cultured human keratinocytes 4 days after grafting to a full-thickness wound on the dorsum of a athymic nude mouse. Collagen-VII immunohistochemistry and propidium-iodide counterstain, ×200.

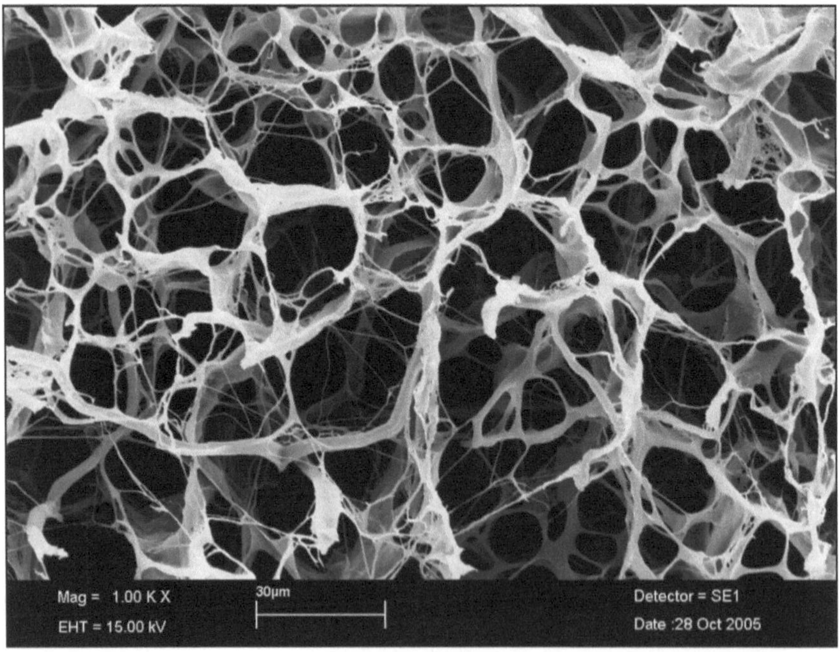

Fig. 6. Environmental electron microscopy of native MatriDerm™.

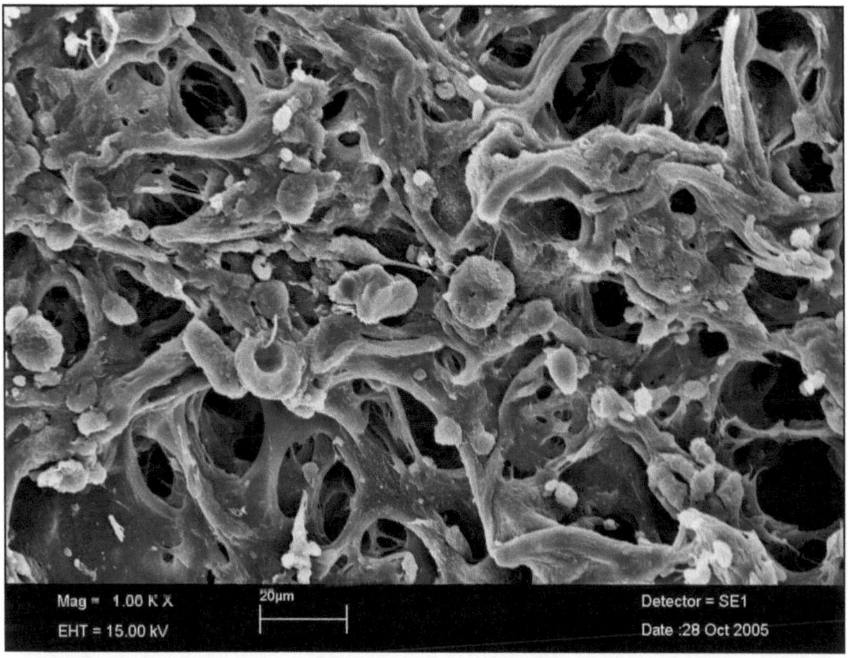

Fig. 7. Environmental electron microscopy of MatriDerm™ seeded with cultured human keratinocytes.

5. Future Perspectives

Nowadays, socioeconomic aspects complicate the treatment modalities not only of burn patients. Nevertheless, further progress in the development and clinical application of tissue-engineered "artificial skin" is needed. It may lead to improvements in skin reconstitution while at the same time overcoming limits of donor site morbidity. The improvement of cosmesis, the regaining of skin functions including sensitivity, sweat glands, dermal appendages, and normal pigmentation in combination with invisible scarring are still problems in skin tissue engineering that need to be solved.

It will still need a lot of effort to attain these goals. A daily growing number of dermal substitutes pave the way for the perfect material to achieve these aims. But reality is still different. Further helpful improvements can be the additional genetical modification of cells and/or dermal substrates to accelerate the cell growth and the healing potential of the transplanted material, which will allow the use of an off-the-shelf epidermal replacement *(2)*. Furthermore, the potential of multipotent adult stem cells may well become a way to solve

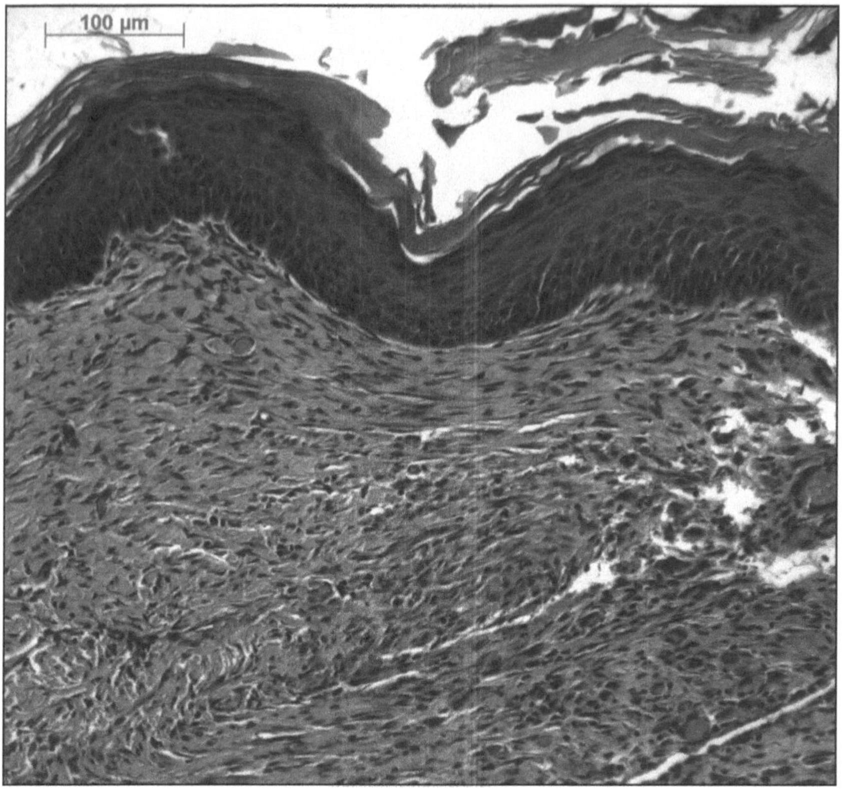

Fig. 8. Murine epithelium 28 days after transplantation to a full-thickness wound on the dorsum of a athymic nude mouse with MatriDerm™ as dermal component.

the mentioned problems—once some of the still remaining technical problems have been solved.

As long as the anatomy and physiology of engineered skin substitutes improves, they will become more similar to native skin autografts. Improvement of skin substitutes will result from inclusion of additional cell types (e.g., endothelial cells, and melanocytes) and from modifications of culture media and scaffolds. Skin-substitute materials may be able to stimulate regeneration rather than repair, and tissue-engineered skin may match the quality of split-skin autografts, our present gold standard ***(3)***.

The aim of surgeons is to help burn patients with the best quality of skin within the shortest time possible. As tissue engineers, we have not yet obtained the ultimate universal skin product.

References

1. Rheinwald, J. G. and Green H. (1975) Serial cultivation of strains of human epidermal keratinocytes: the formation of keratinizing colonies from single cells. *Cell* **6**, 331–344.
2. Jones, I., Currie, L., and Martin, R. (2002) A guide to biological skin substitutes. *Br. J. Plast. Surg.* **55**, 185–193.
3. Boyce, S. T. (2001) Design principles for composition and performance of cultured skin substitutes. *Burns* **27**, 523–533.
4. Ljunggren, C. A. (1898) Von der Fähigkeit des Hautepithels, ausserhalb des Organismus sein Leben zu erhalten, mit Berücksichtigung der Transplantation. *Deutsch Z. Chir.* **47**, 608–615.
5. Carrel, A. and Burrows, M. T. (1910) Cultivation of adult tissues and organs outside the body. *JAMA* 1379–1384.
6. Hadda, S. (1912) Die Kultur lebender Zellen. *Klein Wschr.* **49**, 11–19.
7. Kreibich, K. (1914) Kultur erwachsener Haut auf festem Nährboden. *Arch. Dermatol. Syph.* **120**, 168–178.
8. Mangoldt, V. F. (1895) Die Überhäutung von Wundflächen und Wundhöhlen durch Epithelaussaat, eine neue Methode der Transplantation. *Deut. Med. Wschr.* 798–799.
9. Poumay, Y. and Pittelkow, M. R. (1995) Cell density and culture factors regulate keratinocyte commitment to differentiation and expression of suprabasal K1/K10 keratins. *J. Invest. Dermatol.* **104**(2), 271–276.
10. Pellegrini, G., Ranno, R., Stracuzzi, G., Bondanza, S., Guerra, L., Zambruno, G., Micali, G., and de Luca, M. (1999) The control of epidermal stem cells (holoclones) in the treatment of massive full-thickness burns with autologous keratinocytes cultured on fibrin. *Transplantation* **68**, 868–879.
11. Gallico, G. G., O'Connor, N. E., Compton, C. C., Kehinde, O., and Green, H. (1984) Permanent coverage of large burn wounds with autologous cultured epithelium. *N. Engl. J. Med.* **311**, 448–451.
12. Compton, C. C., Gill, J. M., Bradford, D. A., Regauer, S., Gallico, G. G., and O'Connor, N. E. (1989) Skin regenerated from cultured epithelial autografts on full-thickness burn wounds from 6 days to 5 years after grafting. *Lab. Invest.* **60**, 600–612.
13. Odessey, R. (1992) Addendum: multicenter experience with cultured epidermal autograft for treatment of burns. *J. Burn Care Rehabil.* **13**, 174–180.
14. Cuono, C., Langdon, R., and McGuire, J. (1986) Use of cultured epidermal autografts and dermal allografts as skin replacement after burn injury. *Lancet* **1**, 1123–1124.
15. Rue, L. W., Cioffi, W. G., McManus, W. F., and Pruitt, B. A. (1993) Wound closure and outcome in extensively burned patients treated with cultured autologous keratinocytes. *J. Trauma* **34**, 662–668.

16. Herndon, D. N. and Rutan, R. L. (1992) Comparison of cultured epidermal autograft and massive excision with serial autografting plus homograft overlay. *J. Burn Care Rehabil.* **13**, 154–157.
17. Woodley, D. T., Peterson, H. D., Herzog, S. R., Stricklin, G. P., Burgeson, R. E., Briggaman, R. A., Cronce, D. J., and O'Keefe, E. J. (1988) Burn wounds resurfaced by cultured epidermal autografts show abnormal reconstitution of anchoring fibrils. *JAMA* **259**, 2566–2571.
18. Putland, M., Snelling, C. F. T., MacDonald, I., and Tron, V. A. (1995) Histologic comparison of cultured epithelial autograft and meshed expanded split-thickness skin graft. *J. Burn Care Rehabil.* **16**, 627–640.
19. Poumay, Y., Roland, I. H., Leclerq-Smekens, M., and Lelouop, R. (1994) Basal detachment of the epidermis using dispase: tissue spatial organization and fate of integrin α6β4 and hemidesmosomes. *J. Invest. Dermatol.* **102**, 111–117.
20. Leary, T., Jones, L., Appleby, M., Blight, A., Parkinson, K., and Stanley, M. (1991) Epidermal keratinocyte self-renewal is dependent upon dermal integrity. *J. Invest. Dermatol.* **99**, 422.
21. Rennekampff, H. O., Kiessig, V., and Hansbrough, J. F. (1996) Research review: current concepts in the development of cultured skin substitutes. *J. Surg. Res.* **62**, 288–295.
22. Rennekampff, H. O., Kiessig, V., Griffey, S., Greenleaf, G., and Hansbrough, J. F. (1997) Acellular human dermis promotes cultured keratinocyte engraftment. *J. Burn Care Rehabil.* **18**, 535–544.
23. Loss, M., Wedler, V., Künzi, W., Meuli-Simmen, C., and Meyer, V. E. (2000) Artificial skin, split-thickness autograft and cultured autologous keratinocytes combined to treat a severe burn injury of 93% of TBSA. *Burns* **26**, 644–652.
24. Stark, G. B. and Kaiser, H. W. (1994) Cologne burn centre experiences with glycerol-preserved allogenic skin: part II: combination with autologous cultured keratinocytes. *Burns* **20**, 34–38.
25. Stark, G. B., Kaiser, H. W., Horch, R., Kopp, and J., Spilker, G. (1995) Cultured autologous keratinocytes suspended in fibrin glue (KFGS) with allogenic overgraft for definitive burn wound coverage. *Eur. J. Plast. Surg.* **18**, 267–271.
26. Supp, D. M. and Boyce, S. T. (2005) Engineered skin substitutes: practices and potentials. *Clin. Dermatol.* **23**, 403–412.
27. Lee, Y. S., Yuspa, S. H., and Dlugosz, A. A. (1998) Differentiation of cultured human epidermal keratinocytes at high cell densities is mediated by endogenous activation of the protein kinase C signaling pathway. *J. Invest. Dermatol.* **111**(5), 762–766.
28. Juhasz, I., Murphy, G. F., Yan, H. C., Herlyn, M., and Albelda, S. M. (1993) Regulation of extracellular matrix proteins and integrin cell substratum adhesion receptors on epithelium during cutaneous human wound healing in vivo. *Am. J. Pathol.* **143**, 1458–1469.
29. Wood, F. (2001) The first 7 years of the west Australian skin culture laboratory, in *Cultured Human Keratinocytes and Tissue Engineered Skin Substitutes*, 1st

edition (Horch, R. E., Munster, A. M., and Achauer, B. M., eds), Thieme, Stuttgart, pp. 275–283.

30. Fraulin, F. O. G., Bahoric, A., Harrop, A. R., Hiruki, T., Clarke, H. M. (1998) Autotransplantation of epithelial cells in the pig via an aerosol vehicle. *J. Burn Care Rehabil.* **19**, 337–345.
31. Adant, J. P., Detroz, B., D'Silva, M., Natowitz, L., Ledoux, M., Pestiaux, and B., Leclercq, P. (1993) Skin grafting with fibrin glue in burns. *Eur. J. Plast. Surg.* **16**, 292–297.
32. Currie, L. J., Sharpe, J. R., and Martin, R. (2001) The use of fibrin glue in skin grafts and tissue-engineered skin replacements: a review. *Plast. Reconstr. Surg.* **108**, 1713–1726.
33. Clark, R. A. F., Lanigan, J. M., Dellapelle, P., Manseau, E., Dvorak, H. F., and Colvin, R. B. (1982) Fibronectin and fibrin provide a provisional matrix for epidermal cell migration during wound reepithelialization. *J. Invest. Dermatol.* **79**, 264–269.
34. Geer, D. J., Swartz, D. D., and Andreadis, S. T. (2002) Fibrin promotes migration in a three-dimensional in vitro model of wound regeneration. *Tiss. Eng.* **8**(5), 787–798.
35. Horch, R., Bannasch, H., Kopp, J., Andree, C., and Stark, G. B. (1998) Single-cell-suspensions of cultured human keratinocytes in fibrin-glue reconstitute the epidermis. *Cell Transplant.* **7**(3), 309–317.
36. Bannasch, H., Horch, R. E., Tanczos, E., and Stark, G. B. (2000) Behandlung chronischer Wunden mit kultivierten autologen Keratinozyten als Suspension in Fibrinkleber. *Zentralbl. Chir.* **125**(1), 79–81.
37. Hansbrough, J. F., Morgan, J., Greenleaf, G., Parikh, M., Nolte, C., and Wilkins, L. (1994) Evaluation of graftskin composite grafts on full-thickness wounds on athymic mice. *J. Burn Care Rehabil.* **15**(4), 346–353.
38. Boyce, S. T., Goretsky, M. J., Greenhalgh, D. G., Kagan, R. J., Rieman, M. T., and Warden, G. D. (1995) Comparative assessment of cultured skin substitutes and native skin autograft for treatment of full-thickness burns. *Ann. Surg.* **222**, 743–752.
39. Medalie, D. A., Eming, S. A., Tompkins, R. G., Yarmush, M. L., Krueger, G. G., and Morgan, J. R. (1996) Evaluation of human skin reconstituted from composite grafts of cultured keratinocytes and human acellular dermis transplanted to athymic mice. *J. Invest. Dermatol.* **107**, 121–127.
40. Kopp, J., Bannasch, H., Andree, C., and Stark, G. B. (1996) Kultivierte Keratinozyten auf einem Silikon-Kollagen-Matrix-Träger zur Deckung von Vollhautdefekten. *Langenbecks Arch. Chir.* (Suppl. I), 299.
41. Ronfard, V., Broly, H., Mitchell, V., Galizia, J. P., Hochart, D., Chambon, E., Pellerin, P., and Huart, J. J. (1991) Use of human keratinocytes cultured on fibrin glue in the treatment of burn wounds. *Burns* **17**, 181–184.

42. Bohnert, A., Hornung, J., Mackenzie, I. C., and Fusenig, N. E. (1986) Epithelial-mesenchymal interactions control basement membrane production and differentiation in cultured and transplanted mouse keratinocytes. *Cell Tissue Res.* **244**, 413–429.
43. Cooper, M. L., Andree, C., Hansbrough, J. F., Zapata-Sirvent, R. L., and Spielvogel, R. (1993) Direct comparison of a cultured composite skin substitute containing human keratinocytes and fibroblasts to an epidermal sheet graft containing human keratinocytes on athymic mice. *J. Invest. Dermatol.* **101**, 811–819.
44. Livesey, S. A., Herndon, D. N., Holloyak, M. A., Atkinson, Y. H., and Nag, A. (1995) Transplanted acellular allograft dermal matrix. Potential as a template for the reconstruction of viable dermis. *Transplantation* **60**, 1–9.
45. Horch, R. E., Kopp, J., Kneser, U., Beier, J., and Bach, A. D. (2005) Tissue engineering of cultured skin substitutes. *J. Cell Mol. Med.* **9**, 592–608.
46. Boyce, S. T., Kagan, R. J., Yakuboff, K. P., Meyer, N. A., Rieman M. T., Greenhalgh, D. G., and Warden, G. D. (2002) Cultured skin substitutes reduce donor skin harvesting for closure of excised, full thickness burns. *Ann. Surg.* **2**, 269–279.
47. Wainwright, D. J. (1995) Use of an acellular dermal matrix (AlloDerm) in the management of full-thickness burns. *Burns* **21**, 243–248.
48. Andreassi, L. (1992) History of keratinocytes cultivation. *Burns* **18**, S2–S4.

11

Small Blood Vessel Engineering

Patrick Au, Josh Tam, Dai Fukumura, and Rakesh K. Jain

Summary

Tissue engineering has attracted wide interest as a potential method to alleviate the shortage of transplantable organs *(1)*. To date, almost all of the successfully engineered tissues/organs have relatively thin and/or avascular structures [e.g., skin *(2)*, cartilage *(3)*, and bladder *(4)*], where postimplantation vascularization from the host (angiogenesis) is sufficient to meet the implant's demand for oxygen and nutrients. Vascularization remains a critical obstacle impeding attempts to engineer thicker, metabolically demanding organs, such as heart and liver. One approach in vascularizing an engineered tissue is to add the cellular components of blood vessels (endothelial and perivascular cells) directly to the tissue-engineered construct. We have shown that coimplanting endothelial cells and perivascular cells in a scaffold in vivo can lead to the formation of a vascular network that anastomoses to the host circulatory system. The engineered vessels are stable and functional, and they persist for more than 1 year in vivo. This approach may potentially lead to the creation of a well-vascularized-engineered tissue.

Key Words: Angiogenesis; Endothelial cells; Pericyte; Vessel maturation; Vessel engineering; Retroviral transduction; Intravital microscopy.

1. Introduction

The circulatory system plays an important role in maintaining homeostasis in the body. Its key function is the delivery of nutrients and oxygen and the removal of metabolic waste products. The blood circulation also serves as a conduit for transporting biological signals among different tissues (e.g., growth factors, and hormones). Thus, the development of a vascular network within an engineered tissue is critically important for it to be integrated into the host tissue *(5)*. A number of approaches have been used in attempts to vascularize engineered tissues. Most of these approaches can be grouped into two

From: *Methods in Molecular Medicine, 2nd ed.: Tissue Engineering*
Edited by: H. Hauser and M. Fussenegger © Humana Press Inc., Totowa, NJ

broad categories: (1) growth factor-based approaches where angiogenic growth factors, such as members of the vascular endothelial growth factor (VEGF) or fibroblast growth factor (FGF) families, are used to induce angiogenesis from the host or (2) cell-based approaches where cellular components of a blood vessel, such as endothelial cells and perivascular cells, are seeded into scaffolds directly. To date, attempts at using individual angiogenic factors such as VEGF and basic FGF (bFGF) for the induction of angiogenesis have led to the formation of immature vessels that are aberrant, hemorrhagic, and regress upon growth factor withdrawal *(6)*.

Cell-based approaches may potentially overcome some of these problems and limitations. An engineered scaffold preseeded with endothelial cells eliminates the time delay and the need for specific chemoattractants associated with host endothelial cell recruitment and avoids the need to fine-tune the dose and release schedule of the growth factors involved in angiogenesis *(7)*. We have recently shown that coimplantation of human umbilical cord vein endothelial cells (HUVECs) or human embryonic stem cell derived endothelial cells with 10T1/2 cells, a murine mesenchymal precursor cell, in a collagen gel scaffold can produce long-lasting and functional microvessels in vivo *(8,9)*. Intravital microscopic observation of endothelial cells prelabeled with enhanced green fluorescence protein (EGFP) allows for the examination of the dynamics of vessels formation and evaluation of vessel functions including permeability and vessel contractility.

2. Materials

2.1. Cell Culture and Retrovirus Production

1. Endothelial growth medium (EGM) (Cambrex, East Rutherford, NJ) supplemented with bovine brain extract (BBE, Cambrex).
2. Eagle's basal medium (BME) (Invitrogen, Carlsbad, CA) with 2 mM L-glutamine supplemented with 10% fetal bovine serum (FBS, HyClone, Ogden, UT).
3. Dulbecco's modified Eagle's medium (DMEM) (Invitrogen) supplemented with 10% FBS (HyClone).
4. Opti-MEM I® reduced-serum medium (Invitrogen).
5. Phosphate-buffered saline (PBS) (Sigma, St. Louis, MO).
6. Solution of 0.05% trypsin with ethylenediaminetetraacetic acid (EDTA)-4Na in PBS (Invitrogen).
7. Solution of 0.2% gelatin (Sigma).
8. Solution of polybrene in PBS at 8 μg/ml (Sigma).
9. Lipofectamine 2000 (Invitrogen).
10. 0.45-mm filters (Whatman, Florham Park, NJ).

11. HUVECs (Center for Excellence in Vascular Biology in Brigham & Women's Hospital, Boston, MA).
12. 10T1/2 cells (ATCC, Manassas, VA).
13. 293T/17 cells (ATCC).

2.2. Collagen Gel

1. Collagen type I derived from rat tail tendon (BD Biosciences, Bedford, MA).
2. Fibronectin from human plasma, 0.1% solution in PBS (Sigma).
3. 4-(2-Hydroxyethyl)piperazine-1-ethanesulfonic acid (HEPES).
4. 1N solution of sodium hydroxide (NaOH).
5. 25 mM HEPES-buffered EGM.
6. pH paper.

2.3. Implantation of Collagen Gel

1. Skin punch 4 mm (Roboz Surgical, Rockville, MD).
2. Forceps (Roboz Surgical).
3. Dissection microscope.
4. Coe tray plastic powder (GC America Inc., Alsip, IL).

2.4. Intravital Microscopy

1. Tetramethylrhodamine-conjugated dextran (Molecular Probes, Eugene, OR).
2. PE10 polyethylene tubing (Clay Adams, Sparks, MD).
3. 1-ml syringe (Beckon Dickinson, Franklin Lakes, NJ).
4. 30-G needle (Beckon Dickinson).
5. Econo-Pac 10-DG disposable chromatography column (Bio-Rad Laboratories, Hercules, CA).
6. Tetramethylrhodamine-conjugated bovine serum albumin (BSA) (Molecular Probes).
7. Endothelin-1 (ET-1) in PBS at 100 nM concentration (Sigma).

3. Methods

Tissue engineering has traditionally focused on the design of scaffolds and bioreactors and the evaluation of in vitro function. Less emphasis has been placed on evaluating the tissue's in vivo integration and function. When an engineered tissue is implanted in vivo, it is often examined only by histology at the time of excision. Little information is obtained on the in vivo physiology and function of the tissue. One way to address this limitation is by intravital microscopy ***(10)***. Intravital microscopy allows for direct visualization of the implanted cells and/or tissue. In the case of blood vessel engineering, intravital microscopy provides us with the ability to observe the dynamics of vessel formation in the same animal and to measure physiological parameters in real time.

To visualize the implanted cells in vivo, HUVECs are first transduced with retrovirus to stably express EGFP. Retrovirus-mediated transduction of HUVECs is highly efficient (over 90% of HUVECs expressed EGFP after two to three rounds of transduction in a typical experiment). The EGFP-labeled HUVECs are then mixed together with 10T1/2 in a collagen scaffold. Precise spatial arrangement of HUVECs and 10T1/2 is not necessary as the cells self-assemble into vascular structures within the collagen scaffold. The tissue-engineered construct is then implanted in vivo on the pial surface in a mouse cranial window model. The EGFP-labeled endothelial cells are observed at different timepoints for their self-assembly into blood-perfused vessels. Perfused engineered blood vessels can be visualized by injection of a blood tracer [rhodamine (Rho)-conjugated dextran] through the mouse tail vein. The engineered vessels can be further characterized for their functions, including vessel permeability and contractility. Vessel permeability provides a particularly good measurement of the maturity of a blood vessel. This is because immature blood vessels, such as those inside a tumor, are leaky and hemorrhagic ***(11)***. Obviously, as such vessels have poor function, they are not suitable for use in tissue engineering. Vascular permeability is determined by injecting tetramethylrhodamine-conjugated BSA through tail vein and measuring the fluorescence intensity of the tissue. The tissue fluorescence intensity correlates with the quantity of Rho–BSA that has leaked out of the vessel ***(12)***. Finally, an optimal-engineered vessel should not only be able to carry blood, but it should also be able to respond to physiological stimuli. For example, the property of normal arterioles is the ability to control systemic blood pressure by modulating their vascular resistance upon stimulation. An engineered blood vessel can be evaluated for this function by stimulating it with vasoactive agents (vasoconstrictor such as ET-1) and measuring its contractile response.

3.1. Retrovirus Production

1. The 293 packaging cells are cultured in 100-mm tissue-culture dishes with DMEM supplemented with 10% FBS. The packaging cells should be passaged 2 days prior to transfection at 1:4 split. On the day of transfection, the 293 cells should be approximately 80–90% confluent (*see* **note 1**).
2. On the day of transfection, the 293 cells are given fresh media (10 ml DMEM 10% FBS). The packaging cells are then transiently transfected with the retrovirus vector encoding EGFP, gag/pol, and VSVG using Lipofectamine 2000. The plasmids (5 μg Vesicular stomatitis virus glycoprotein (VSVG), 7 μg gag/pol, and 15 μg retroviral vector) are mixed together in 1.5 ml Opti-MEM in a 15-ml centrifuge tube. In a

separate tube, 40 μl lipofectamine 2000 is added to 1.5 ml Opti-MEM, and it is allowed to incubate at room temperature for 5 min. After 5 min of incubation, the two tubes of solution are mixed together and allowed to further incubate for 20 min at room temperature. The transfection mixture is then added dropwise to the dish of packaging cells. To ensure uniform mixing, the dish is gently agitated, and it is then placed in a 37° C incubator overnight (*see* **note 2**).

3. After overnight incubation, the 293ET cells are washed gently three times with 10 ml PBS. It is important not to pipette directly on the cells, because 293 cells tend to detach easily. After washing with PBS, 10 ml fresh DMEM 10% FBS is added to the dish of cells, which is then placed in the incubator for overnight culture (*see* **note 3**).
4. The next day, the supernatant containing retrovirus is collected. Fresh media is added to the dish, and the cells are allowed to incubate overnight.
5. The supernatant is spun at (3500 × g) for 5 min to remove any cell debris. It is then passed through a 0.45-mm filter (Whatman) and is either used immediately for infection or kept at −80° C (*see* **note 4**).
6. The supernatant is collected from the packaging cells for three more times.

3.2. Transduction of HUVECs

1. HUVECs are maintained on a gelatin-coated flask. To coat the flask, 2 ml 0.2% solution of gelatin is added and allowed to cover the bottom surface of the flask for 10–15 min at room temperature. The gelatin solution is then removed by aspiration, and the flask is used immediately.
2. One day prior to transduction, the HUVECs are passaged at 1:3 to 1:4 split. It is important for cells to be subconfluent when transduced by retrovirus.
3. For transduction of HUVECs, supernatant containing EGFP retrovirus is first diluted at 1:1 with EGM. Polybrene is then added at final concentration of 8 μg/ml to enhance transduction efficiency. The HUVECs are incubated with the retrovirus supernatant for 4 h in a 37° C incubator. After 4 h, the retrovirus supernatant is replaced with fresh EGM, and the cells are allowed to recover overnight. The transduction procedure is repeated the next day. The expression of EGFP in the transduced HUVECs is confirmed with a fluorescence microscope. Usually, over 90% of HUVECs express EGFP after two to three rounds of transduction (*see* **Fig. 1**, **note 5**).

3.3. Preparation of Cranial Window

1. For the surgical procedures, animals (25–30 g) are anesthetized with s.c. injection of a cocktail of 90 mg Ketamine (Parke-Davis, Morris Plains, NJ) and 9 mg Xylazine (Fermenta, Kansas City, MO) per kilogram body weight. All surgical procedures are performed under aseptic conditions in a specially adapted laminar flow hood, with

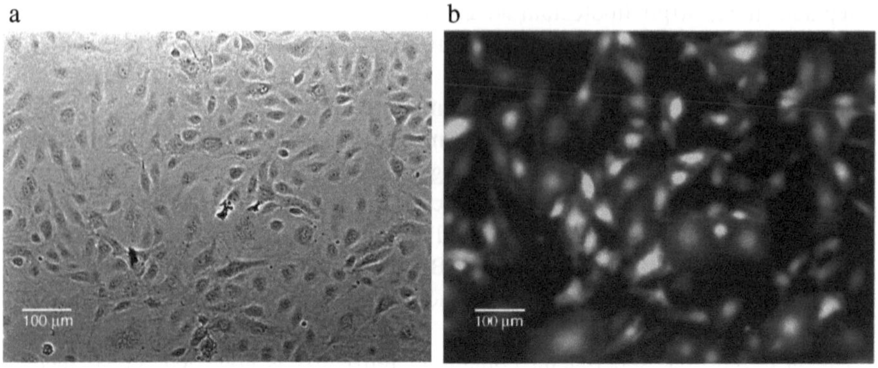

Fig. 1. Endothelial cells were transduced with retrovirus encoding enhanced green fluorescence protein (EGFP). After two to three rounds of transduction, more than 90% of the cells are positive for EGFP (a, bright field and b, EGFP).

all equipments being steam, gas, or chemically sterilized. During surgery, the body temperature of the animals is kept constant at 36–37° C by means of a heating pad. All mice are housed in individual microisolator cages after window implantation. All manipulations are done in laminar flow hoods.

2. The head of an animal is fixed by a stereotactic apparatus. The skin on top of the frontal and parietal regions of the skull is cleaned with antimicrobial (Bacteriostatic, 0.9% sodium chloride; Abbott Laboratories, North Chicago, IL) solution.
3. A longitudinal incision of the skin is made between the occiput and forehead. Then the skin is cut in a circular manner on top of the skull, and the periosteum underneath is scraped off to the temporal crests. A 6-mm circle is drawn over the frontal and parietal regions of the skull bilaterally. Using a high-speed air-turbine drill (CH4201S; Champion Dental Products, Placentia, CA) with a burr tip, 0.5 mm in diameter, a groove is made on the margin of the drawn circle. This groove is made thinner by cautious and continuous drilling of the groove until the bone flap becomes loose. Cold saline is applied during the drilling process to avoid thermal injury of the cortical regions.
4. Using a malis dissector, the bone flap is separated from the dura mater. After the removal of bone flap, gelfoam is placed on the cutting edge to stop the bleeding. The dura mater is then removed from the pia mater underneath by fine scissors except at the area where the dura mater adheres to the sagittal sinus. The pia mater is continuously kept moist with physiological saline. The window is sealed with an 8-mm cover glass, which will be glued to the bone with histocompatible cyanoacrylate glue mixed with Coe tray plastic powder. Give buphrenorphine after surgery. The mouse is allowed to recover for 7 days.

3.4. Preparation of Collagen Scaffold

1. All reagents for preparing the collagen scaffold should be kept on ice to prevent spontaneous polymerization of the collagen solution.
2. To prepare 1 ml solution of collagen scaffold, 1.5 mg rat tail type 1 collagen and 90 μg human plasma fibronectin are added to a 15-ml centrifuge tube kept on ice. HEPES-buffered EGM (25 mM HEPES) is then added to the mixture to bring the total volume to 1 ml. The pH of the collagen–fibronectin solution is adjusted to 7.4 by the addition of 1N NaOH, and it is confirmed with pH paper. The solution is kept on ice until it is used.
3. EGFP-labeled HUVEC and 10T1/2 cells are trypsinized and counted. 10T1/2 cells are maintained in BME with 10% FBS. It is important to keep cells from being overgrown, and the cells are passaged at 80–90% confluence (*see* **note 6**).
4. About 800,000 EGFP-labeled HUVECs and 200,000 10T1/2 cells are added to a 15-ml centrifuge tube. The cells are then centrifuged for 5 min at 1200 × rpm. The supernatant is removed and the cells are resuspended in the collagen–fibronectin solution. The cell suspension is pipetted up and down a few times to make sure the cells are well mixed in the solution. The cell suspension is then added into one well of a 12-well plate, and the cell-culture plate is placed in an incubator at 37° C and 5% CO_2. This step should be done quickly to prevent the cells from settling to the bottom of the plate before the collagen solution can polymerize (*see* **note 7**).
5. After incubation for half an hour, the collagen solution would polymerize to form a solid construct. One milliliter of warmed EGM is then added to the well. The collagen scaffold is cultured for 1 day in vitro to allow the cells to acclimate to the collagen scaffold before it is implanted into a cranial window (*see* **note 8**).

3.5. Implantation of Collagen Scaffold

1. The cover glass is removed from a cranial window by first drilling a space in the glue with a 26-G needle syringe. The needle is drilled down to the space between the cover glass and the skull. Once the needle tip is placed underneath the cover glass, the needle is lifted up to remove the cover glass. The cover glass often shatters during its removal, and the glass fragments are carefully removed from the cranium with fine forceps. The pia mater is kept moist with PBS.
2. A skin puncher (4-mm diameter) is applied to the collagen scaffold to create circular pieces that are 4 mm in diameter. Using a fine forceps, a circular piece of collagen scaffold is carefully placed on top of the pia mater. It is important to gently lift the collagen pieces from the well without damaging them.
3. The window is sealed with a new 8-mm cover slip, which is glued to the bone with a mixture of histocompatible cyanoacrylate glue and Coe tray plastic power. The

animal is kept warm at 36–37° C during the recovery from anesthesia by means of a heating pad.

3.6. Tail Vein Injection of Rho-Conjugated Dextran

1. To determine the functionality of engineered vessels, tail vein injection of tetramethylrhodamine–dextran (2M MW: red) is performed. Tetramethylrhodamine-conjugated dextran is first dissolved in PBS at 10 mg/ml. The solution is then filtered through a $0.22-\mu$m filter to remove any undissolved aggregates.
2. A 1-ml syringe is used to fill the Rho–dextran solution, and the syringe is capped off with a 30-G needle.
3. Using pliers, the metal tip of a 30-G needle is bent and broken off from the plastic housing. The blunt end of the needle is then inserted into a PE10 tubing around 4 inches in length. The other end of the PE10 tubing is connected to the Rho–dextran-filled syringe.
4. The sharp end of the 30-G needle is used to the cannulate the dorsal vein of a mouse tail. Back pressure is applied on the syringe to make sure that blood can flow freely in the cannula (*see* **note 9**).
5. Hundred microliters of Rho–dextran is injected slowly through the tail vein. If there is resistance to injection, it is likely a result of a blood clot in the needle or a displacement of the needle away from the cannulated blood vessel, and the vessel needs to be recannulated.
6. The mouse is then imaged with multiphoton laser-scanning microscopy to visualize and quantify the morphological changes of EGFP–HUVECs and 10T1/2 cells at different timepoint. The images are analyzed with the ImageJ image-processing program for measuring vascular density. In our experience, blood flow in the engineered vessels begins around day 7 and the number of perfused vessels increases and plateaus at around 3 weeks. The engineered vessels are stable and functional, and we have observed engineered vessels that have persisted for 1 year (*see* **Fig.** 2).

3.7. Vascular Permeability Measurement

1. Tetramethylrhodamine-labeled BSA is diluted in PBS to 1% solution.
2. Unconjugated Rho is removed by passing the Rho–BSA solution through an Econo-Pac size exclusion column.
3. Similar to the procedure for injecting Rho–dextran, a mouse is injected with 100 μl Rho–BSA through the tail vein.
4. The mouse is then imaged with single-photon fluorescence intravital microscopy.
5. An area of engineered vessels is selected and an image of the vessels is taken. Fluorescence intensity of the selected area of tissue is then measured every 2 min for a total of 20 min by a photomultiplier using a ×20 objective lens.
6. The total volume and surface area of the engineered vessels in the image taken earlier are analyzed and measured in the ImageJ image-processing program. The effective vascular permeability (P) is calculated as follows: $P = (1 - HT)\, V/S\, [1/(I_0 - I_b) \times$

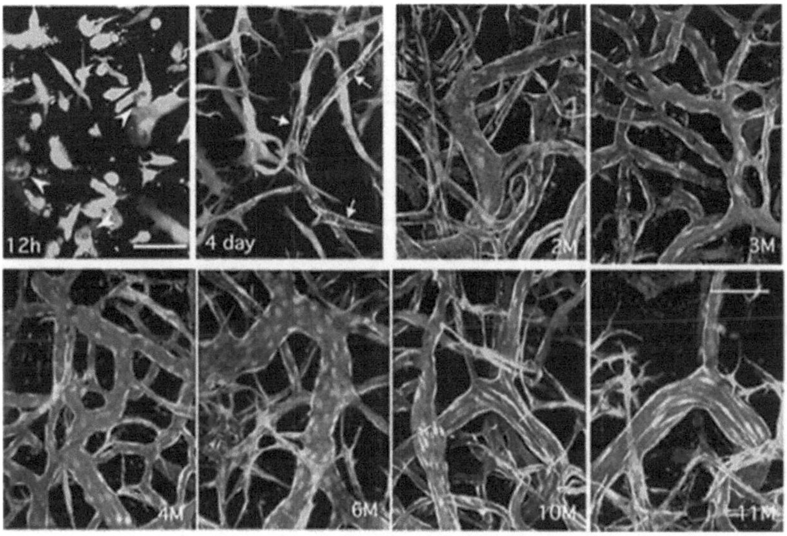

Fig. 2. Top left 2 panels : Initial morphological changes in the 3-D constructs in vivo. EGFP-HUVECs (green) and 10T1/2 cells were seeded in fibronectin–type 1 collagen gel. One day after incubation, the 3-D constructs were implanted into transparent cranial windows in severe combined immunodeficient mice. Before the 3-D constructs became perfused, EGFP labelled HUVECs (green) formed a mesh-like network. Immediately after implantation, the HUVECs exhibited round or spindle-shape morphologies (12 hour). Intracellular vacuoles were also observed in some HUVECs (arrow heads). Thereafter, HUVECs formed long interconnected tubes with many branches (Day 4). Large vacuoles in the tubes resemble the lumen of capillaries (arrows). Similar structural changes were reported in an in vitro 3-D angiogenesis assay (13). There were no apparent differences between the HUVEC – alone group (not shown) and the co-implantation (HUVEC + IOT1/2) group during these early morphological changes, scale bar: 50 μm. Top right 2 panels and bottom row, 2 months to 11 months after co-implantation of EGFP labelled HUVECs (green) and 10T1/2 cells, engineered blood vessels were visualized by multi-photon laser scanning microscopy following intravenous infection injection of tetramethylrhodamine-dextran that allowed visualization of perfused vessels (red.) scale bar: 50 μm (8).

$dI/\mathrm{d}t + 1/K]$, where I is the average fluorescence intensity of the whole image, I_0 is the value of I immediately after the filling of all vessels by Rho–BSA, and I_b is the background fluorescence intensity. HT is the average hematocrit. V and S are the total volume and surface area of vessels within the tissue volume covered by the surface image, respectively. The time constant of BSA plasma clearance (K) is 9.1×10^3 s (*see* **Table** 1).

Table 1
Vascular Permeability of Tissue-Engineered Vessels and Other Vessels in the Mouse Cranial Window

HUVEC + 10T1/2 *(8)*	Control cranial window *(14)*	Tumors *(14,15)*[a]	Type I collagen gel with VEGF *(16,17)*
$1.33 \pm 0.47 \times 10^{-7}$ cm/s	0.3–0.6 $\times 10^{-7}$ cm/s[b]	2.9–3.9 $\times 10^{-7}$ cm/s[b]	2.5–4.9 $\times 10^{-7}$ cm/s[b]

HUVEC, human umbilical cord vein endothelial cell; VEGF, vascular endothelial growth factor.
Vascular permeability of engineered vessels was determined at day 36 when the perfused vessels in coimplantation construct are relatively stable. Data are expressed as mean ± SEM ($n = 6$each).
[a]includes LS174T human colon adenocarcinoma, U87 human glioma, and MCaIV murine mammary carcinoma *(14–17)*.
[b]Range of average values from literature.

3.8. Arteriolar Contractility Assay

1. The cover glass of a cranial window is carefully removed, and the surface of the brain is superfused with warm PBS to maintain a moist surface.
2. For vessel contrast enhancement, 100 μl Rho–dextran (10 mg/ml) is injected through tail vein. Arterioles and venules are distinguished by their morphology and flow

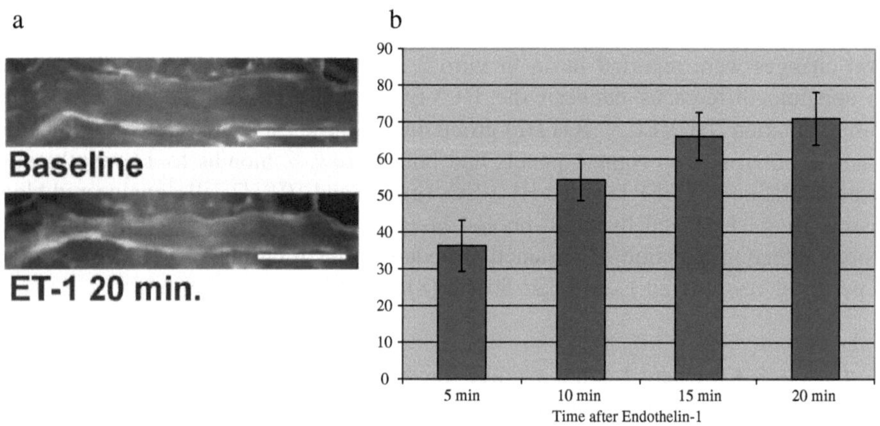

Fig. 3. Engineered vessels were superfused with 100 nM of ET-1, a vasoconstrictive agent. The engineered vessels contracted upon stimulation with endothelin-1 (a) and the contraction was time dependent (b) (scale bar, 100 μ m). *(8)*

pattern in vivo. Arterioles branch out from larger vessels and have faster flow rates and smaller diameters. Venules, on the contrary, merge into larger vessels and have slower flow rates and larger diameters.

3. An arteriolar engineered vessel is selected, and an image of the vessel is taken with two-photon laser-scanning microscopy using a ×20 water-immersion objective.
4. After the baseline measurement, the superfusate is replaced with 100 nM ET-1 in PBS. The same engineered vessel is imaged repeatedly for 20 min to monitor its contractile response to ET-1 (*see* **Fig.** 3).

Notes

1. The density of 293 cells is a critical parameter. Transfection with lipofectamine with too low a cell density will result in a high number of cell death, whereas if cell density is too high, the 293 cells tend to detach from the cell-culture dish.
2. Instead of lipofectamine, other methods of transfection, such as calcium chloride, can also be used.
3. The packaging cells should be washed three times with PBS carefully to remove any residual lipofectamine-containing plasmids. We have had experience where the residual lipofectamine is transferred to the HUVECs culture, causing the HUVECs cells to fuse together because of the ectopic expression of VSVG on their cell membrane.
4. If the viral supernatant is not used immediately, it can be stored in a −80° C freezer and used within 3 months. The viral supernatant is stored in 5-ml aliquots to minimize the cycle of freeze–thaw.
5. If the virus titer is low, spin infection can potentially enhance the infectivity of the retrovirus. Cells are plated in a 6-well plate at subconfluency. Two milliliters viral supernatant is added to the well along with polybrene at 8 μg/ml. The plate is then centrifuged for 45 min at room temperature at 650 × g. After centrifugation, the plate is incubated for 4 h at 37° C before exchange with fresh medium.
6. We usually use HUVECs before passage 5 (approximately 15 population doublings) and 10T1/2 cells before passage 15.
7. We used the HUVEC versus 10T1/2 cell ratio of 4:1 based on our preliminary studies. When we use a higher concentration of 10T1/2 cells, the implanted gels often shrink because of overgrowth of 10T1/2 cells, and the onset of perfusion is delayed. Conversely, when we use a lower concentration of 10T1/2 cells, viability and capillary formation are poor in long-term culture (approximately 2 weeks; in vitro).
8. If the collage gel solution does not polymerize within half an hour, it is likely that the pH of the solution is not at 7.4. Another potential cause is that the collagen type I is beyond its expiration date and has become degraded.
9. It is important to begin the cannulation from the tip of a mouse tail especially if repeated injection is required because the vessel cannot be cannulated distal to

where it has been previously cannulated. The dorsal vein can be better visualized by wiping the tail with 70% ethanol, warming the tail in warm water, or using a rubber band to tie off the proximal end of the tail.

Acknowledgments

This work was supported by a Program Project Grant (P01CA80134) and Bioengineering Research Partnership Grant (R01 CA85140) to R. K. J. and D. F., an R01 Grant (CA96915) to D. F., and American Heart Association Pre-doctoral Fellowship to P. A.

References

1. Vacanti, J. P. and Langer, R. (1999) Tissue engineering: the design and fabrication of living replacement devices for surgical reconstruction and transplantation. *Lancet* **354**(Suppl. 1), SI32–SI34.
2. Compton, C. C., Butler, C. E., Yannas, I. V., Warland, G., and Orgill, D. P. (1998) Organized skin structure is regenerated in vivo from collagen-GAG matrices seeded with autologous keratinocytes. *J. Invest. Dermatol.* **110**, 908–916.
3. Vunjak-Novakovic, G., Obradovic, B., Martin, I., Bursac, P. M., Langer, R., and Freed, L. E. (1998) Dynamic cell seeding of polymer scaffolds for cartilage tissue engineering. *Biotechnol. Prog.* **14**, 193–202.
4. Atala, A., Bauer, S. B., Soker, S., Yoo, J. J., and Retik, A. B. (2006) Tissue-engineered autologous bladders for patients needing cystoplasty. *Lancet* **367**, 1241–1246.
5. Jain, R. K., Au, P., Tam, J., Duda, D. G., and Fukumura, D. (2005) Engineering vascularized tissue. *Nat. Biotechnol.* **23**, 821–823.
6. Blau, H. M. and Banfi, A. (2001) The well-tempered vessel. *Nat. Med.* **7**, 532–534.
7. Levenberg, S., Rouwkema, J., Macdonald, M., Garfein, E. S., Kohane, D. S., Darland, D. C., Marini, R., van Blitterswijk, C. A., Mulligan, R. C., D'Amore, P. A., and Langer, R. (2005) Engineering vascularized skeletal muscle tissue. *Nat. Biotechnol.* **23**,879–884.
8. Koike, N., Fukumura, D., Gralla, O., Au, P., Schechner, J. S., and Jain, R. K. (2004) Tissue engineering: creation of long-lasting blood vessels.*Nature* **428**, 138–139.
9. Wang, Z. Z., Au, P., Chen, T., Shao, Y., Daheeon, L. M., Bai, H., Arzigian, M., Fukumura, D., Jain, R. K., Scadden, D. T. (2007) Endothelial cells derived from embryonic stem cells form durable blood vessels in vivo. Nat. *Biotechnol.* **25**, 317–318.
10. Jain R. K., Booth, M. F., Padera T. P., Munn L. L., Fukumura D., Brown, E. (in press) Applications of Non-Linear Intravital Microscopy, in *Handbook of Biological Nonlinear Optical Microscopy* (So, P., Masters, B., eds), Oxford University Press, Oxford, UK.
11. Jain, R. K. (2003) Molecular regulation of vessel maturation. *Nat. Med.* **9**, 685–693.

12. Yuan, F., Leunig, M., Berk, D. A., and Jain, R. K. (1993) Microvascular permeability of albumin, vascular surface area, and vascular volume measured in human adenocarcinoma LS174T using dorsal chamber in SCID mice. *Microvasc. Res.* **45**, 269–289.
13. Yang, S., Grahm, J., Kahn, J. W., Schwartz, E. A., Gereitsen, M. E., (1999) Functional coles for PECAM-1 (C031) and VE-Cadherin (CD144) in tube assembly and Lumen formation in three-dimensional collagen gels. *Am J. Pathol.* **155**, 887–895.
14. Monsky, W. L., Fukumura, D., Gohongi, T., Ancukiewcz, M., Weich, H. A., Torchilin, V. P., Yuan, F., and Jain, R. K. (1999) Augmentation of transvascular transport of macromolecules and nanoparticles in tumors using vascular endothelial growth factor. *Cancer Res.* **59**, 4129–4135.
15. Yuan, F., Salehi, H. A., Boucher, Y., Vasthare, U. S., Tuma, R. F., and Jain, R. K. (1994) Vascular permeability and microcirculation of gliomas and mammary carcinomas transplanted in rat and mouse cranial windows.*Cancer Res.* **54**, 4564–4568.
16. Dellian, M., Witwer, B. P., Salehi, H. A., Yuan, F., and Jain, R. K. (1996) Quantitation and physiological characterization of angiogenic vessels in mice: effect of basic fibroblast growth factor, vascular endothelial growth factor/vascular permeability factor, and host microenvironment. *Am. J. Pathol.* **149**, 59–71.
17. Fukumura, D., Gohongi, T., Kadambi, A., Izumi, Y., Ang, J., Yun, C. O., Buerk, D. G., Huang, P. L., and Jain, R. K., Huang, P. L., and Jain, R. K. (2001) Predominant role of endothelial nitric oxide synthase in vascular endothelial growth factor-induced angiogenesis and vascular permeability.*Proc. Natl. Acad. Sci. U. S. A.* **98**, 2604–2609.

12

Artificial Pancreas to Treat Type 1 Diabetes Mellitus

Riccardo Calafiore and Giuseppe Basta

Summary

Substitution of diseased organ/tissues with totally artificial machines or transplantable biohybrid devices where functionally competent cells are enveloped within immunoprotective artificial membranes could represent one of the future goals in medicine. In particular, artificial or, closer to feasibility, biohybrid artificial pancreas (BHAP) could replace the function of pancreatic islet β-cells that have been destroyed by autoimmunity, thereby obviating the need to treat patients with type 1 diabetes mellitus (T1DM) with multiple daily insulin injections. State-of-the-art diabetes therapy and perspectives in the use of BHAP, with special regard to islet-cell-containing microcapsules fabricated with alginate-based polymers, including applications to experimental animal models according to different chemical procedures, are reviewed. Special emphasis has been given to preparation methods, immunoprotection strategies, and biocompatibility of the islet-cell-containing microbarriers, as well as to approaches to ameliorate these features. Currently available BHAP prototypes have been critically reviewed to define expectations about the next generation devices targeting the final cure of T1DM.

Key Words: Diabetes mellitus; Hyperglycemia; Transplantation; Cell therapy; Pharmacologic immunosuppression.

1. Introduction

1.1. Definitions

The artificial pancreas is a machine that while constantly measuring blood glucose levels by an artificial glucose sensor is able to release minute-by-minute insulin for remission of glucose excursion within physiological values. In this respect, the machine works like a normal pancreas. Downstream the glucose sensor, electronic signals drive an insulin infusion pump through a computer algorithm that controls insulin delivery appropriate for the detected glucose levels. However, despite the intense study over the 1980s and 1990s,

From: *Methods in Molecular Medicine, 2nd ed.: Tissue Engineering*
Edited by: H. Hauser and M. Fussenegger © Humana Press Inc., Totowa, NJ

in an attempt to adapt the machine to the needs of patients with type 1 (insulin-dependent) diabetes mellitus (T1DM), eventually trying to miniaturize a big apparatus into a cardiac pacemaker-like device, the artificial pancreas has not yet become a clinical reality, due to a series of technical pitfalls. Overcoming the chemical/physical/engineering problems, possibly resulting in stable application of the machine to live beings, could still require many years of study to come.

Easier and certainly closer to practical applicability is an artificial pancreas that is powered by live, insulin-secreting pancreatic cells obtained from humans or possibly better animals, whether may these cells be primary or genetically engineered or finally pancreatic or extrapancreatic stem cells. Such a device, named "biohybrid artificial pancreas" (BHAP), should consist of a hybrid machinery, made of artificial micromembranes or macromembranes enveloping live insulin-secreting cells that usually but not necessarily are whole pancreatic islets of Langerhans, to protect the tissue upon transplantation into diabetic recipients, from the host's immune response. The BHAP that could represent a possible strategy to operatively apply an artificial organ to the therapy of T1DM falls in the field of tissue engineering. This is an interdisciplinary approach that couples engineering principles to life sciences in an effort to develop biohybrid substitutes that maintain, replace, restore, or anyhow improve the targeted organ/tissue function ***(1)***. Cells/tissues entrusted with only mechanical/biobarrier tasks, such as skin, are quite different from others that are associated with refined systems for bioactive molecule biosynthesis and delivery, such as the endocrine pancreas (insulin, glucagons, etc.), or thyroid, or liver, and so forth. It is all but demonstrated that such tissues once disconnected from their natural matrix are capable of retaining their sophisticated functional competence, especially if embodied within artificial materials. These not only should be associated with a tissue/artificial membrane interface that primarily is harmless to both the incorporated cells and the host, but they also are supposed to play an active role in regulating transmembrane molecule trafficking that is indispensable for both function of the enveloped tissue and harmless integration with the host that needs tissue replacement. Special care is to be taken in selecting either artificial materials or live tissues, as main components of the BHAP, in compliance to physiological competence and general safety principles.

BHAP could apply to transplant for cell therapy of insulin-dependent diabetes, whether originally be it type 1 (main indication) or type 2 or another type of disease (*see* **Table 1**). Because BHAP incorporates live insulin-secreting cells, within selective permeable and biocompatible artificial membranes, the device would provide for continuous insulin delivery, under regulated biofeedback by the extracellular (blood and interstitial tissue) glucose levels.

Table 1
Ethiologic Classification of Diabetes Mellitus

Type 1 diabetes (β-cell destruction leading to absolute insulin deficiency)
Immune-mediated
Idiopathic
Type 2 diabetes (may range from predominantly insulin resistance with relative insulin deficiency to a predominantly secretory defect with insulin resistance)
Other specific types
Genetic defects of β-cell function
Genetic defects in insulin action
Diseases of the endocrine pancreas
Endocrine diseases
Drug-induced or chemical-induced
Infections
Uncommon forms of immune-mediated diabetes
Other genetic syndromes sometimes associated with diabetes

1.2. Diabetes Mellitus

Diabetes mellitus is a group of metabolic diseases (*see* **Table 1**) characterized by hyperglycemia resulting from defects in insulin secretion, insulin action, or both. The chronic hyperglycemia of diabetes is associated with long-term damage, dysfunction, and failure of various organs, especially the eyes, kidneys, nerves, heart, and blood vessels.

Several pathogenic processes are involved in the development of diabetes. These range from autoimmune destruction of the pancreatic islet β-cells with consequent total insulin deficiency to abnormalities that result in resistance to insulin action.

Symptoms of marked hyperglycemia include polyuria, polydipsia, weight loss, sometimes polyphagia, and blurred vision. Impairment of growth and susceptibility to certain infections may also accompany chronic hyperglycemia. Acute life-threatening consequences of uncontrolled diabetes are hyperglycemia with ketoacidosis or the nonketotic hyperosmolar syndrome.

The vast majority of cases of diabetes fall into two broad ethiopathogenetic categories. In one category, T1DM, the cause is absolute deficiency of insulin secretion. Individuals at increased risk of developing this type of diabetes can often be identified by serological evidence of an autoimmune pathologic process occurring in the pancreatic islet cells by genetic markers. In the other, much more prevalent category, type 2 diabetes mellitus (T2DM), the cause is a combination of resistance to insulin action and an inadequate compensatory insulin secretory response.

1.3. Type 1 Diabetes Mellitus

This form of diabetes, which accounts for 5–10% of those with diabetes, results from cellular-mediated autoimmune destruction of the β-cells of the pancreas. Markers of β-cell immune destruction include islet cell autoantibodies, autoantibodies to insulin, autoantibodies to glutamic acid decarboxylase (GAD65), and autoantibodies to tyrosine phosphatases (IA-2 and IA-2β). One and usually more of these autoantibodies are present in 85–90% of individuals when fasting hyperglycemia is initially detected. Also, the disease has strong human leukocyte antigen (HLA) associations, with linkage to the *DQA* and *DQB* genes, and it is influenced by the *DRB* genes. These *HLA-DR/DQ* alleles can be either predisposing or protective.

In this form of diabetes, the rate of β-cell destruction is quite variable, being rapid in some individuals (mainly infants and children) and slow in others (mainly adults). Some patients, particularly children and adolescents, may present with ketoacidosis as the first manifestation of the disease. Others have modest fasting hyperglycemia that can rapidly change to severe hyperglycemia and/or ketoacidosis in the presence of infection or other stress. Still others, particularly adults, may retain residual β-cell function sufficient to prevent ketoacidosis for many years; such individuals eventually become dependent on insulin for survival and are at risk of ketoacidosis. At this latter stage of the disease, there is little or no insulin secretion, as manifested by low or undetectable levels of plasma C-peptide (*see* **Fig. 1**).

1.4. Epidemics and Clinical Features of Diabetes Mellitus

Diabetes mellitus is quite a common endocrine disorder (*see* **Table 2**) characterized by metabolic derangements, namely high blood glucose levels that if not rapidly or anyway properly treated may lead to acute (ketoacidosis and coma) or chronic complications (cardiovascular disease, renal insufficiency, premature blindness, and disabling neuropathic disease) that overall severely hamper life expectations of the affected patients. It is estimated that over 150 million people cumulatively suffer for T1DM and T2DM, but the number is destined to raise to over 300 million people by 2025 (*see* **Fig. 2**) *(2)*. Insulin dependency, a constant feature of T1DM, is associated with inevitable attending of choir to strictly monitor blood glucose levels by finger-prick capillary blood sampling, several times a day to assess how many insulin injections are necessary to

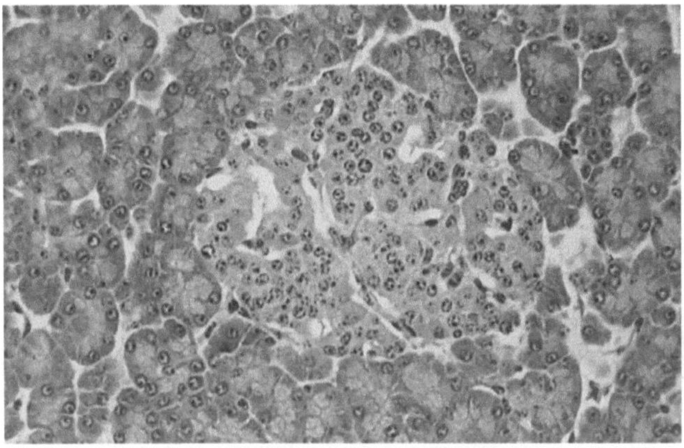

Fig. 1. Histologic section of human pancreas upon fixation and staining with hematoxylin & eosin (H&E) showing significant decrease of islet β-cell population (light microscopy ×40).

maintain a fair daily mean blood glucose control. However, even the most accurate insulin therapy regimen does not grant for preventing occurrence of such complications, thereby reflecting the fact that it is almost impossible to reproduce or mimic stimulus-coupled minute-by-minute physiological insulin release of the normal islet β-cells.

1.5. Insulin Therapy Regimens for the Therapy of Diabetes

According to guidelines promulgated by the American Diabetes Association *(3)*, the injection of insulin is essential for management of patients with T1DM but may be needed by patients with T2DM for intermittent or

Table 2
Frequency of Type 1 Diabetes Mellitus (T1DM) Worldwide

Country	Male/female	Frequency/100,000 people
Finland	1 : 1	28.6
Sweden	1 : 0.9	20.8
Norway	1 : 1.15	17.6
USA	1 : 1.10	14.7
Scotland	1 : 1.09	13.8
Israel		4.2
Japan	1 : 0.7	0.8

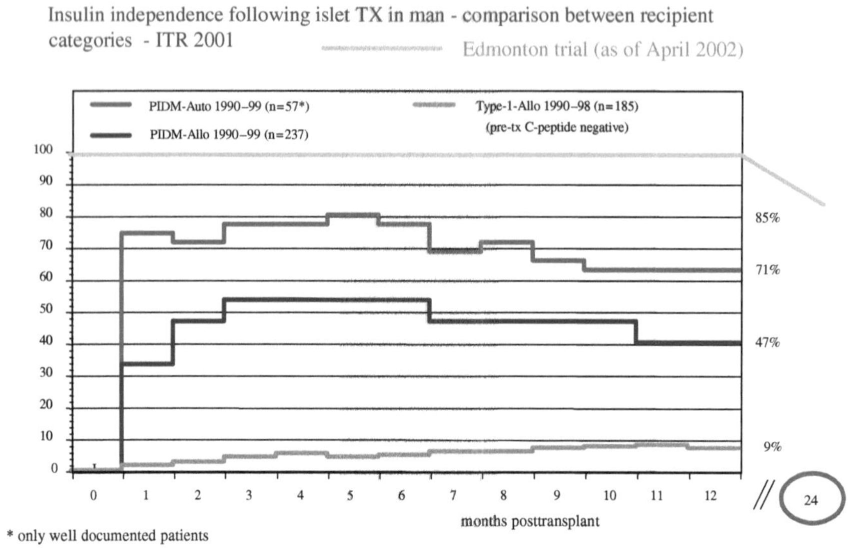

Fig. 2. Graph indicating the outcome of human islet allografts in the pre-Edmonton versus Edmonton protocol era (International Islet Registry, 2005). Insulin independence following islet TX in man—comparison between recipient categories (ITR, 2001). *, only well-documented patients.

continuous glycemic control. Today, new insulin molecules are available, that are associated with very flexible pharmacokinetics and pharmacodynamics, with regard to both short-acting (i.e., insulin *lispro*, insulin *aspart*, and insulin *glulisine*) and long-acting (i.e., insulin *glargine* and insulin *detemir*) preparations. These molecules have resulted in considerable improvement of metabolic control in patients with T1DM or in those with insulin-dependent T2DM. The Diabetes Control Clinical Trial data group has inequivocally demonstrated that the most effective approach to try to prevent/attenuate though not completely eliminate the risk of contracting secondary complications of T1DM is to apply intensive insulin therapy regimens, based on at least four daily administration of both short-active and long-active insulin preparations, including the new analogic insulin molecules. This remains the best way to maintain a balanced blood glucose control that results in improved values of glycated hemoglobin (HbA1c), a well-known long-term marker of metabolic control (the higher the HbA1c values, the worse the extent of metabolic control) although at the expense of strict blood glucose monitoring to decide, every day, the insulin schedule. Intensive insulin therapy regimens have certainly helped achievement of better daily mean blood glucose levels in patients with T1DM, but on the contrary, they are associated with two major drawbacks: (1) treated patients are exposed to higher risk of hypoglycemia, which positively correlates with decrease of HbA1c levels and (2) the improved metabolic control may reduce but not eliminate the risk of development of secondary complications of the disease. Last but not least, compliance of those T1DM patients undertaking intensive insulin regimens with both strict blood glucose monitoring and multiple insulin injection schedule daily is scarce, especially in the younger segments of the diabetic population.

1.6. Biological Systems for Endocrine Pancreatic Substitution

1.6.1. Human Pancreatic and Islet Transplantation

In light of serious concerns and technical pitfalls still associated with either "open" or "closed loop"-like artificial systems for insulin delivery, the T1DM problems could be solved only if the original physiological model, namely viable and functional insulin producing β-cells, was fully replaced. Such a goal might theoretically be accomplished by transplanting the endocrine pancreas as either whole pancreas or isolated human islet cells *(4)*. Both the former and the latter enabled reversal of hyperglycemia for even long periods of time in generally immunosuppressed patients with T1DM *(5)*. However, mandatory general immunosuppression dictates careful selection of the recipients, among those with brittle T1DM where glycemic control cannot be achieved by standard conventional or intensive insulin regimen protocols.

In the specific instance of human islets, separated and purified from cadaveric donor pancreases for allograft, potential risks associated with general immunosuppression should be weighed against higher frequency of secondary complications associated with "conventional" control of blood glucose by intensive insulin therapy regimens. Undoubtedly, islet unlike whole pancreas transplantation is a minimally invasive procedure, conducted under local anesthesia. Moreover, although before 2000 the majority of human islet allograft trials reported by the International Islet Transplant Registry (ITR) *(6)* conducted in generally immunosuppressed T1DM patients had fully functioned in 9% of the instances (remission of hyperglycemia, withdrawal of exogenous insulin) for maximum 1 year, important clinical results were achieved in 2000 and later by the Edmonton protocol (*see* **Fig. 3**), devised at the University of Alberta, Canada. According to this protocol, where changes in grafted islet mass and quality/patient, and cyclosporine/steroid-free immunosuppressive therapy regimens, resulted in clear clinical success, 100% of the transplanted patients

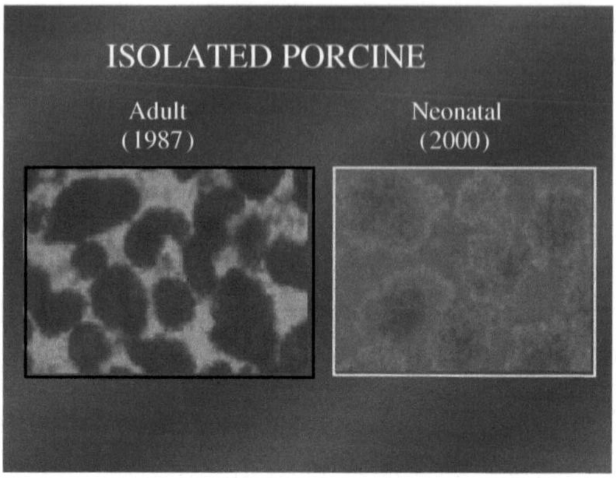

Fig. 3. Photomicrographs showing adult (left panel) and neonatal (right panel) porcine islets under light microscopy with (adult) or without (neonatal) diphenylthiocarbazone staining (×10).

(7/7) were insulin-free at 1 year posttransplant, with gradient decline of success (25% off-insulin patients, 5 years posttransplant). The lessons from the Edmonton protocol taught us that in the best conditions, with respect to both high and pure islet mass (not less than 10,000 islets equivalents/kg patient b.w.) and more tolerable immunosuppressive schedules, human islet allografts may do the job of reversing T1DM for long periods of time. This can very well be maintained as "proof of principle." However, single donor organs do not yield, in the majority of the instances, those many islets as required to produce full metabolic effects for remission of hyperglycemia. Quite often, two to three pancreases per recipient are required, which further exacerbates the restricted availability of human donor tissue that could never fulfill demand for transplant of all diabetic patients who would benefit from the procedure *(7)*. Hence, alternate, nonhuman insulin-secreting cell/tissue sources should be developed (*see* **Fig. 4**). For instance, it is current opinion that either adult or neonatal *(8)* porcine islets might successfully replace human donor tissue. In fact, pork insulin would be acceptable to humans, and it also would be possible to access *specific pathogen free* (SPF) pig-breeding stocks complying with standard safety requirements. These would include common porcine infectious contamination (*see* **Fig. 5**), with the exclusion of porcine endogenous retroviruses (PERVs), whose potential noxious effects on humans, incidently, are actually being revisited *(9)*. This being the case, it is possible that conventional or advanced protocols for the recipient's immunosuppression are on gestation, another approach is coming after many years of study into a sharper focus that

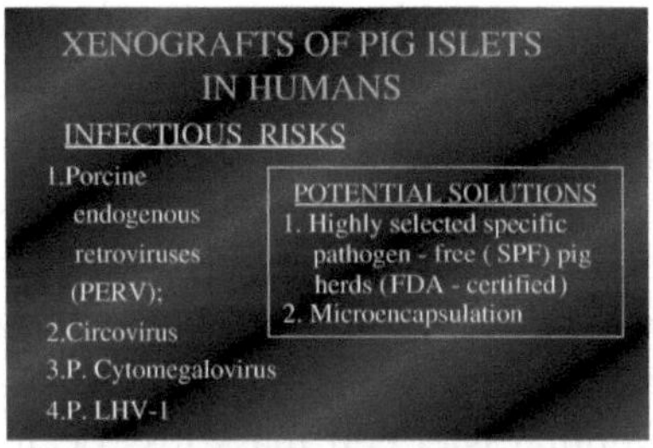

Fig. 4. Xenografts of pig islets in humans. *Fishman JA. Patience C Xenotransplantation: infections risk revisited Am. J. Transpl. 4(9): 1383-90, 2004.*

are active for allograft immunoprotection may not apply to xenografts (i.e., pig islets to human).

1.6.2. BHAP: General Features and Issues

In principle, as mentioned elsewhere, BHAP consists of isolated islets or, possibly in the future, engineered/nonengineered or stem-derived, insulin-secreting cells that are enveloped within a device (macro/micro) that is

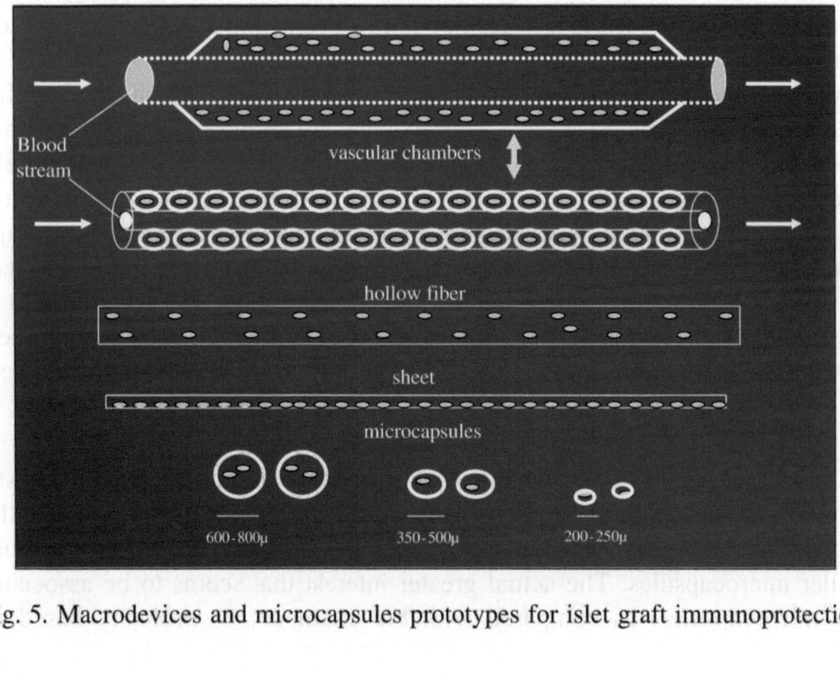

Fig. 5. Macrodevices and microcapsules prototypes for islet graft immunoprotection.

composed of biocompatible and selective permeable membranes. Of the two main BHAP constituents, the live cells and the artificial membranes, the former should provide for long-term insulin source that upon secretion should diffuse out of the latter preserving adequate physiological kinetics to prevent blood glucose fluctuations over the normal range. It is well known that adult islet cells are associated with poor mitotic index (1–2%), meaning that they virtually do not expand in vitro or in vivo, if not minimally, under physiological conditions. This implies that any eventual islet cell mass loss occurring within the device could not be efficiently replaced by newborn cells. Although some mitotic figures have been detected within microencapsulated islets *(9)*, these may not suffice to secure sufficient islet β-cell turnover, as observed in the normal endocrine pancreas. In terms of environmental biology of the enveloped islets, regardless of the employed enveloping membranes, the embodied unlike naked transplanted islets cannot benefit from direct neovascularization/innervation, and their survival solely depends on passive diffusion of gas, solutes, and soluble factors, with variations associated with the selected implant site. As for this specific point, it is obvious that sites where oxygen tension and blood supply are better would be preferable. Hence for instance, subcutaneous tissue would not be very suitable for islet-containing devices because of poor oxygen supply. Other sites, where oxygen and nutrients would be near excellent, such as the liver or spleen, that also are very close to physiological site of insulin action unfortunately cannot hold device implant, despite sporadic attempts *(10)*.

As far as the device shell is concerned, several problems of physical configuration geometries and size have been extensively studied over the 1980s and 1990s, but the feeling is that more has to be done and unfolded in this field. In fact, the device's configuration and size are instrumental not only to secure sufficient nutrient supply but also to grant immunoselection that is a major feature of a BHAP. It has been increasingly made clear that the window between the two main properties, adequate filtration/permeability and selectiveness to prevent ingress to unwanted cells/molecules (i.e., immunity, etc.), is extremely narrow. The right compromise to fulfill both may have not accomplished yet.

1.6.2.1. Morphology/Configuration Types

The idea of physically protecting the transplanted islet tissue from the host's immune system has been devised many decades ago, with large, differently-shaped, diffusion chambers prevailing over the actually more popular, much smaller microcapsules. The actual greater interest that seems to be associated with microcapsules, as compared with big chambers, is probably based on better results achieved by the former, in terms of either survival/functional performance of the chamber's content or biocompatibility of the implanted device. However, this may not represent the final word on this field. Increasing evidence can now be marshaled upon the fact that microcapsules could be grafted into a prevascularized bed, starting from an original idea ***(11)*** that had aimed to create a compromise between the two main types of immunoprotection devices, microcapsules deposited in a macrochamber (*see* **Fig. 6**).

1.6.3. Macrodevices

Many of the past-designed "macro"-devices for islet transplant immunoprotection today retain only historical value, whether would they be represented by vascular chambers, directly anastomosed to blood vessels (ultrafiltration chambers) ***(12)***, or by diffusion hollow fibers, associated with several physical configurations (planar, spherical, etc.), positioned either subcutaneously or intraperitoneally, and operating by simple passive exchange/diffusion of metabolites/oxygen/nutrients.

A wide range of biomaterials has been employed for fabrication of devices in an attempt to provide the islet grafts with immunoprotection. Some of them (*see* **Table 3**) are still being used, such as polyacrylonitrile–polyvinylchloride

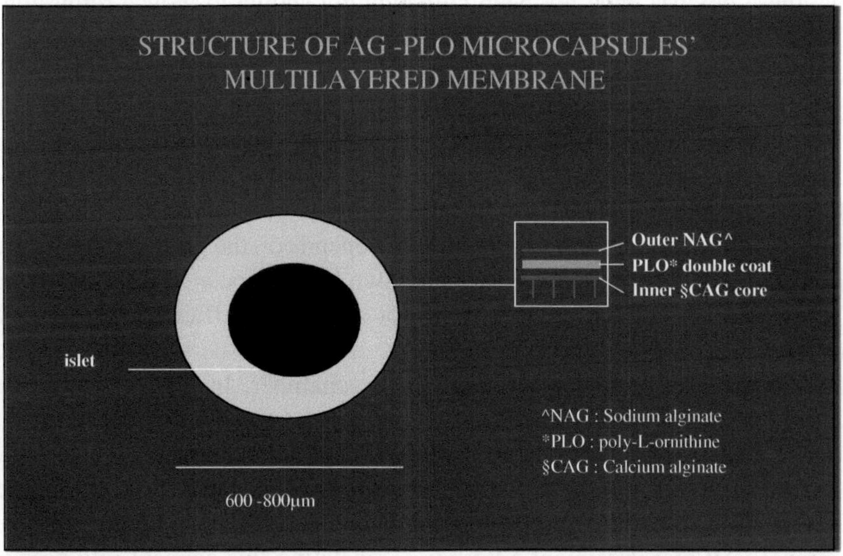

Fig. 6. Schematic representation of cross-section of alginate/poly-L-ornithine microcapsules (University of Perugia).

Table 3
Epidemiologic Features of Diabetes Mellitus

All diabetes and IGT	2003	2025
Total world population (billions)	6.3	8.0
Adult population (20–79 years) (billions)	3.8	5.3
Number of people with diabetes (20–79 years) (billions)	194	333
World diabetes prevalence (20–79 years) (%)	5.1	6.3
Number of people with IGT (20–79 years) (billions)	314	472
IGT prevalence (20–79 years) (billions)	8.2	9.0

(PAN-PVC), while others have progressively disappeared. Hollow fibers have been essentially employed for subcutaneous ***(13)*** or intraperitoneal ***(14)*** islet grafts. The fibers were composed of a selective permeable macromembrane (nominal molecular weight cut-off – MWCO – <100 kD) that while aimed at preventing access to immune cells or antibodies would allow transit of metabolically active substances, indispensable for survival and function of the embodied islets. The membranes were considered sufficiently biocompatible to circumvent foreign body tissue reactivity. Although successful in a few rodent trials, these devices have not been convincingly shown to induce remission of diabetes in higher mammalians. Ease of implant and retrieval procedure was uncoupled with sufficient embodied islets' survival rate, probably because of the device's disproportional surface/volume ratio with respect to islet mass content. This resulted, in the majority of cases, in limited cell nutrient supply and consequently in central necrosis of the enveloped cell/tissue. Such technical pitfalls were particularly sharp in large animals as compared with rodents. Moreover, although nutrient supply largely depends on the graft site (i.e., subcutaneous tissue and peritoneal cavity are known to be associated with low oxygen tension), and size of the devices does not allow for sufficient site selection, other factors could deserve consideration. A major issue with macrodevices relates to the islet-seeding procedure in the chamber. In fact, it was evident that only if islets were embedded within a gel (usually but not limited to alginate) matrix ***(13)***, they would survive; on the contrary, "loose" islets would die fast. Unfortunately, the islet density loading capacity in gel matrices was very low, which impaired accomplishment of metabolically sufficient functional islet mass/recipient's kg b.w. Translating this concept into practice, although rodents responded to graft of relatively small gel-embedded islet-containing "straws" ***(15)***, large animals would require unbearably big devices to obtain comparable metabolic effects. Of course, the same would translate to humans, where it has been estimated that reversal of hyperglycemia would be associated with unbearably large fibers filled with alginate-embedded islets. As elsewhere mentioned, prevascularizing the device, prior to islet entrapment, has resulted in interesting preliminary results in rodents ***(16)***. Whether these findings will ever apply to large-size mammalians remain to be assessed in the future.

As an alternative to prevascularized membranes, another approach to improve microdevice's environmental conditions could consist of direct blood supply to device through an arterio-vein (A-V) vascular graft (vascular prosthesis). Although in the past the vascular chambers were equipped with selective membranes, in charge of immunoprotecting the islet grafts, a potential new approach could be to deposit microencapsulated islets in the device so that an improved nutrient supply could be coupled with immunoprotection afforded by the microcapsules. The "vascular" graft of microencapsulated islets (*see* **Fig. 6**) while brilliant in principle, and preliminarily associated with results that were convincing in diabetic dogs ***(17)***, although not in humans, where only partial and transient remission of hyperglycemia was achieved ***(18)*** (in this setting the islets had been microencapsulated prior to loading in the vascular prosthesis), has remained, so far, an open issue. In particular, safety issues with regard to risk of the device's clotting/breakage and subsequent thrombosis should be carefully scrutinized prior to adapting this approach to humans.

In general, taking into consideration the risk/benefit balance, macrodevices may offer some advantages over other physical islet graft immunoprotection strategies, namely (1) retrievability, which is quite easy because of the device's size and predictable graft site and (2) physical configuration, which is today quite versatile, in an attempt to optimize oxygen/nutrient diffusion and physiological insulin kinetics. On the contrary, some issues also are apparent: (1) biocompatibility, which requires continuous search for new materials; (2) breakage risk, which grows bigger with device's size; and (3) surgical maneuvers for graft, which could still be invasive.

1.6.4. Microcapsules

Unlike macrodevices, microcapsules hold the advantage of offering far better volume/surface ratio, which means faster diffusion kinetics of hormone and nutrients. Microcapsules are easy to handle and graft into usually but not limited to the peritoneal cavity, within a minor injection procedure, under local anesthesia. Unlike reported elsewhere ***(19)***, conventional-size microcapsules (CSMs) are not suitable for intrahepatic injection, as usually done with naked islets (in generally immunosuppressed patients with T1DM), because they would completely plug the intrahepatic porta vein network. However, conformal microcapsules (CMs) fabricated in our laboratory (*see* **section 3.2.2**) could eventually fit the portal tree and then be used for intrahepatic graft. For conventional microcapsules, alternative to the peritoneal cavity, other sites in the mesenteric area that surely are better vascularized should be examined, and they are in progress in our laboratory.

We believe, at this juncture, that the only immunoprotection devices that have gained over time sufficient experimental evidence and confidence are microcapsules, especially if formulated with alginate-based biopolymers, which have been complexed or not with polyaminoacids (i.e., polylysine and polyornithine). This is why larger attention will be devoted here to this microcapsules prototype.

We have been able to develop, after 20 years of work in the field, a method for fabrication of highly purified alginate/poly-L-ornithine (AG/PLO) microcapsules for immunoprotection of either islet allografts or islet xenografts with no recipient's immunosuppression. Pharmaceutical grade 1.6% Sodium Alginate (NAG) (Stern Italia, Milan, Italy) was highly purified so as to lower its intrinsic endotoxin content down to 7 EU/g.

1.7. Alginate-Based Microcapsules: State-of-the-Art

1.7.1. General Considerations

Certainly, microcapsules constitute a selective permeable immunobarrier that is the most important protection shield from the islet graft-directed immune destruction pathways. Hence, microcapsules must be fabricated according to state-of-the-art procedures and chemical composition. Although many have been the polymers time by time proposed for microencapsulation of islet grafts (*see* **Table 4**) *(20–25)*, many of them ultimately have failed to provide for convincing immunoprotective barrier. Others have not been supported by a sufficient body of either in vitro or in vivo experimental studies and have been almost abandoned. Undoubtedly, the only polymer approved for fabricating microcapsules for human use is alginate *(26)*.

Extracted from brown seaweeds (*see* **Table 5**), the alginic saccharidic polymers (guluronic and mannuronic acid block patterns) since the time of the first successful reports from Lim and Sun *(27)* have gained progressive popularity for fabrication of microcapsules for transplant purposes. Unfortunately, standard parameters for encapsulation have not been made available to all laboratories challenged in this field. As a result, in the absence of comparable

Table 4
Biomaterials for Fabrication of Macrodevices

Polyacrylonitrile–polyvinylchloride
Acrylates
Cellulose nitrate
Cellulose-mixed esters
Polytetrafluoroethylene
Alginates
Polyethylene glycol

criteria, some positive data have been overwhelmed with negative results. In general, a major issue has been represented by the unavailability of ultrapurified alginate *(28)*. This being the major component of microcapsules, endotoxin-free and pyrogen-free criteria should be thoroughly fulfilled. But this has not been the case with the majority of laboratories. Another issue is storage of the alginate solutions: if improperly stored, the basic polymer could undergo acid hydrolysis with loss of physicochemical properties and consequential fabrication of altered microcapsules. Moreover, many laboratories have been caught in disputes about whether high mannuronic (M) or high guluronic (G) algin derivatives would be associated with better gel microbeads to be subsequently overlayered with polyaminoacids [PLO or poly-L-lysine (PLL) or other chemical compounds]. Advantageous gelling properties of high G over high M, coupled with lower immunogenicity of the former, as compared with the latter, were brought about by some groups *(29)* but not confirmed by our as well as other laboratories. Issues have been raised from the past to present days about whether small/medium microcapsules or CMs *(30)* would better fulfill needs

Table 5
Most Common Biopolymers for Fabrication of Microcapsules

Alginates with no polyaminoacidic coating
Barium
Calcium/barium
Agarose
Polyethylene glycol
Chitosan
Hydroxyethyl-methacrylate-methyl-methacrylate
Copolymers of acrylonitrile

for efficient and safe transplantation. Having developed different capsule sizes, on the basis of the same chemical formulation (AG/PLO) although with small stoichiometric adjustments, we have gathered enough experience to believe that large capsules (over 800μm in equatorial diameter) should be abandoned because of unfavorable volume/surface ratio that restricts biochemical exchange and diffusion rates. Special care should be taken in assessing uniformity and homogeneity of the alginate gel distribution, whether an alginate gel bead may represent the whole finite capsule ***(31)*** or rather just a core that needs overlayering with other polymers (i.e., alginate-polylysine or polyornithine) ***(32)***. With regard to "uncoated" alginate beads, prevalently made of barium alginate, these have been proven to represent immunocompetent barriers for islet allografts but not xenografts, by inhibiting contact with the immune cells involved with allograft rejection (*see* **section 1.6.1**), thereby limiting the applicability of such formulation. On the contrary, AG/PLO and alginate/PLL have been shown to enable function of islet xenografts in diabetic animal models. This has been clearly demonstrated, among others, by our group with full reversal of hyperglycemia upon microencapsulated Sertoli cell preconditioned neonatal porcine islet cell clusters in nonobese diabetic (NOD) mice with spontaneous autoimmune diabetes ***(33)*** (*see* **section 3.6.3**). This animal model is extremely close to human type 1 diabetes thereby raising hope that the same system could one day apply to patients with T1DM. It is noteworthy to mention that whereas PLL alginate gel bead coating has been deemed to provoke some immune response in recipient animals, PLO has been shown in our laboratory and recently confirmed ***(19)*** to be free of any inflammatory reaction when co-reacted with alginate hydrogels and implanted in either rodents or higher mammalians, man included ***(20)***.

2. Materials

2.1. Alginates

1. NAG, a naturally occurring polysaccharide from seaweed (marine brown algae), is used in the encapsulation process. It is extracted from raw material sources by Monsanto-Kelco (20N Wacker Dr, Chicago, IL, USA) and is prepared in a powdered ultrapure form according to specifications provided by the manufacturer.
2. Composition: Chemically, alginate is a linear copolymer of two uronic acids: D-mannuronic acid and L-guluronic acid linked by 1,4 glycosidic bonds. The two monomers are arranged in homopolymeric blocks, M-blocks and G-blocks, as well as by sequences containing both monomers, MG-blocks. The proportions and sequential arrangement of the uronic acids depend on the species of

algae and the specific algal tissue (stem or leaf) from which the alginate is extracted. The chemical composition of alginate is described in terms of its fractions (F_G or F_M), proportion of polyguluronic acids (Pgas) and polymannuronic acids (Pmas), and the M/G ratio. The source of the raw material (i.e., the algal species) largely determines the compositional structure of the alginate, as summarized in Table 5.

3. Method of production: Kelco has over 50 years experience in the production of alginate for commercial application. The company harvests *Macrocystis pyrifera* along the western coasts of North and South America, where large forests of these brown algae grow naturally in fairly shallow waters (between 2 and 40 m deep). Information regarding the manufacturing process is provided in the original Kelco document entitled "Technical Information – Alginate" (*Kelco Algin/Hydrophilic Derivatives of Alginic Acid for Scientific Water Control*, 2nd ed. Appendix II and III). Kelco manufactures under good manufacture practice (GMP) guidelines, and detailed information on their quality control procedures at the Chicago facility producing ultrapure alginate is provided in Appendix VIII and IX of the document. The alginate production process includes the following steps:
   Seaweed harvest → Washing → Alginate extraction → Filtration (0.2-μm filter) → Precipitation → Drying.
4. Characterization of the ultrapure alginate includes the following:

   i. Viscosity analysis – values are given for a 2% aqueous alginate solution measured on a Brookfield Viscometer RTV at 25°C and 60 rpm.
   ii. Mannuronic/guluronic acid content – obtained by 400 mHz HNMR spectral analysis.
   iii. pH.
   iv. Protein content.
   v. Heavy metal and salt content.

5. Specifications: Ultrapure alginate (Kelco LV) is custom produced by Monsanto-Kelco according to the well-established specifications (*see* **Table** 6).
6. Acceptance criteria: Powdered ultrapure alginate is accepted if it complies with the specifications outlined above. The mannuronic acid and guluronic acid content and the chemical composition of the alginate are determined by HNMR spectral analysis.
7. Purification: The multiple filtration process conducted at the University of Perugia on the raw material employs positively charged filters that remove any lipopolysaccharide (LP) content. Consequently, there is no trace of this endotoxin in the final preparation.
8. Assay of purity: The limulus amebocyte lysate (LAL) test is performed to detect endotoxin levels. This test is performed at both the University of Perugia and the Diatranz Ltd's laboratories (SOP Q209) to confirm the absence of endotoxin.

Table 6
Typical Alginate Composition from Various Seaweeds

Algae species	F_G (%)	F_M (%)	Pgas	Pmas	M/G ratio
Macrocystis pyrifera	39	61	17.7	40.6	1.56
Laminaria digitalia	41	59	12.7	60.5	1.45
Laminaria hyperborea	69	31	18.5	61.2	0.45

9. Storage: Storage is in endotoxin-free polystyrene bottles at 6°C with no light exposure.

2.2. *Reagents*

1. 1.6% NAG (pyrogen-free and endotoxin-free).
2. 10 kD polyethylene glycol (PEG).
3. 70 kD Dextran (DEX).
4. 0.05 and 0.01% PLO.
5. 0.05% NAG.

2.3. *Other Reagents*

1. 1.2% $CaCl_2$: 12 g anhydrous $CaCl_2$ in 1 l distilled deionized H_2O.
2. 1.6% isotonic NAG.
3. 0.1% isotonic NAG: 1:10 dilution of the mother solution (5 ml 1.6% alginate solution + 45 ml 0.9% NaCl solution).
4. 0.12% PLO: 120 mg PLO powder in 100 ml NaCl 0.9% solution.
5. 0.06% PLO: 1:2 dilution of the PLO mother solution (25 ml 1.6% alginate solution + 25 ml NaCl 0.9% solution).
6. 55 mM Na citrate: 1618 g Na citrate in 100 ml distilled/deionized H_2O.
7. 27 mM ethylenediaminetetraacetic acid (EDTA): 1005 g EDTA in 100 ml distilled deionized H_2O.
8. 0.12% PIPES: 60 mg Pipes in 50 ml sterile 0.9% NaCl solution.
   All the employed reagents must be sterile.

3. Methods

3.1. *Preparative Procedure*

1. Carefully wash the islet pellet twice in saline to eliminate any possible contaminating proteins (if tissue pellet is less than 1 ml, it is advisable to use 15-ml Falcon plastic tubes).

A homogenous suspension composed of islets–saline–NAG at 2:1:20 ratio (flexible, depending upon purity of the islet preparation, which mandates careful use of alginate in the dilution suspension process) is prepared.

2. Setup a peristaltic pump (i.e., Gilson 2 or similar) as follows:
   i. Dial 250.
   ii. Air flow: 5 l/min.
   iii. Needle tip distance from collection bath: 3 cm.
3. Schedule a "dry run" with saline initially so as to make sure that the system is operative.

The collection solution is composed of 200 ml 1.2% $CaCl_2$ in a 250-ml glass beaker.

4. Begin the extrusion process and collect the formed microspheres, after waiting for all the microbeads to deposit in the beaker bottom.
5. Leave the capsules in the beaker for 5 min in $CaCl_2$ or alternatively for 10 min in 0.1% Pipes, which will strengthen the capsule's chemical core.
6. Replace 100 ml $CaCl_2$ solution with saline and wait for the capsules to sediment in the beaker bottom and wait for 5 min.
7. Aspirate the capsules by pipette and transfer them to 50-ml Falcon plastic tubes and wash with saline twice.
8. Aspirate off the supernatant and add 0.12% PLO at a volume that is double the capsules' packed volume (mark this level on the tube) and manually rock the tube for 10 min.
9. Wash with saline and aspirate off the supernatant.
10. Wash with 0.1% Pipes, pH 7, and rock the tube for 3 min (same volume as that of PLO) and thereafter aspirate off the supernatant.
11. Wash twice with saline to remove unbound PLO and aspirate off the supernatant.
12. Add 0.1% NAG, same volume as above, and rock the tube for 6 min and aspirate off the supernatant.
13. Wash with saline twice to remove unbound NAG.
14. Add 55 nM Na citrate or 27 mM EDTA, same volume as above, and rock the tube for 6 min; aspirate off the supernatant.
15. Carefully wash twice with saline.
16. Resuspend the final capsule preparation in tissue-culture medium.

According to the original Lim and Sun's method ***(25)***, microcapsules were prepared by premixing the islet cell suspension with a solution of NAG, a polysaccharide extracted from brown seaweeds, thereby pumping the alginate–islet complex through a microdroplet generator in the presence of air-shear forces (*see* **Fig. 3**). Upon collection in a $CaCl_2$ bath, the islet-containing microdroplets immediately turned into calcium-alginate gel microbeads. These were sequentially coated with aminoacidic polycations, such as PLL or, uniquely in our laboratory, PLO, at appropriate molar concentrations. Finally, the microspheres were overlayered with highly purified NAG that provided the multilayered permselective membrane with biocompatibility (*see* **Fig. 7**). These CSMs, measuring an average 700–800 μm in equatorial diameter, have undergone several adjustments, including those originated in our own laboratory (*(34)*, *35*).

Microcapsules may be fabricated following other chemical procedures while still retaining an alginate shell. Other capsule types may be may be prepared with different biopolymers although experience gathered with alginates has been, so far, unsurmounted.

3.1.1. Main Features of Microcapsules and Encapsulated Islet Graft-Related Problems

Several concepts need to be outlined with regard to mechanisms that may be able to preserve functional longevity, viability, and morphologic integrity of the microencapsulated islet grafts.

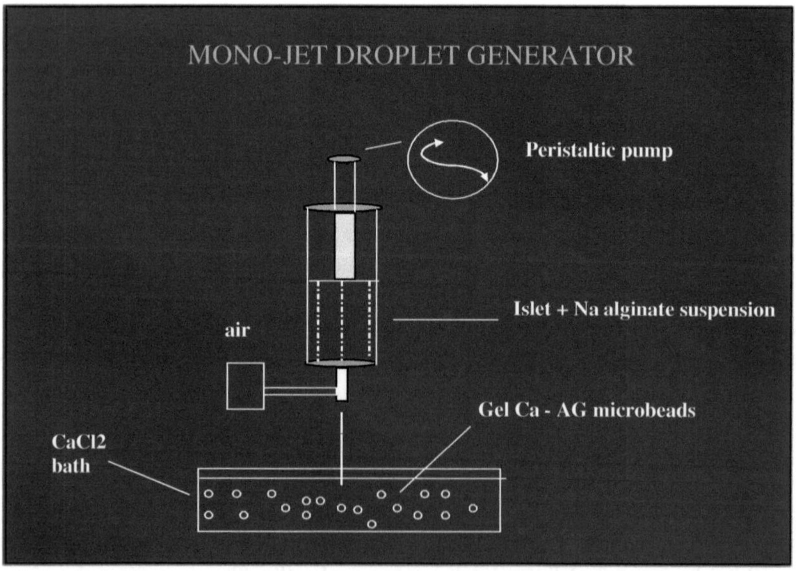

Fig. 7. Jet-head devise for fabrication of alginate-based microcapsules (University of Perugia).

3.2. Morphology and Size

3.2.1. Conventional-Size Microcapsules

A major issue that has not always been properly addressed by the laboratories involved with microencapsulation research pertains to fabrication of round-shaped, monodisperse, and smooth-surfaced microcapsules, regardless of the employed fabricative polymer. Owing to different technical apparatus, time by time employed by different laboratories, microcapsules have seldom resulted in regular shape and size. With respect to alginate microcapsules, according to the original recipe of Lim and Sun, the NAG/islet cell suspension was extruded through a microdroplet generator (*see* **Fig. 8**), where mechanical pressure and air-shearing forces drove formation of the gel microbeads, to be subsequently coated by multilayered PLL and NAG. By finely tuning all the physical parameters involved in the procedure (suspension flow rate and air flow rate) and using a microdroplet divalent cationic collection bath, properly formulated (usually $CaCl_2$), the capsules came out very regular and uniform in size and shape. A variable here was represented by the capsule's content, because of impure islet preparations, where the islets were contaminated with

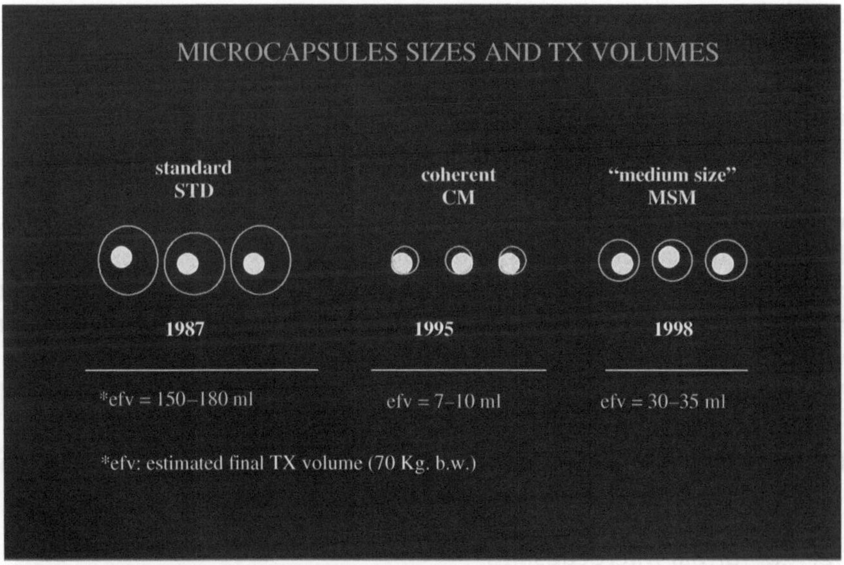

Fig. 8. Schematic representation of differently-sized alginate/poly-L-ornithine microcapsules (University of Perugia).

bigger exocrine/ductal tissue chunks. In this instance, gel microbeads were irregularly shaped, and sometimes broken, with the contained tissue protruding through the capsular membrane ***(36)***. On the contrary, uniform and pure islet cell preparations would be associated with regular-shaped capsules. This generation of capsules measured an average 800 μm in equatorial diameter, and they appeared to be too large for transplant of microencapsulated islets into small-size or, even worse, large-size diabetic animals. Here came the possibility to simply adjust the apparatus by adding static electricity ***(19)***. In particular, a droplet charged with high static voltage is suspended from a needle and attracted to a collecting vessel with opposing polarity. Upon overcoming the voltage threshold potential, the droplets move from the needle to the collection vessel. While microcapsules usually form reproducibly, in terms of size and shape, the former resulting considerably smaller as compared with tradition microdroplet generator, the apparatus may be unsafe because of the high voltage generated while any eventual damage to the microcapsule's islet cell content, theoretically absent because of the static nature of electricity, still needs to be assessed. In many instances, length of the encapsulation procedure is a negative factor. In fact, the suspended islet cell mass is exposed to chemical concentrations and physical/electric forces that may damage the cell viability and functional properties. This problem for instance applies to PEG microcapsules fabricated by laser beam-driven photopolimerization, which is a lengthy process ***(37)***. Moreover, the laser process must be triggered by initiators that may be unsafe because of their potential mutagenic properties. Staying with alginate-based microcapsules, and the traditional microdroplet generator, small changes in air flow and suspension flow rate, associated with regulation of the basic alginate polymer temperature, could help to prepare smaller capsules as compared with the initial 800-μm spheres, decreasing in size to 400–500 μm with no need for particular stressful treatment. We have named these last generation microcapsules, formulated in our laboratory with AG/PLO, "medium-size microcapsules" (MSMs). They are associated with favorable properties, in terms of volume/surface ratio, reasonable final transplant volume in either lower or higher mammalians with diabetes, and are currently used in our laboratory for transplant preclinical and pilot clinical trials.

The possibility of fabricating tiny microcapsules that tightly envelop each islet with virtually no dead space in between deserves special mention.

3.2.2. Conformal Microcapsules

Increasing interest is being shown on microcapsules that while retaining immunoprotective properties are composed by an extremely thin film-like membrane that tightly adheres to the islet surface. A major advantage of this approach would be the final transplant volume of the conformal encapsulated islets, with the possibility of widening the umbrella of the potential graft sites. We have devised a method ***(24)*** for fabrication of coherent microcapsules that is based on original fabricative principles of PEG.

3.3. Preparative Procedure

CMs are fabricated according to an original method developed in our laboratory (*see* **Fig. 9**) *(38)*. PEG, DEX, and NAG were transferred to a plastic tube and vigorously shaken at a 1:1:1 volume ratio until fine microdroplet emulsion was formed. The highly purified islet tissue, derived from rat or neonatal or human pancreas, was added to the emulsion and subsequently rocked on a rotating plate to facilitate the individual islet engulfment into each microdroplet. The emulsion containing the islets was gently poured in a $CaCl_2$ bath, under continuous stirring, until complete microdroplet gelling. The microbeads were then sequentially coated with 0.05 and 0.01% PLO and finally with 0.05% NAG, which overall formed a permselective membrane. The islet-containing CMs were culture-maintained in HAM F12 supplemented with 10% fetal calf serum and antibiotics at 37°C and 95% air/CO2.

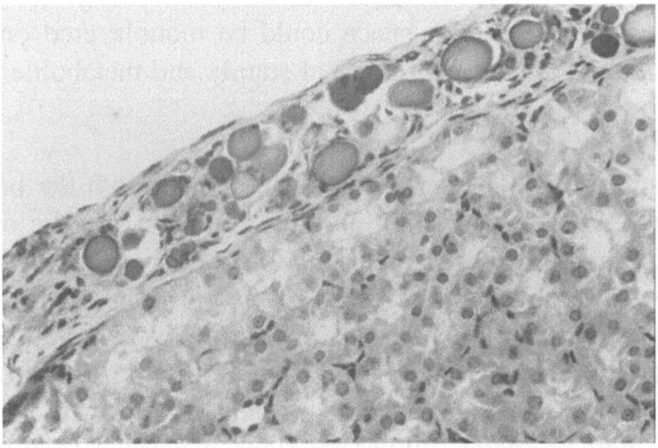

Fig. 9. Histologic section of subcapsular renal region containing empty microcapsules at 30 days posttransplant, showing no signs of inflammatory cell reaction (H&E, ×10) (University of Perugia).

3.4. Transplant Sites for Microcapsule

The original microcapsules devised by Lim and Sun were designed to be grafted intraperitoneally, given their large size that would lead to huge final graft volume per recipient. At that time, only mice and rats had been tested with intraperitoneal encapsulated islet grafts. However, problems have arisen when shifting the graft models from rodents to large-size mammalians. In fact, aside of past reports from our own *(39)* and other few laboratories *(40)*, no reproducible data were drawn from these animal models. The same applied to microencapsulated porcine islet xenografts into diabetic primates that apparently have succeeded only in one case *(41)* but failed elsewhere *(42)*. It is possible that the peritoneal cavity does not represent the best site for encapsulated islet graft, especially in high mammalians, where the volume/surface ratio may be unfavorable in terms of biochemical exchange, also in light of the low environmental oxygen tension *(43)*. In effect, our studies in diabetic rodents were associated with encapsulated islet cell overload to achieve the recipient's normoglycemia upon transplant. But the same would not be feasible in large-size animals and ultimately humans where too many islets would be required, should these recipients be assimilated to "big rats." Recent attention seems to be devoted to prevascularized beds where (possibly) small-size microcapsules containing the islets could be deposited. The beds could be positioned in the mesenteric area, thereby being closer to the liver, the first site of insulin action, with the advantage of being both removable, in case of need, and replenishable with microencapsulated fresh islets in case of functional exhaustion of the former graft. Of course, this kind of approach would ideally see the application of CMs that occupying narrow space could be monolayered on the vascular membrane, with great benefit for nutrient supply and metabolite exchange.

3.5. Biocompatibility

Probably, any artificial material that is introduced into the body somehow is "foreign" and as such may elicit an either immune-specific or not specific inflammatory response. However, introduction of new materials *(44)* and refinements in already established polymers for fabrication of macrodevices and microdevices has considerably helped surmounting many problems. As a consequence, most devices will not elicit a major inflammatory response when implanted "empty." At maximum, some fibroblast overgrowth will outlayer the body's surface with no major cellular reactions. Nevertheless, it is increasingly emerging that size, geometry, and site of implant may play different roles in the control of the host's response. Some sites, such as the peritoneal cavity or subcutaneous tissue, react very vigorously regardless of nature of the grafted materials. It is enough to say that even autologous blood may elicit an inflammatory response if leaking in the peritoneum. Consensus is being gathered on the fact that the smaller is the object the better is the host's tolerance. Although neovascularized membranes are being tested representing a novel version of the old macrodevices, increasing attention is being devoted to microcapsules, with regard to those formulated with alginate-derived polymers. In a way, this looks obvious, because alginate-based capsules, as mentioned elsewhere, are the only capsule prototypes that have been extensively studied. The challenge here is the availability of highly purified, pyrogen-free and endotoxin-free alginic acid fabricated according to US Pharmacopeia and National Formulary recommendations. We have established that during 1980s and 1990s work on microencapsulation an efficient system that has been gradually improved over time, for multiple sequential filtration, dialyzation, and solution of pharmaceutical grade raw powder of M-enriched NAG. Empty microcapsules fabricated with this alginate and overlayered with PLO and an outer alginate film have been extensively proven to avoid any host's inflammatory cell reaction in rodents or dogs or primates and ultimately humans (*see* **Fig. 10**) ***(20)***. Although both conventional and conformal capsules have been made available by our

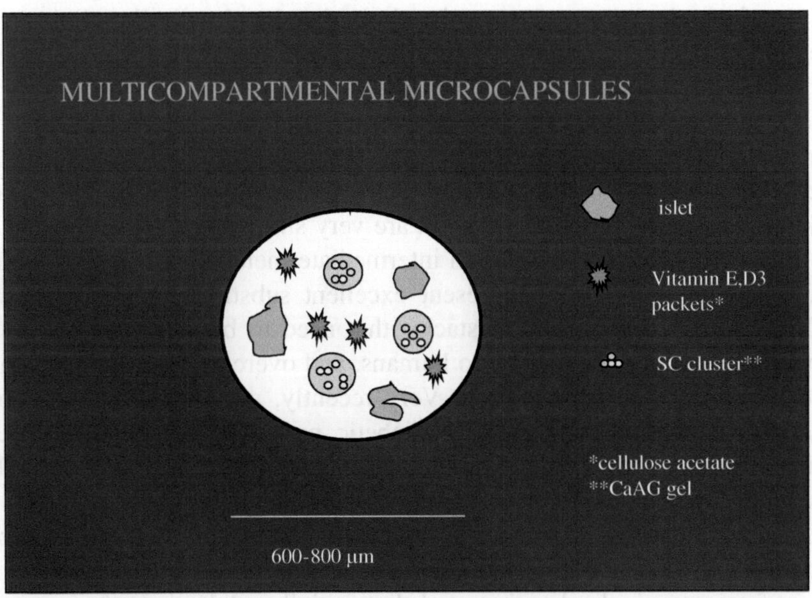

Fig. 10. Schematic representation of multicompartmental microcapsules (University of Perugia).

original procedures, we have primarily devoted our attention to MSMs that have been proven to be associated in our system with superior immunodefense capability *(45)*, as compared with the other types. MSMs do not occupy large spaces while still able to prevent immune rejection in both islet allografts and islet xenografts in diabetic animal models.

3.6. Cell Sources

The restricted availability of cadaveric human donor organs makes it unrealistic that patients with T1DM will be cured by human islet grafts. This even more applies to microencapsulated islet grafts, because lessons from rodents have clearly shown that the microencapsulated islet graft system needs to be overloaded to achieve reasonable metabolic results. In particular, if 5000 microencapsulated islet equivalents (IEQ) are necessary to reverse diabetes in a mouse weighing 20–25 g, we should expect that to induce remission of hyperglycemia in T1DM patients, weighing an average 70 kg, a few millions of human islets would be required. Unfortunately, we know that in the best conditions a single pancreas permits retrieval of about 300–50,000 IEQ. Hence, grafting many organs into a single recipient, should microcapsules be used for immunoprotection, could never be fulfilled given human donor organ scarcity. As a proof of principle, we have uniquely initiated a closed pilot clinical trial, under surveillance and control by the Italian Ministry of Health (19383, PRE 805), where microencapsulated human islets are being grafted intraperitoneally into 10 patients with T1DM and no general immunosuppression. However, other cell sources are required. Since the beginning of our research, we had turned our attention to adult porcine islets as human islet β-cell surrogates. We know that pork and human insulin are very similar, molecularly speaking. Also in general, porcine and human intermediate metabolism are very similar. Hence, porcine islets could represent excellent substitutes for patients with T1DM, despite a few serious obstacles that need to be surmounted. First of all, porcine islets are xenogeneic to humans, and overcoming a specie immune barrier certainly is a difficult task. Very recently, pig islet grafts have been shown to reverse hyperglycemia in diabetic primates that had been totally immunosuppressed by the co-stimulatory pathway blockade (*46*, *47*). This stays as an excellent proof of principle, closing the circle about potential use of pig islets into humans. However, general immunosuppression may expose the patients to unbearable risks. Here comes help from microcapsules. We have in fact extensively demonstrated that initially adult, thereafter neonatal, porcine islets separated and purified in our laboratory *(27)* may enable reversal of hyperglycemia either in diabetic rodents or in diabetic large-size animals.

We have fabricated especially composite microcapsules that permit access to complementary immunoprotection strategies (*see* **section 3.6.4**) and allow acceptance of the porcine islet xenografts. Another possible restraint on the way to applying microencapsulated porcine islet xenograft to humans is the issue associated with PERV potential transmission between species ***(48)***. PERV transmission has been up and down within the international scientific community, as a major barrier to porcine islet xenografts, although no clear proofs about potential harmful effects have been gathered. Interestingly, the PERV issue has been recently revisited by the same authors who had brought it up in 1999, discounting the possible harmful effects in humans ***(49)***. Moreover, retrospective studies conducted after years in patients who had been grafted with porcine islet tissue have never shown signs of PERV transmission in the patients (***50, 51***). Finally, microcapsules might per se represent a diffusion barrier for PERVs ***(52)***, although macrodevices do not seem to be associated with similar advantage ***(53)***.

3.6.1. Microencapsulated Islet Grafts: Principles, Basic, and Specific Problems

3.6.1.1. Not Encapsulated Islet Cell Graft-directed Allogeneic Immune Reactivity

When "naked" islets are grafted into either humans or animal models of diabetes, a series of immune responses are triggered that vary with regard to allogeneic or xenogeneic (discordant/nondiscordant) nature of donor tissue.

In general, staying with allografts (i.e., human-to-human), at least two orders of immune responses should be taken into account:

1. Immune rejection.
2. Recurrence of autoimmune diabetes.

A well-recognized role here is played by cellular immunity ***(54)***. T-lymphocytes (Tc) sense foreign antigens as peptides that are presented in association with major histocompatibility complex (MHC) molecules. In particular, CD8 Tc recognize antigens presented in association with class I MHC molecules, whereas CD4 Tc generally recognize antigens presented with class II MHC molecules. However, two signals are required for Tc activation: signal 1 is provided by Tc receptor/antigen binding; signal 2, "co-stimulator," is provided by nonantigen-specific signals from an antigen-presenting cell (APC) (i.e., macrophage, dendritic cell, etc.). The process defined as "co-stimulation" is composed of a complex cell surface receptor–ligand interactions network, which has been partly clarified in recent years (*see* **Fig. 6**). Typical participants in such interactions are CD80/CD86 molecules, expressed by APCs that bind CD28 co-receptor molecule on Tc; interactions between CD40 on APC and CD40L on Tc also result in important co-stimulatory signals. The role played by co-stimulatory molecules in inducing islet allograft rejection is shown by observations where blocking CD40–CD40L interactions in vivo could result in indefinite islet allograft survival *(55)*. It is worth to retain a classic distinction between a direct (or donor MHC-restricted) antigen presentation where Tc engage native MHC molecules directly on the surface of donor-derived APC and indirect (or host MHC-restricted) antigen presentation pathway in which donor-derived antigens are captured by recipient APCs, processed and reexposed, in association with recipient MHC molecules.

An issue that deserves careful attention in islet cell transplantation in T1DM consists of autoimmune recurrence of the disease on the transplanted tissue. Recurring disease has been thought to destroy the transplanted islet tissue in either humans or animal models of autoimmune diabetes. Tc-dependent autoimmune islet injury behaves similarly to xenograft-mediated response, in that islet damage can be triggered by autoreactive CD4 Tc, which are specific for islet antigens that have been processed and presented by APCs, similarly to the indirect pathway immune response process. Although CD8 Tc may contribute to developing the islet graft-directed autoimmune destruction, the latter clearly appears to be a CD4 Tc-mediated event.

Overall strategies so far employed to prevent islet allograft-directed immunity have included both general immunosuppressive agents, as mentioned elsewhere, and molecular mechanisms to block the co-stimulator pathway, such as blocking CD40–CD40L and B7–B28 interactions. The former has been usually applied to human islet allografts by using different pharmacological agents. The latter has been preliminarily studied in lower and higher mammal animal models of diabetes (*see* **Fig. 11**) *(56)*.

3.6.2. Not Encapsulated Islet Cell Graft-Directed Xenogeneic Immune Reactivity

In the case of xenogeneic islet transplantation, both cellular and humoral immunity may induce rapid destruction of the heterologous tissue.

In particular, hyperacute rejection results in rapid destruction of vascular-connected tissue/organs grafted across xenogeneic immune barriers, by virtue of action of the host's natural antibodies. These hook up to antigens located on the surface of endothelial cells within the xenograft, thereby activating complement with subsequent graft destruction *(57)*. The most important target of these antibodies is the terminal carbohydrate galactose α(1,3)-galactose (α-Gal). The presence of high-level α-Gal in all adult porcine endothelial cells does account for the immediate activation of a heterologous host's immune system. α-Gal expression seems to be lower in neonatal pig islets that should confer them a relative higher extent of protection when implanted in xenogeneic hosts. Among several possible strategies that actually are available for immunoprotection of xenogeneic discordant grafts, on the way to potentially using pig tissue into humans, generation of transgenic pigs that express human complement-regulatory proteins might be important ***(51)***. These proteins, CD 55 (DAF, decay-accelerating factor), CD 46 (MCP-1, membrane cofactor protein), and CD 59 (protectin), do not prevent anti-αGal antibody binding to endothelial cells (thus complement activation). These factors rather inhibit subsequent steps of the complement cascade finally resulting in cell lysis prevention. Results so far achieved with transgenically modified organs, and by extension, with pancreatic islet cells separated from the modified pancreases actually are under investigation.

Besides acute, humoral-mediated rejection pathways, islet xenograft rejection is less dependent on donor APC than allograft rejection. In fact, depletion of passenger leukocytes is not associated with the beneficial effects described for allografts. Xenograft rejection may occur in the absence of CD 8 Tc, whereas CD 4 Tc are predominantly involved with the immune destruction process, which reflects the steps already described for the indirect allograft rejection pathway ***(58)***.

In conclusion, islet xenograft rejection is an extremely rapid and efficient process that involves both the humoral and cellular arms of the immune system, and it continues, so far, to be difficult to contrast with the actually available means. Introduction of recent innovative general immunosuppressive protocol regimens has been preliminarily shown able to control xeno-islet rejection across discordant species barriers ***(40)***, although these observations need to be confirmed. Finally, just very recently, another report has shown the ability of co-stimulatory pathway (B7-CD28) blockade in allowing long-term survival and function of pocrine islet xenografts into primates with total pancreatectomy-induced diabetes ***(41)***. Use of special types of BHAP could dramatically improve the outcome of islet xenografts.

3.6.3. Encapsulated Islet Graft-Directed Destruction Pathways

The lessons learnt from "naked" islet allografts or xenografts in diabetic recipients may help to clarify the role and efficacy of microencapsulation in the natural history of the enveloped islet graft-directed immune attack and destruction. The original principle of "immunoisolation" was based on physical separation of the grafted islet cell antigens from the host's immune system. In this way, the antigen presentation necessary for triggering the islet allograft-directed immune destruction would be circumvented. Hence, physical isolation by preventing physical cell-to-cell contact between graft and the host's immune system could prevent activation of CD 8 Tc subsets, thereby obviating the encapsulated islet allograft-directed immune attack.

As mentioned elsewhere, CD 4 Tc subsets play a major role in the xenograft immune destruction process, within the indirect pathway of the host's immune response. A similar immune attack pattern could apply to grafted microencapsulated islet cells. Either small molecules (shed antigens and so forth), leaking out of the capsule's membrane, or part of the membrane's biopolymers could elicit an inflammatory response resulting in the host's macrophage recruitment. Macrophages would serve as APCs where the engulfed antigens are processed, thereby presenting to CD4 Tc subsets. The activated Tc could act either themselves or by activation of CD 8 Tc and/or secreting cytokines [interleukin-2 (IL-2), IL-5, and interferon-γ (IFN-γ)] that stimulate macrophages to produce additional cytokines [IL-1, tumor necrosis factor (TNF), IL-6, histamine, prostaglandins, and leukotrienes] that are associated with proinflammatory roles ***(59)***. An important role here is played by IL-8. Recent in vitro studies showed that IL-8 may be associated with multiple effects on neutrophils, with respect to shape changes, lysosome enzymes release, induction of respiratory burst, generation of superoxides and hydrogen peroxide, generation of bioactive lipids, and increase in adhesion molecule expression ***(53)***. IL-8 may induce in vitro and in vivo chemotaxis of CD4+ or CD8+ human peripheral blood Tc. In fact, intradermal injection of human IL-8 causes a rapid concentration-dependent neutrophil infiltration in all animal species so far examined. Likewise, subcutaneous injection of IL-8 also causes later Tc migration into the injection site. Moreover, LPs, which may contaminate collagenase batches for pancreatic digestion and islet isolation, may synergistically potentiate IL-8 action. IL-8 production is further potentiated by TNF-α released from monocytes. It then is apparent that specific immune response and not specific inflammatory reaction often are tightly intertwined and somehow connected by cytokines that are in charge of a dangerous crosstalk targeting the enveloped islet grafts.

Whether of strictly immune or at least initially not specific inflammatory nature and thereafter specific immune, the final common way will be pericapsular inflammatory cell tissue infiltration. In the instance, bioincompatible materials are employed for fabrication of the microcapsule fibroblast, and foreign body tissue reaction cell overgrowth of the microspheres will prevail. On the contrary, if the capsular immunobarrier competence is insufficient, prevalent immune cells (CD4, macrophage, neutrophils, etc.) will appear as a major component of the pericapsular tissue cell infiltrate. Consequently, regardless of the triggering event, inflammatory cell tissue infiltrate build-up around the capsules will dramatically impair diffusion of nutrients, gas, and solutes through the membrane. As result, the encapsulated islet cells will "chock-off" rapidly, thereby losing their viability and function.

3.6.4. Complementary Immunoprotection Strategies

Because microcapsules represent immunoisolatory and biocompatible barriers, one would expect that they are able to prevent either islet allograft-directed or islet xenograft-directed immune destruction. However, apparently this goal may be achieved with allografts, as proven by great number of reports worldwide, but less easily with xenografts. Some authors claim that microcapsules, with regard to the alginate-based ones, cannot protect islet xenografts at all. Hence, some of them have proposed that encapsulated islet xenograft recipients additionally are treated with short-term or smaller dose courses of general immunosuppressants. We believe that this strategy would considerably weaken the concept of physical immunoprotection that the capsules per se define and constitute. Rather, our belief is that can be improved and eventually potentiated microcapsules' immunobarrier competence by introducing the concept of composite microcapsules. Our idea has been to create multiple compartments (*see* **Fig. 12**) within conventional microcapsules, where islets take room together with agents that hold either anti-inflammatory properties ***(53)*** or anti-oxidizing effects ***(60)***. These solutions have shown to considerably improve both acceptance of the encapsulated islet grafts and viability of the enveloped islet cells. Another strategy developed in our laboratory, on the same wavelength, has been to preculture microencapsulated neonatal pig islets on homologous prepubertal Sertoli cell monolayers, taking advantage of the many growth factors, anti-apoptotic and anti-oxidizing agents, and finally immunomodulatory molecules secreted by these cells. As an important outcome, we have uniquely shown that NOD mice with spontaneous, overt autoimmune diabetes reversed hyperglycemia upon microencapsulated neonatal porcine islet xenografts, precultured with Sertoli cells for extraordinarily long periods of time ***(27)***. If applicable to higher mammalians, the procedure could represent a major breakthrough for encapsulated islet grafts.

3.6.5. Lights and Shadows With Microencapsulated Islet Grafts

Most of either allogeneic or xenogeneic islet-containing AG-PLL/PLO microcapsules have been usually implanted intraperitoneally into diabetic hosts that had not been generally immunosuppressed. Unfortunately, the lack of standardized materials and procedures has hampered impartial assessment of the majority of the results communicated by different laboratories. In general, diabetic rodent studies were associated with greater and more consistent success, in terms of full remission of hyperglycemia, following microencapsulated islet transplant (TX), as compared with large-size mammal trials ***(61)***. Nevertheless, the transplanted rodents carrying spontaneous autoimmune diabetes (NOD mice and Bio Breeding (BB) rats) have shown lower remission rate in comparison with mice and rats, where diabetes had been artificially induced by streptozotocin ***(62,63)***. In summary, the most experienced laboratories indeed have provided convincing evidence that microencapsulated islet grafts may work satisfactorily in small-size animals. On the contrary, very few reports documented reversal of hyperglycemia coupled with exogenous insulin withdrawal in dogs with spontaneous or pancreatectomy-induced diabetes, grafted with intraperitoneal microencapsulated canine or porcine islets (*see* **Table 7**). In these animals, full reversal of hyperglycemia also was accomplished in our laboratory, when the encapsulated islets had been embodied within a special vascular chamber, directly anastomosed to blood vessels ***(33)***. The canine background had subsequently permitted us to apply vascular grafts of encapsulated human islets into nonimmunosuppressed patients with T1DM. These were associated in the early 1990s with only partial and transient metabolic results American Society for Artificial Internal Organs (ASAIO). Preliminary success was reported in one patient with T1DM ***(64)*** grafted with intraperitoneal AG/PLL-microencapsulated human islets, but interpretation of the data was clouded by the fact that the recipient was under immunosuppression with cyclosporine due to previous kidney graft. An important exception to this uncertain trend was represented, at that time, by a trial of AG/PLL-encapsulated porcine islet xenograft into spontaneously diabetic, nonimmunosuppressed monkeys. In these animals, full remission of hyperglycemia and exogenous insulin withdrawal were achieved in all recipients and sustained in some of them for extraordinarily long periods of time, throughout 3 years of post-TX follow-up ***(35)***. This striking, still unmatched result seems to demonstrate that under the best conditions, microcapsules may constitute an effective, biocompatible, and immunoselective physical barrier, which enables immunoprotection of islet xenografts with no host's general immunosuppression.

Table 7
Chemistry of Alginate for Microencapsulation

1. Viscosity: 2% – 100–300 cps (Brookfield 25°C, speed 3, 60 rpm)
2. pH: 6.4–8.0
3. Protein content: <0.5%
4. Filtration: through 0.2-μm filter
5. Endotoxin level—measured by LAL test: 39 EU/g (any level below 100 EU/g in this test is considered endotoxin-free)
6. Molecular weight: 120,000–190,000 kD
7. Mannuronic acid (M) content: M fraction (F_M), 61%
8. Guluronic acid (G) content: G fraction (F_G), 39%.
9. Viscosity: 2% – 100–300 cps (Brookfield 25°C, speed 3, 60 rpm)
10. pH: 6.4–8.0
11. Protein content: <0.5%
12. Filtration: through 0.2-μm filter

Chemical analysis: Ca, <100 ppm; Cu, <40 ppm; Fe, <60 ppm; Hg, <40 ppb; Mg, <40 ppm; Zn, <40 ppm; Pb, <50 ppm; Si, <10 ppm; Mn, <10 ppm; Sr, <40 ppm; As, <100 ppb.

Preliminary results of physically immunoprotected islet allografts/xenografts have so far been quite disappointing, in terms of reproducible and consistent evidence of function, quantifiable as a full reversal of hyperglycemia.

The trials of encapsulated islet cell allografts or xenografts in nonimmunosuppressed diabetic recipients have shown so far mixed results in the majority of the instances unable to confirm reproducible effectiveness of this approach. Many of these issues derive from the lack of uniform strategic plans to fabricate consistent microcapsules by the scientific community that is involved with this research. It remains unclear if alginate-based microcapsules, whether they be overcoated with different polyaminoacidic formulations or rather with no outer coats (i.e., cross-linked barium alginate), are the best chemical formulation for immunoisolatory biomembranes. Indeed, the vast majority of graft studies thus far communicated in the literature derive from alginate microcapsules. Of course, search for new biopolymers to make high-performance microcapsules should be pursued. In particular, the membrane's durability, immunoselectiveness, and biocompatibility are issues that would deserve the highest attention. Moreover, to surmount technical problems that have hampered clear-cut success of encapsulated islet grafts, another issue should be addressed, with regard to design/configuration of the microcapsules. This is in light of increasing success of artificial scaffold ***(65)*** where the islets are deposited under much better nutrition supply conditions. Hence, should microencapsulated islets still be implanted as "loose" microspheres? Should they rather be incorporated within laminar prevascularized membranes to implement nutrient/oxygen supply and improve metabolic insulin kinetics? Are the capsules themselves useful or should they be replaced by artificial scaffolds? These are the still unanswered questions that need to be addressed in the near future to achieve success in this field.

3.7. Outlook

In light of all observations that have been made, bioartificial pancreas shows continuous progress, thanks to recent technical achievements that pertain to both the "artificial" and the "biological" components of the device. If porcine islet xenografts have been recently proven to survive and function till the induction of full reversal of hyperglycemia in totally immunosuppressed diabetic monkeys, it is more than a hope to believe that similar results may be accomplished by microencapsulated neonatal porcine islets (eventually precultured in Sertoli cells) with the advantage that the recipients should not undergo any general immunosuppression and its possible severe side effects. Work is actually being in progress in our laboratory to this end using composite microcapsules that have performed very well in rodents. Hence, while our ongoing pilot clinical trial with microencapsulated human islets into nonimmunosuppressed patients with T1DM is the first controlled study of capsules graft in humans, proofing the principle that microcapsules may be safely and effectively employed in patients, we expect a different future for this type of bioartificial pancreas setting. In the short-term/mid-term, when ban to operative use of pig islets in humans is likely to be lifted, microencapsulation of neonatal pig islets could be the best way to apply the system to patients with T1DM. There would be no restrictions for insulin-producing cell/tissue procurement that actually limits progress of human islet-cell-grafting procedures, with the PERV issue gradually being overcome.

Finally, if studies for the identification and collection of functional totipotent and/or pluripotent/multipotent stem cells that are able to transdifferentiate into insulin-producing β-like cells are successful, another powerful cell source could flank porcine islets for microencapsulation and transplantation for the final cure of T1DM.

Notes

1. 1.6% Alginate solution for microcapsule fabrication should be kept at 4°C wrapped in aluminium foil to prevent light exposure until use.
2. Operative use of alginate should take place within 4–5 min of pulling the solution out of the cold room.

3. Distance from the microdroplet generator's tip and the surface of the $CaCl_2$ collection solution should never exceed 3.5 cm.
4. While preparing CMs, the ratio alginate : DEX : PEG should be 1:1:1.
5. The gelled microbeads should not stay in $CaCl_2$ longer than 5 min to avoid walls chemical unwanted reactions.
6. Final washes of microcapsules should always follow this sequence: saline for 5 min, Hank's balanced salt solution for 5 min, and final culture medium for 5 min.
7. The following timing should be strictly observed: islets in culture for 24 h – encapsulation – additional 24 h in culture transplantation.

References

1. Colton, C. K. (1995) Implantable biohybrid artificial organ. *Cell Transplant.* **4**, 415–436.
2. Williams, R. (1998) The burden of diabetes in the next millennium. *Diabetes Rev. Int.* **7**(3), 21–27.
3. American Diabetes Association (position statement) (2005) Standard of medical care in diabetes, *Diabetes Care* **28**(Suppl. 1), S4–S36.
4. Calafiore, R. (1997) Perspectives in pancreatic and islet cell transplantation for the therapy of IDDM, *Diabetes Care* **20**, 889–896.
5. Ryan, E. A., Bigam, D., and Shapiro, A. M. (2006) Current indications for pancreas or islet transplant. *Diabetes Obes. Metab.* **8**, 1–7.
6. International Islet Transplant Registry (1999) *Newsletter 8*, 1.
7. Shapiro, A. M., Lakey, J. R., Paty, B. W., Senior, P. A., Bigam, D. L., and Ryan, E. A. (2005) Strategic opportunities in clinical islet transplantation. *Transplantation* **79**, 1304–1307.
8. Korbutt, G. S., Elliott, J. F., Ao, Z., Smith, D. K., Warnock, G. L., and Rajotte, R. V. (1996) Large scale isolation, growth, and function of porcine neonatal islet cells. *J. Clin. Invest.* **97**, 2119–2129.
9. Hunkeler, D., Sun, A. M., Rajotte, R., Gill, R., Calafiore, R., and Morel, P. (1999) Bioartificial organs and acceptable risk. *Nat. Biotechnol.* **17**, 1045.
10. Toso, C., Mathe, Z., Morel, P., Oberholzer, J., Bosco, D., Sainz-Vidal, D., Hunkeler, D., Buhler, L. H., Wandrey, C., and Berney, T. (2005) Effect of microcapsule composition and short-term immunosuppression on intraportal biocompatibility. *Cell Transplant.* **14**, 159–167.
11. Orive, G., Hernandez, R. M., Gascon, A. R., Calafiore, R., Chang, T. M., de Vos, P., Hortelano, G., Hunkeler, D., Lacik, I., Shapiro, A. M., and Pedraz, J. L. (2003) Cell encapsulation: promise and progress. *Nat. Med.* **9**, 104–107.
12. Petruzzo, P., Pibiri, L., De Giudici, M. A., Basta, G., Calafiore, R., Falorni, A., Brunetti, P., and Brotzu, G. (1991) Xenotransplantation of microencapsulated pancreatic islets contained in a vascular prosthesis: preliminary results. *Transpl. Int.* **4**, 200–204.

13. Lacy, P. E., Hegre, O. D., Gerasimidi-Vazeou, A., Gentile, F. T., and Dionne, K. E. (1991) Subcutaneous xenografts of rat islets in acrylic copolymer capsules maintain normoglycemia in diabetic mice. *Science* **254**, 1782–1784.
14. Lanza, R. P., Beyer, A. M., Staruk, J. E., and Chick, W. L. (1993) Biohybrid artificial pancreas: long-term function of discordant islet xenografts in streptozotocin-diabetic rats. *Transplantation* **56,** 1067–1072.
15. Lanza, R. P., Borland, K. M., Staruk, J. E., Appel, M. C., Solomon, B. A., and Chick, W. L. (1992) Transplantation of encapsulated canine islets into spontaneously diabetic BB/Wor rats without immunosuppression. *Endocrinology* **131**, 637–642.
16. de Vos, P., Hillebrands, J. L., De Haan, B. J., Strubbe, J. H., and Van Schilfgaarde, R. (1997) Efficacy of a prevascularized expanded polytetrafluoroethylene solid support system as a transplantation site for pancreatic islets. *Transplantation* **63**, 824–830.
17. Maki, T., Otsu, I., O'Neil, J. J., Dunleavy, K., Mullon, C. J., Solomon, B. A., and Monaco, A. P. (1996) Treatment of diabetes by xenogeneic islets without immunosuppression: use of a vascularized bioartificial pancreas. *Diabetes* **45**, 342–347.
18. Calafiore, R. (1992) Transplantation of microencapsulated pancreatic human islets for the therapy of diabetes mellitus: a preliminary report. *ASAIO J.* **38**, 34–37.
19. Kizilel, S., Garfinkel, M., and Opara, E. (2005) The bioartificial pancreas: progress and challenges. *Diabetes Technol. Ther.* **7**, 968–985.
20. Duvivier-Kali, V. F., Omer, A., Parent, R. J., O'Neill, and Weir, G. C. (2001) Complete protection of islets against allorejection and autoimmunity by a simple barium alginate membrane. *Diabetes* **50**, 1698–1705.
21. Iwata, H., Amemiya, H., Matsuda, T., Takano, H., Hayashi, R., and Akutsu, T. (1989) Evaluation of microencapsulated islets in agarose gel as bioartificial pancreas by studies of hormone secretion in culture and by xenotransplantation. *Diabetes* **38**(Suppl. 1), 224–225.
22. Cruise, G. N., Hegre, O. D., Lamberti, F. V., Hager, S. R., Hill, R., Scharp, D. S., and Hubbell, J. A. (1999) In vitro and in vivo performance of porcine islets encapsulated in interfacially photopolymerized poly(ethylene glycol) diacrylate membranes. *Cell Transplant.* **8**, 293–306.
23. Zielinski, B. A. and Aebischer, P. (1994) Chitosan as a matrix for mammalian cell encapsulation. *Biomaterials* **15**, 1049–1056.
24. Dawson, R. M., Broughton, R. L., Stevenson, W. T., and Sefton, M. V. (1987) Microencapsulation of CHO cells in a hydroxyethyl-methacrylate methyl methacrylate copolymer. *Biomaterials* **8**, 360–366.
25. Kessler, L., Pinget, M., Aprahaim, M., Dejardin, P., and Damge, C. (1991) In vitro and in vivo studies of the properties of an artificial membrane for pancreatic islet encapsulation. *Horm. Metab. Res.* **23**, 312–317.

26. Calafiore, R., Basta, G., Luca, G., Lemmi, A., Montanucci, M. P., Calabrese, G., Racanicchi, L., Mancuso, F., and Brunetti, P. (2006) Microencapsulated pancreatic islet allografts into nonimmunosuppressed patients with type 1 diabetes: first two cases. *Diabetes Care* **29**, 137–138.
27. Lim, F. and Sun, A. M. (1980) Microencapsulated islets as bioartificial endocrine pancreas. *Science* **210**, 908–910.
28. Hunkeler, D., Prokop, A., Powers, A., Haralson, M., DiMari, S., and Wang, T. (1997) A screening of polymers as biomaterials for cell encapsulation. *Polymer News* **22**, 232.
29. Draget, K. I., Skjak-Braek, G., and Smidsrod, O. (1997) Alginate based new materials. *Int. J. Biol. Macromol.* **21**, 47–55.
30. Calafiore, R., Basta, G., Osticioli, L., Luca, G., Tortoioli, C., and Brunetti, P. (1996) Coherent microcapsules for pancreatic islet transplantation: a new approach for bioartificial pancreas. *Transplant. Proc.* **28**, 812–813.
31. Duvivier-Kali, V. F., Omer, A., Lopez-Avalos, M. D., O'Neil, J. J., and Weir, G. C. (2004) Survival of microencapsulated adult pig islets in mice in spite of an antibody response. *Am. J. Transplant.* **4**, 1991–2000.
32. Strand, B. L., Gaserod, O., Kulseng, B., Espevik, T., and Skjak-Baek, G. (2002) Alginate-polylysine-alginate microcapsules: effect of size reduction on capsule properties. *J. Microencapsul.* **19**, 615–630.
33. Luca, G., Nastruzzi, C., Calvitti, M., Becchetti, E., Baroni, T., Neri, L. M., Capitani, S., Basta, G., Brunetti, P., and Calafiore, R. (2005) Accelerated functional maturation of isolated neonatal porcine cell clusters: in vitro and in vivo results in NOD mice. *Cell Transplant.* **14**, 249–261.
34. Calafiore, R., Basta, G., Falorni, A., Calcinaro, F., Pietropaolo, M., and Brunetti, P. (1992) A method for the large-scale production of microencapsulated islets: in vitro and in vivo results. *Diabetes Nutr. Metab.* **5**, 23.
35. Lanza, R. P. (1995) Encapsulation technologies. *Tissue Eng.* **1**, 18.
36. de Vos, P., Wolters, G. H., and van Schilfgaarde, R. (1994) Possible relationship between fibrotic overgrowth of alginate-polylysine-alginate microencapsulated pancreatic islets and the microcapsule integrity. *Transplant. Proc.* **26**, 782–783.
37. Sawhney, A., Pathak, C. P., and Hubbell, J. A. (1994) Modification of Langerhans islet surfaces with immunoprotective poly(ethylene glycol) coatings. *Biotech. Bioeng.* **44**, 383.
38. Calafiore, R., Basta, G., Osticioli, L., Luca, G., Tortoioli, C., and Brunetti, P. (1996) Coherent microcapsules for pancreatic islet transplantation: a new approach for bioartificial pancreas. *Transplant. Proc.* **28**, 812–813.
39. Brunetti, P., Basta, G., Faloerni, A., Calcinaro, F., Pietropaolo, M., and Calafiore, R. (1991) Immunoprotection of pancreatic islet grafts within artificial microcapsules. *Int. J. Artif. Organs* **14**, 789–791.

40. Soon-Shiong, P., Feldman, E., Nelson, R., Komtebedde, J., Smidsrod, O., Skjak-Braek, G., Espevik, T., Heintz, R., and Lee, M. (1992) Successful reversal of spontaneous diabetes in dogs by intraperitoneal microencapsulated islets. *Transplantation* **54**, 769–774.
41. Sun, X., Ma, X., Zhou, D., Vacek, I., and Sun, A. M. (1996) Normalization of diabetes in spontaneously diabetic cynomologous monkeys by xenografts of microencapsulated porcine islets without immunosuppression. *J. Clin. Invest.* **98**,1417–1422.
42. Elliott, R. B., Escobar, L., Tan, P. L., Garkavenko, O., Calafiore, R., Basta, P., Vasconcellos, A. V., Emerich, D. F., Thanos, C., and Bambra, C. (2005) Intraperitoneal alginate-encapsulated neonatal porcine islets in a placebo-controlled study with 16 diabetic cynomolgus primates. *Transplant. Proc.* **37**, 3505–3508.
43. Wu, H., Avgoustiniatos, E. S., Swette, L., Bonner-Weir, S., Weir, G. C., and Colton, C. K. (1999) In situ electrochemical oxygen generation with an immunoisolation device. *Ann. N. Y. Acad. Sci.* **875**, 105–125.
44. Desai, T. A., Chu, W. H., Rasi, G., Sinibaldi-Vallebona, P., Guarino, E., and Ferrari, M. (1999) Microfabricated biocapsules provide short-term immunoisolation of insulinoma xenografts. *Biomed. Microdevices* **1**, 131–138.
45. Basta, G., Sarchielli, P., Luca, G., Racanicchi, L., Nastruzzi, C., Guido, L., Mancuso, F., Macchiarulo, G., Calabrese, G., Brunetti, P., and Calafiore, R. (2004) Optimized parameters for microencapsulation of pancreatic islet cells: an in vitro study clueing on islet graft immunoprotection in type 1 diabetes mellitus. *Transplant. Immunol.* **13**, 289–296.
46. Hering, B. J., Wijkstrom, M., Graham, M. L., Hardstedt, M., Aasheim, T. C., Jie, T., Ansite, J. D., Nakano, M., Cheng, J., Li, W., Moran, K., Christians, U., Finnegan, C., Mills, C. D., Sutherland, D. E., Bansal-Pakala, P., Murtaugh, M. P., Kirchhof, N., and Schuurman, H. J. (2006) Prolonged diabetes reversal after intraportal xenotransplantation of wild-type porcine islets in immunosuppressed nonhuman primates. *Nat. Med.* **12**, 301–303.
47. Cardona, K., Korbutt, G. S., Milas, Z., Lyon, J., Cano, J., Jiang, W., Bello-Laborn, H., Hacquoil, B., Strobert, E., Gangappa, S., Weber, C. J., Pearson, T. C., Rajotte, R. V., and Larsen, C. P. (2006) Long-term survival of neonatal porcine islets in nonhuman primates by targeting costimulation pathways. *Nat. Med.* **12**, 304–306.
48. Patience, C., Takeuchi, Y., and Weiss, R. A. (1997) Infection of human cells by an endogenous retrovirus of pigs. *Nat. Med.* **3**, 282–286.
49. Fishman, J. A. and Patience, C. (2004) Xenotransplantation: infectious risk revisited. *Am. J. Transplant.* **4**, 1383–1390.
50. Lindeborg, E., Kumagai-Braesch, M., Tibell, A., Christensson, B., and Moller, E. (2004) Biological activity of pig islet-cell reactive IgG antibodies in xenotransplanted diabetic patients. *Xenotransplantation* **11**, 457–470.

51. Valdes-Gonzalez, R. A., Dorantes, L. M., Garibay, G. N., Bracho-Blanchet, E., Mendez, A. J., Davila-Perez, R., Elliott, R. B., Teran, L., and White, D. J. (2005) Xenotransplantation of porcine neonatal islets of Langerhans and Sertoli cells: a 4-year study. *Eur. J. Endocrinol.* **153**, 419–427.
52. Elliott, R. B., Escobar, L., Garkavenko, O., Croxson, M. C., Schroeder, B. A., McGregor, M., Ferguson, G., Beckman, N., and Ferguson, S. (2000) No evidence of infection with porcine endogenous retrovirus in recipients of encapsulated porcine islet xenografts. *Cell Transplant.* **9**, 895–901.
53. Pakhomov, O., Martignat, L., Honiger, J., Clemenceau, B., Sai, P., and Darquy, S. (2005) AN69 hollow fiber membrane will reduce but not abolish the risk of transmission of porcine endogenous retroviruses. *Cell Transplant.* **14**, 749–756.
54. Gill, R. G. and Wolf, L. (1995) Immunobiology of cellular transplantation. *Cell Transplant.* **4**, 361–370.
55. Grewal, I. S. and Flavell, R. A. (1998) CD40 and CD154 in cell-mediated immunity. *Ann. Rev. Immunol.* **16**, 111–135.
56. Weber, C. J., Safley, S., Hagler, M., and Kapp, J. (1999) Evaluation of graft-host response for various tissue resources and animal models. *Ann. N. Y. Acad. Sci.* **875**, 233–254.
57. Rayat, G. R., Rajotte, R. V., and Korbutt, G. S. (1999) Potential application of neonatal porcine islets as treatment for type 1 diabetes: a review. *Ann. N. Y. Acad. Sci.* **875**, 175–188.
58. Gill, R. G. (1999) Antigen presentation pathways for immunity to islet transplants: relevance to immunoisolation. *Ann. N. Y. Acad. Sci.* **875**, 255–260.
59. Ricci, M., Blasi, P., Giovagnoli, S., Rossi, C., Macchiarulo, G., Luca, G., Basta, G., and Calafiore, R. (2005) Ketoprofen controlled release from composite microcapsules for cell encapsulation: effect on post-transplant acute inflammation. *J. Control Release* **107**, 395–407.
60. Luca, G., Nastruzzi, C., Basta, G., Brozzetti, A., Saturni, A., Mughetti, D., Ricci, M., Rossi, C., Brunetti, P., and Calafiore, R. (2000) Effects of anti-oxidizing vitamins on in vitro cultured porcine neonatal pancreatic islet cells. *Diabetes Nutr. Metab.* **13**, 301–307.
61. Lanza, R. P., Ecker, D. M., Kuhtreiber, W. M., Marsh, J. P., Ringeling, J., and Chick, W. L. (1999) Transplantation of islets using microencapsulation: studies in diabetic rodents and dogs. *J. Mol. Med.* **77**, 206–210.
62. Fan, M. Y., Lum, Z. P., Fu, X. W., Levesque, L., Tai, I. T., and Sun, A. M. (1990) Reversal of diabetes in BB rats by transplantation of encapsulated pancreatic islets. *Diabetes* **39**, 519–522.
63. Lum, Z. P., Tai, I. T., Krestow, M., Norton, J., Vacek, I., Sun, A. M. (1991) Prolonged reversal of diabetic state in NOD mice by xenografts of microencapsulated rat islets. *Diabetes* **40**, 1511–1516.

64. Soon-Shiong, P., Heintz, R. E., Merideth, N., Yao, Q. X., Yao, Z., Zheng, T., Murphy, M., Moloney, M. K., Schmehl, M., and Harris, M. (1994) Insulin-independence in a type I diabetic patient after encapsulated islet transplantation. *Lancet* **343**, 950–951.
65. Dufour, J. M., Rajotte, R. V., Zimmerman, M., Rezania, A., Kin, T., Dixon, D. E., and Korbutt, G. (2005) Development of an ectopic site for islet transplantation, using biodegradable scaffolds. *Tissue Eng.* **11**, 1323–1331.

13

Human Articular Chondrocytes Culture

Andrea Barbero and Ivan Martin

Summary

In this chapter, we describe a strategy to expand adult human articular chondrocytes in monolayer while maintaining high post-expansion chondrogenic capacity. This method relies on the use of a specific growth factor cocktail during the two-dimensional (2D) propagation of chondrocytes. In addition, a protocol to isolate and grow single-colony-derived strains of chondrocytes is presented as a model to study the biology of different chondrocyte subpopulations.

Key Words: Chondrocyte; Cartilage; Growth factor; Cell expansion; Clone.

1. Introduction

Human articular chondrocytes have been extensively used during the past two decades to repair articular cartilage defects with good clinical results and until now they remain the favorite source for the cell-based treatment of cartilage lesions. Chondrocyte-based cartilage repair techniques, including autologous cartilage implantation (ACI) *(1)*, the more recent alternative matrix-mediated ACI (MACI) *(2)*, or the grafting of tissue engineered (TE) cartilaginous constructs *(3)*, require that autologous articular chondrocytes isolated from a small biopsy are efficiently expanded prior to being grafted in the defect.

Finding conditions that permit fast amplification of chondrocytes while maintaining their capacity to generate cartilaginous tissue has been and is the objective of different research groups. The main obstacle for an efficient propagation of chondrocyte derived from the fact that their in vitro expansion is intrinsically associated with cellular de-differentiation (***4***, ***5***) and reduced ability to re-differentiate.

From: *Methods in Molecular Medicine, 2nd ed.: Tissue Engineering*
Edited by: H. Hauser and M. Fussenegger © Humana Press Inc., Totowa, NJ

De-differentiation occurs when chondrocytes are cultured under conditions allowing them to attach and spread on a two-dimensional (2D) surface. In this environment, chondrocytes gradually lose their spherical shape and acquire an elongated fibroblast-like morphology. These morphologic alterations are accompanied by profound biochemical changes, as indicated by the reduction or total loss of synthesis of aggrecan and type II collagen (cartilage-specific proteins) and the increase in synthesis of versican and type I collagen (proteins associated with an undifferentiated mesenchymal cell phenotype) ***(4, 6–8)***. De-differentiated chondrocytes have the capacity to re-differentiate when transferred into an environment supporting a spherical morphology; however, their original phenotype is not fully re-acquired (***4***, ***9***).

In an effort to overcome this problem, different strategies have been proposed to expand chondrocytes under conditions maintaining their original phenotype. They include the culture of chondrocytes within 3D gels or scaffolds, encapsulated in alginate beads or at the surface of microcarrier beads ***(10–14)***. These culture techniques allow chondrocytes to remain round and continue to express cartilage-specific genes but are limited in the extent of proliferation achieved ***(15)***.

One alternative approach to obtain a large number of chondrocytes capable to generate cartilaginous tissues, in contrast to the above-described strategy to maintain the chondrocytic phenotype, consists of keeping cells in a more "plastic" state, thus enhancing their de-differentiation and also their ability to re-differentiate. Indeed, we demonstrated that medium supplementation with specific growth factors during monolayer expansion of human adult articular chondrocytes [in particular, the combination of transforming growth factor beta-1 (TGFβ-1), fibroblast growth factor-2 (FGF-2), and platelet-derived growth factor-BB (PDGF-BB) (TFP)] increases cell proliferation rate, accelerates the process of cell de-differentiation, and—most importantly—enhances the re-differentiation capacity of the expanded cells (*see* **Fig. 1**) (***16***, ***17***).

A striking characteristic of human articular chondrocytes, possibly undermining standardized clinical outcome of chondrocyte-based cartilage repair techniques, is the large variability in proliferation rate and post-expansion chondrogenic capacity of populations from different donors, even if expanded with the selected growth factor combination (***17***). This could be explained by the presence of different fractions of subpopulations with differential chondrogenic ability in cartilage biopsies (*see* **Fig. 2**) (***16***) and prompts for fundamental studies of human articular chondrocytes at the clonal level.

On the basis of these considerations, in this chapter, we will describe the methods (i) to expand human articular chondrocytes in monolayer with growth

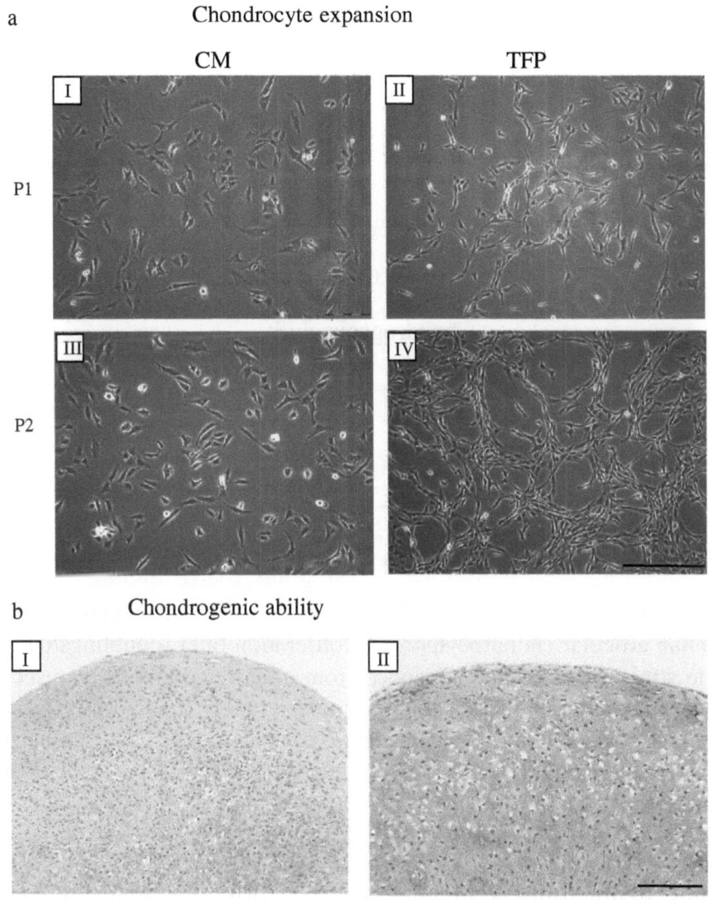

Fig. 1. Expansion and chondrogenic capacity of human articular chondrocytes cultured in monolayer without or with growth factors. (a) Phase-contrast microscopical appearance of human articular chondrocytes at first passage, P1 (I and II), or second passage P2 (III and IV) of expansion in complete medium, CM (I and III), or CM supplemented with the growth factors Transforming Growth Factor beta-1 (TGFβ -1)/ Fibroblast Growth Factor-2 (FGF-2)/Platelet-Derived Growth Factor-BB (PDGF-BB), TFP medium (II and IV) (for the expansion of chondrocytes, *see* **section 3.2.**). Cell spreading is enhanced and accelerated by TFP. Generally, chondrocytes cultured in TFP medium tend to grow in structured patterns. Bar = 100 μm. (b) Sulfate glycosaminoglycans (GAG) staining (Safranin O) of representative pellets generated by chondrocytes expanded in CM (I) or TFP (II) medium after 2 weeks' culture in chondrogenic medium (ChM) (*see* **section 3.4.**). Cells expanded in TFP medium generate more cartilaginous tissues as compared with cells expanded in CM. Bar = 100 μm.

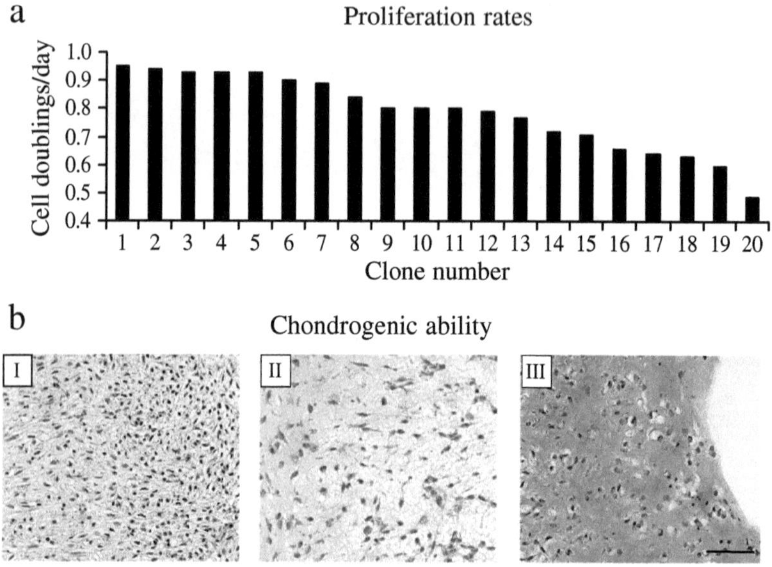

Fig. 2. Expansion and chondrogenic capacity of different clonal-derived populations of human articular chondrocytes. (a) Proliferation rates (doublings/day) of clonal chondrocyte strains isolated and expanded from a cartilage biopsy (*see* **section 3.3.**). The proliferation rate is largely variable among clones. (b) Safranin O staining of selected pellets generated by chondrocytes expanded in TFP. The tissue generation capacity is largely variable among clones, from null (I) to low (II) or high (III). Bar = 100 μm.

factors, thus maintaining a higher cartilage formation capacity, and (ii) to expand single-colony-derived strains of chondrocytes, as a model to study the biology of different chondrocyte subpopulations. A 3D pellet culture technique will also be described as a standard and simple assay to determine the post-expansion chondrogenic capacity of chondrocytes.

2. Materials

2.1. Isolation of Human Articular Chondrocytes

1. Sterile Ca^{2+}/Mg^{2+}-free phosphate-buffered saline (PBS): 0.14 M NaCl, 2.7 mM KCl, 1.5 mM KH_2PO_4, pH 7.2, Gibco-BRL®, cat. no. 14200. Store at 4°C.
2. Penicillin/streptomycin/glutamine (PSG) mixture: penicillin G (10,000 units/ml), streptomycin sulfate (10, 000 μg/ml), and L-glutamine (29.2 mg/ml; Gibco-BRL®, cat. no. 1291232). Store at −20°C in 5 ml aliquots.

3. Cartilage biopsy tissues harvested and kept in PBS containing 10% (v/v) PSG mixture.
4. Sterilized scalpel blades and forceps.
5. Sterile 50-ml centrifuge tubes.
6. Tissue culture Petri dishes (21 and/or 56 cm^2) and tissue culture flasks (75 and/or 150 cm^2).
7. 0.2-μm syringe filter (Nalgene®, cat. no. 190-2520) and 10-ml sterile syringes.
8. Complete medium (CM): Dulbecco's modified Eagle's medium (DMEM; formula containing 4.5 mg/ml D-glucose and 0.1 mM nonessential amino acids, Gibco-BRL®, cat. no. 10938-025), with 10 mM HEPES buffer (Gibco-BRL®, cat. no. 15630-056), 1 mM sodium pyruvate (Gibco-BRL®, cat. no. 11360-039), 1% (v/v) PSG mixture, 10% (v/v) fetal bovine serum (FBS; Gibco-BRL®, cat. no. 10270-106). Store at 4°C (*see* **note 1**).
9. Collagenase type II (>300 U/mg; Bioconcept Worthington, cat. no. LS4176). Store at 4°C.
10. 100-μm nylon filter (BD Falcon™, cat. no. 352360).
11. Orbital shaker mounted within a temperature-controlled 37°C incubator with an atmosphere of 95% air/5% CO_2, 95% humidity.
12. Class II laminar air hood.

2.2. Monolayer Expansion of Human Articular Chondrocytes

1. TGFβ-1 (R&D Systems, cat. no. 240-B). Store at −20°C. Prepare a stock solution of 10 μg/ml in filter-sterilized PBS containing 4 mM HCl and 1 mg/ml human serum albumin (HSA) and store at −20°C in 100 μl aliquots. Prepare a working solution of 100 ng/ml in filter-sterilized DMEM containing 1.25 mg/l HSA and store at 4°C for not longer than 2 weeks.
2. FGF-2 (R&D Systems, cat. no. 233-FB). Store at −20°C. Prepare a stock solution of 10 μg/ml in filter-sterilized PBS containing 1 mM dithiothreitol (DTT, GIBCO-BRL®, cat. no. 15508) and 1 mg/ml HSA and store at −20°C in 100 μl aliquots. Prepare a working solution of 1 μg/ml in filter-sterilized DMEM containing 1.25 mg/l HSA and store at 4°C for not longer than 2 weeks.
3. PDGF-BB (R&D Systems, cat. no. 220-BB). Store at −20°C. Prepare a stock solution of 10 μg/ml in filter-sterilized PBS containing 4 mM HCl and 1 mg/ml HSA and store at −20°C in 100 μl aliquots. Prepare a working solution of 1 μg/ml in filter-sterilized DMEM containing 1.25 mg/l HSA and store at 4°C for not longer than 2 weeks.
4. TFP medium: add 10 μl/ml working solution TGFβ-1, 5 μl/ml working solution FGF-2, and 10 μl/ml working solution PDGF-BB to CM (*see* **section 2.1.**) immediately prior to use. The final concentration of the growth factors are 1 ng/ml TGFβ-1, 5 ng/ml FGF-2, and 10 ng/ml PDGF-BB. Prepare the medium in the required amount and do not store.
5. PBS (*see* **section 2.1.**).

6. 15 mg/ml collagenase solution: dissolve the powder (*see* **section 2.1.**) in PBS and filter sterilize the solution. Store at −20°C in 900 μl aliquots.
7. Trypsin–ethylenediaminetetraacetic acid (EDTA) solution: 0.05 g/l trypsin/0.2 g/l EDTA in Ca^{2+}/Mg^{2+}-free Hanks' balanced salt solution (HBSS; Gibco-BRL®, cat. no. 25300-54).
8. Sterile tubes, tissue culture recipients, sterilizing filters (*see* **section 2.1.**).
9. Trypan blue solution: 0.4% trypan blue in 0.81% sodium chloride and 0.06% potassium phosphate (SIGMA®, cat. no. T8154).
10. Hemocytometer (Marienfeld-superior®, cat. no. 6300-30).
11. Inverted microscope.
12. Class II laminar air hood.

2.3. Clonal Analysis

1. CM and TFP medium (*see* **sections 2.1.** and **2.2.**).
2. Collagenase type II and trypsin–EDTA solutions (*see* **section 2.1.**).
3. Sterile tubes, sterilizing filters (*see* **section 2.1.**).
4. 96- and 12-well plates.
5. Trypan blue solution, hemocytometer.
6. Inverted microscope.
7. Class II laminar air hood.

2.4. Pellet Culture of Human Articular Chondrocytes (AHAC)

1. DMEM; PSG mixture; sodium pyruvate; HEPES buffer (*see* **section 2.1.**).
2. TGFβ-1 (*see* **section 2.2.**).
3. Dexamethasone (SIGMA®, cat. no. D-2915). Store at room temperature. Prepare a 10^{-5} M solution in DMEM, filter sterilize, and store the solution at −20°C in 1 ml aliquot. When thawed, store at 4°C for no longer than 1 week.
4. L-Ascorbic acid 2-phosphate (SIGMA®, cat. no. A-8960). Store at room temperature. Prepare a 10 mM solution in DMEM, filter sterilize, and store the solution at −20°C in 500 μl aliquots, protect to light. When thawed, store at 4°C for no longer than 2 days.
5. ITS^{+1}: 1 mg/ml insulin, 0.55 mg/ml transferrin, 0.55 μg/ml sodium selenite, 50 mg/ml bovine serum albumin, 470 μg/ml linoleic acid (SIGMA®, cat. no. I2521). Store at 4°C.
6. Chondrogenic medium (ChM): DMEM supplemented with 10 mM HEPES buffer, 1 mM sodium pyruvate, 1% (v/v) PSG mixture, 1% (v/v) ITS^{+1}, 0.1 mM L-ascorbic acid 2-phosphate, 1.25 mg/ml HSA, 10^{-7} M dexamethasone, and 10 ng/ml TGFβ-1. Store at 4°C (*see* **note 1**).
7. 1.5-ml polypropylene conical tubes (Sarstedt, cat. no. 72.692.05).
8. Orbital shaker (*see* **section 2.1.**).
9. Centrifuge tubes, syringe filters, syringes (*see* **section 2.1.**).
10. Class II laminar air hood.

3. Methods

3.1. Isolation of Human Articular Chondrocytes

1. Weigh the cartilage biopsy and place it in a Petri dish containing sterile PBS and 10% (v/v) PSG mixture. Using a scalpel blade, cut the tissue (*see* **note 2**) into cubes of 1–2-mm side.
2. Prepare a 1.5 mg/ml collagenase solution (collagenase for *cartilage digestion*) by dissolving the powder in DMEM containing 1% (v/v) PSG mixture and 5% (v/v) FBS. Filter sterilize the solution using a 0.2-μm syringe filter. Use the solution freshly prepared.
3. Place the cartilage pieces in a 50-ml centrifuge tube and add 1 ml 1.5 mg/ml collagenase solution per 100 mg tissue. Incubate the tube for 22 h in a humidified 37°C/5%CO_2 incubator on an orbital shaker (30 rpm). The angle of inclination at which the platform rotates is 10°.
4. Filter the digested tissue through a 100-μm nylon filter and centrifuge the resulting cell suspension at 200 × *g* for 4 min.
5. Remove the collagenase solution and wash the isolated chondrocytes by adding 20 ml sterile PBS containing 5% (v/v) FBS. Gently resuspend the cells and centrifuge again at 200 × *g* for 4 min.
6. Remove the surnatant and add 10 ml CM. Resuspend the chondrocytes and count the number of cells (*see* **note 3**).

3.2. Monolayer Expansion of Human Articular Chondrocytes

1. Centrifuge the chondrocytes at 200 × *g* for 4 min and remove the surnatant. Resuspend the cells in TFP medium. Plate chondrocytes in tissue culture flasks at a density of 10^4 cells/cm^2. Incubate the flask in a humidified 37°C/5% CO_2 incubator. Change TFP medium twice a week.
2. Allow the chondrocytes to expand until 80–90% confluency [first passage (P1) chondrocytes]. Remove the expansion medium and wash with PBS.
3. Prepare the collagenase solution for *cell detachment*: dilute one part of 15 mg/ml collagenase solution with four parts of PBS (final concentration of the collagenase will be 3 mg/ml) (*see* **note 4**).
4. Add the 3 mg/ml collagenase solution to the culture flask (0.1 ml solution per cm^2) and incubate the flask in a humidified 37°C/5% CO_2 incubator for 10 min.
5. Collect the 3 mg/ml collagenase solution containing some detached chondrocytes in a 50-ml centrifuge tube.
6. Add trypsin–EDTA solution to the same culture flask (0.1 ml solution per cm^2) and incubate the flask in a humidified 37°C/5% CO_2 incubator for 5 min or until all the cells are detached from the culture flask.
7. Collect the trypsin–EDTA solution with the detached chondrocytes in the tube with the collagenase solution. Wash the flask with CM and collect the washing solution in the same tube. Centrifuge the chondrocytes at 200 × *g* for 4 min.

8. Remove the surnatant and add 10 ml CM, resuspend the chondrocytes, and count the number of cells (*see* **note 3**).
9. Seed the P1 chondrocytes in new flask(s) at the density of 5×10^3 cells/cm^2 in TFP medium and allow the cells to expand until 80–90% confluency (P2 chondrocytes) (*see* **note 5**).
10. Repeat steps 3–8.

3.3. Clonal Analysis

1. Resuspend 10^5 freshly isolated chondrocytes (*see* **section 3.1.**) in 1 ml CM (cell suspension 1: 10^5 cells/ml).
2. Perform a serial dilutions of cell suspension 1 in different tubes as follows: tube 1, add 100 μl cell suspension 1 to 900 μl CM (cell suspension 2: 10^4 cells/ml); tube 2, add 100 μl cell suspension 2 to 900 μl CM (cell suspension 3: 10^3 cells/ml); tube 3, add 100 μl cell suspension 3 to 20 ml TFP medium (cell suspension 4: 5 cells/ml) (*see* **note 6**).
3. Plate 200 μl aliquots of the cell suspension 4 (corresponding to one cell) in each well of a 96-well plate. Incubate the plate in a humidified 37°C/5% CO_2 incubator.
4. To select the wells in which a single cell has adhered, observe carefully the plate under phase-contrast microscope (objective ×10) not earlier than 3 days following seeding and exclude empty wells or wells containing more than one microcolony (colonies containing two to eight cells) (*see* **note 7**).
5. Culture the selected wells in TFP medium (200 μl/well) as described under **section 3.2.** until the chondrocytes reach confluency (*see* **note 8**).
6. Remove the TFP medium and wash the wells with PBS. Detach cells by using collagenase and trypsin solutions as described under **section 3.2**.
7. Collect the detached chondrocytes from each well in a tube, add 6 ml TFP medium, and resuspend well the cells. Plate 2 ml aliquots of the cell suspensions in 3 wells of a 12-well plate. Incubate the plate in a humidified 37°C/5% CO_2 incubator and allow the cells to expand until 80–90% confluency (*see* **note 9**).

3.4. Pellet Culture of AHAC

A pellet culture is an easy model that allows to investigate the chondrogenic capacity of relative small amount of chondrocytes in short time. The system here described has been adapted from that reported for the differentiation of bone marrow-derived mesenchymal progenitor cells ***(18)***.

1. Centrifuge P2 chondrocytes and resuspend the cells in ChM (*see* **section 2.4.**) at 10^6 cells/ml.
2. Aliquot 0.5 ml cell suspension (corresponding to 5×10^5 chondrocytes) in 1.5 ml polypropylene conical tubes and centrifuged cells at $250 \times g$ for 5 min.

3. Incubate the tubes in a humidified 37°C/5%CO_2 incubator for 24 h. After that time, place the tubes onto a 3D orbital shaker at 30 rpm and culture them for 2 weeks with medium changes twice per week (*see* **note 10**). The angle of inclination at which the platform rotates is 10°.

After the pellet culture, the stage of cell differentiation and the quality/quantity of cartilage matrix deposition can be examined using various techniques, including the gene mRNA expression assays, biochemical analyses, and histological and immunohistochemical techniques [see ***(17)*** for specific details].

Notes

1. CM and ChM are stable at 4°C for no longer than 1 month. Before use, warm up media in a 37°C water bath.
2. When dissecting full-depth cartilage, take attention not to cut through to the underlying bone, as the resulting cell culture may become contaminated with osteoblasts.
3. To count the number of chondrocytes, ensure that the cells are well suspended. Put 50 μl cell suspension in one tube and add 50 μl trypan blue solution (dilution factor: 2). Distribute a small drop of cell trypan blue suspension in one chamber (1 mM square) of a hemocytometer. Using a microscope (objective ×10), count the unstained cells present in at least three fields. Calculate the number of chondrocytes per milliliter using the following formula: average number of cells $\times 10^4 \times 2$.
4. The incubation of the chondrocyte layer with collagenase before the trypsin treatment allows to isolate single chondrocytes; if the layers are directly treated with trypsin, cells would detach from the culture dish as aggregates.
5. The total expansion time for the two passages is about 14–16 days. During this time, cells undergo eight to nine population doublings. Starting from 200,000 cells (generally isolable from a cartilage biopsy of less than 100 mg), the number of chondrocytes that can be obtained after two passages in culture would be superior to 50 millions.
6. Culture of human articular chondrocyte clones in the absence of TFP supplements does not allow reproducible expansion: single-colony-derived strains reach only sporadically the level of expansion required for further analysis.
7. The percentage of wells that after 3 days seeding will contain a single microcolony is expected to be around 15–20%.
8. Frequently, cells adhere at the border of the wells; therefore, the arising cell populations are limited to grow and generally are not capable to cover the entire surface. In this case, proceed to the next step when the cells within the colonies become tightly close with each other.
9. The total expansion time differs largely between the different clonal cell populations, ranging 22–35 days. Also note that during this time, cells would undergo a large number of population doublings, typically 17–21.

10. Cell aggregates initially with a shape of flat disks develop into rounded pellets during the culture period.

References

1. Brittberg, M., Lindahl, A., Nilsson, A., Ohlsson, C., Isaksson, O., and Peterson, L. (1994) Treatment of deep cartilage defects in the knee with autologous chondrocyte transplantation. *N. Engl. J. Med.* **331**, 889–895.
2. Bartlett, W., Skinner, J. A., Gooding, C. R., Carrington, R. W., Flanagan, A. M., Briggs, T. W., and Bently, G. (2005) Autologous chondrocyte implantation versus matrix-induced autologous chondrocyte implantation for osteochondral defects of the knee: a prospective, randomised study. *J. Bone Joint Surg. Br.* **87**, 640–645.
3. Marcacci, M., Berruto, M., Brocchetta, D., Delcogliano, A., Ghinelli, D. M., Gobbi, A., Kon, E., Pederzini, L., Rosa, D., Sacchetti, G. L., and Zanasi, S. (2005) Articular cartilage engineering with Hyalograft C: 3-year clinical result. *Clin. Orthop. Relat. Res.* **435**, 96–105.
4. Benya, P. D. and Shaffer, J. D. (1982) Dedifferentiated chondrocytes reexpress the differentiated collagen phenotype when cultured in agarose gels. *Cell* **30**, 215–224.
5. Binette, F., McQuaid, D. P., Haudenschild, D. R., Yaeger, P. C., McPherson, J. M., and Tubo, R. (1998) Expression of a stable articular cartilage phenotype without evidence of hypertrophy by adult human articular chondrocytes in vitro. *J. Orthop. Res.* **16**, 207–216.
6. Mayne, R., Vail, M. S., Mayne, P. M., and Miller, E. J. (1976) Changes in type of collagen synthesized as clones of chick chondrocytes grow and eventually lose division capacity. *Proc. Natl. Acad. Sci. U. S. A.* **73**, 1674–1678.
7. von der Mark, K., Gauss, V., von der Mark, H., and Muller, P. (1977) Relationship between cell shape and type of collagen synthesised as chondrocytes lose their cartilage phenotype in culture. *Nature* **267**, 531–532.
8. Watt, F. M. (1988) Effect of seeding density on stability of the differentiated phenotype of pig articular chondrocytes in culture. *J. Cell Sci.* **89**, 373–378.
9. Bonaventure, J., Kadhom, N., Cohen-Solal, L., Ng, K. H., Bourguignon, J., Lasselin, C., and Freisinger, P. (1994) Reexpression of cartilage-specific genes by dedifferentiated human articular chondrocytes cultured in alginate beads. *Exp. Cell Res.* **212**, 97–104.
10. Gugala, Z. and Gogolewski, S. (2000) In vitro growth and activity of primary chondrocytes on a resorbable polylactide three-dimensional scaffold. *J. Biomed. Mater. Res.* **49**, 183–191.
11. Guo, J. F., Jourdian, G. W., and MacCallum, D. K. (1989) Culture and growth characteristics of chondrocytes encapsulated in alginate beads. *Connect. Tissue Res.* **19**, 277–297.
12. Kimura, T., Yasui, N., Ohsawa, S., and Ono, K. (1984) Chondrocytes embedded in collagen gels maintain cartilage phenotype during long-term cultures. *Clin. Orthop. Relat. Res.* **186**, 231–239.

13. Malda, J., Kreijveld, E., Temenoff. J. S., van Blitterswijk, C. A., and Riesle, J. (2003) Expansion of human nasal chondrocytes on macroporous microcarriers enhances redifferentiation. *Biomaterials* **24**, 5153–5161.
14. Malda, J., van Blitterswijk, C. A., Grojec, M., Martens, D. E., Tramper, J., and Riesle, J. (2003) Expansion of bovine chondrocytes on microcarriers enhances redifferentiation. *Tissue Eng.* **9**, 939–948.
15. Lee, D. A., Bentley, G., and Archer, C. W. (1993) The control of cell division in articular chondrocytes. *Osteoarthritis Cartilage* **1**, 137–146.
16. Barbero, A., Ploegert, S., Heberer, M., and Martin, I. (2003) Plasticity of clonal populations of dedifferentiated adult human articular chondrocytes. *Arthritis Rheum.* **48**, 1315–1325.
17. Barbero, A., Grogan, S. P., Schafer, D., Heberer, M., Mainil-Varlet, P., and Martin, I. (2004) Age related changes in human articular chondrocyte yield, proliferation and post-expansion chondrogenic capacity.*Osteoarthritis Cartilage* **12**, 476–484.
18. Johnstone, B., Hering, T. M., Caplan, A. I., Goldberg, V. M., and Yoo, J. U. (1998) In vitro chondrogenesis of bone marrow-derived mesenchymal progenitor cells. *Exp. Cell Res.* **238**, 265–272.

14

Cardiomyocytes From Human and Mouse Embryonic Stem Cells

Christine Mummery, Marcel A. G. van der Heyden, Teun P. de Boer, Robert Passier, Dorien Ward, Stieneke van den Brink, Marga van Rooijen, and Anja van de Stolpe

Summary

Human and mouse embryonic stem (ES) cells have the potential to differentiate to cardiomyocytes in culture. They are therefore of interest for studying early human and mouse heart development, as well as properties of cardiomyocytes from both species, including their responses to cardiac drugs, and, at some point in the future, may represent a source of transplantable cells for cardiac muscle repair. The differentiation protocols that are effective depend in part on the species from which the ES cell lines were derived, and in part on the individual cell lines and the methods used for their propagation prior to differentiation. Here, several methods for generating and characterizing cardiomyocytes from mouse and human ES cells are described, as well as methods for dissociation of cardiomyocytes into single-cell suspensions which are useful both for characterizing cells by antibody staining and electrophysiological measurements, as well as preparing cells for transplantation into (animal) hearts.

Key Words: Mouse; Human; Embryonic stem cells; Human embryonic stem cell; Cardiomyocyte differentiation; Co-culture; Dissociation.

1. Introduction

Mouse embryonic stem cells (mESCs) have the capacity to form derivatives of the three germ layers in culture. In vitro differentiation usually requires an initial aggregation step to form structures termed "embryoid bodies" (EBs). After a few days of culture under appropriate conditions of cell density, culture medium, and serum supplement, cardiomyocytes form adjacent to an outer

From: *Methods in Molecular Medicine, 2nd ed.: Tissue Engineering*
Edited by: H. Hauser and M. Fussenegger © Humana Press Inc., Totowa, NJ

epithelial layer of the EB, with characteristics of visceral endoderm and basal mesenchymal cells. Formation of cardiomyocytes is readily identifiable by spontaneous contraction of the EBs [reviewed in *(1)*]. Differentiation may be enhanced by culture supplements such as retinoic acid, bone morphogenetic proteins (BMPs) or ascorbic acid, serum omission, or co-culture with endoderm-like cells *(2–4)*. Cardiomyocyte differentiation in mouse EBs (mEBs) recapitulates the programmed expression of cardiac genes observed in the mouse embryo in vivo, both in the kinetics and in the sequence in which genes are upregulated. GATA-4 and Nkx2.5 transcripts appear before mRNAs encoding ANF, MLC-2v, α-MHC, and β-MHC. Sarcomeric proteins are also expressed in a manner similar to that seen in normal myocardial development. The electrical properties and phenotypes of cardiomyocytes derived from mouse EB cultures have been examined in some detail *(1)*. The rate of contraction decreases with differentiation and maturation in culture, as in normal mouse development, and their differentiation as such can be divided into three developmental stages: early (pacemaker-like or primary myocardial-like cells), intermediate, and terminal (atrial-like, ventricular-like, nodal-like, His-like, and Purkinje-like cells) *(5)*. At early stages, the nascent myofibrils are sparse and irregular, but myofibrillar and sarcomeric organization increases with maturation. Functional gap junctions develop between cells, and eventually, their phenotype resembles that of neonatal rat myocytes. Likewise, the electrophysiological properties of mESC-derived cardiomyocytes develop during differentiation in a manner reminiscent of their development in the mouse embryo *(1)*. Fully differentiated mESC-cardiomyocytes are responsive to β-adrenergic stimulation, although early mESC-cardiomyocytes are not *(6)*. They also exhibit many features of excitation–contraction coupling found in isolated fetal or neonatal cardiomyocytes.

The first report of cardiomyocytes derived from human embryonic stem cells (hESCs) *(7)* appeared almost 3 years after hESC were first derived from blastocyst-stage embryos *(8)*. To induce cardiomyocyte differentiation, hESC (cell line H9.2) were dispersed using collagenase IV into small clumps (3–20 cells) and grown for 7–10 days in suspension to form EB-like structures, comparable to mESC but apparently without the distinct outer layer of endoderm cells. After plating these EBs onto gelatine-coated culture dishes, beating areas were first observed in the outgrowths 4 days after plating (i.e., 11–14 days after the start of the differentiation protocol). A maximum in the number of beating areas was observed 20 days after plating (27–30 days of differentiation), with 8.1% of 1884 EBs scored as beating. This spontaneous differentiation to cardiomyocytes in aggregates was also observed by others using different cell lines for example, H1, H7, H9, H9.1, and H9.2 ***(9)***. However, in this report, approximately 70% of the EBs displayed beating areas after 20 days of differentiation. On day 8 of this differentiation protocol (growth in suspension followed by plating in culture dishes), 25% of the EBs was beating. A third group also demonstrated spontaneous derivation of cardiomyocytes from hESC lines H1, H7, H9, and H14, but in this case, 10–25% of the EBs were beating after 30 days of differentiation ***(10)***. The reasons for these apparent differences in efficiency are not clear. In addition, counting beating EBs may not accurately reflect the conversion of hESC to cardiomyocytes because individual EBs may contain significantly different numbers of cardiac cells. Recently, the differentiation of several independent hESC lines BG01 and BG02 has been described ***(11,12,20)***. Following dissociation of hESC by collagenase IV into small clumps, cells were grown for 7 days as EBs and cultured on adherent plates for another 7 days. In this case, immunoreactivity was demonstrated for the cardiac marker cardiac troponin I.

An alternative method for the derivation of cardiomyocytes from hESC was described by Mummery et al. ***(2,13)***. Beating areas were observed following co-culture of hES2 cells with a mouse visceral endoderm-like cell line (END2). Endoderm plays an important role in the differentiation of cardiogenic precursor cells that are present in the adjacent mesoderm in vivo. Earlier co-culture of END2 cells with mouse P19 embryonal carcinoma (EC) cells, a mouse EC cell line with pluripotent differentiation properties, and with mESC already showed that beating areas appeared in aggregated cells and that culture medium conditioned by the END2 cells contained cardiomyogenic activity ***(14)***. For the derivation of cardiomyocytes from hESC, mitotically inactivated END2 cells were seeded on a 12-well plate and co-cultured with the hESC line hES2. This resulted in beating areas in approximately 35% of the wells after 12 days in co-culture ***(13)***.

While these methods appear to be effective, all produce cardiomyocytes at low efficiency. Several potential cardiogenic factors have been tested in hESC. No significant improvement in cardiomyocyte differentiation has been achieved by adding dimethylsulphoxide (DMSO), retinoic acid ***(7,9)***, or BMP-2 ***(13,15)***. It is not clear whether these factors do not play a role in cardiac differentiation of hESC or whether differentiation protocols were not optimal. One factor that has been described as enhancing cardiomyocyte differentiation of hESC is the demethylating agent 5'-deoxyazacytidine. Treatment of hESC aggregates with 5'-deoxyazacytidine induced enhanced cardiomyocyte differentiation and upregulated the expression of cardiac α myosin heavy chain, as determined by real-time reverse transcriptase–polymerase chain reaction (RT–PCR), up to two-fold ***(9)***.

The presence of fetal calf serum (FCS) during differentiation also has important effects on differentiation efficiency. In most reports to date, serum has been present in the culture medium, and differentiation efficiency has been described as being dependent on serum batch. Serum may contain inhibitory factors. For example, Sachinidis et al. ***(16)*** observed a 4.5-fold upregulation in the percentage of beating mEBs after changing to a serum-free differentiation medium. We also recently observed a greater than 20-fold increase in cardiomyocyte yield in hES2—END2 co-cultures in serum-free medium ***(17)***, which was further enhanced by ascorbic acid. A serum-free protocol is described here for both hES2 and hES3 cells.

2. Materials

General requirements: Tissue-culture facility with laminar flow cabinet, CO_2 incubator, inverted phase contrast microscope and stereo (dissecting) microscope, preferably with heated stage.

2.1. mESC Cardiomyocyte Differentiation

2.1.1. mESC Cardiomyocyte Differentiation in EBs

1. mESC lines: E14, D3, and R1 [D3 and R1 can be purchased from the American Type Culture Collection (ATCC) (ATCC cat. no. CRL-1934, SCRC-1011)].
2. Mouse embryonic fibroblast (MEF) feeder cells: MEF-F1 cells are available from commercial sources (ATCC cat. no. SCRC-1045 or SCRC-1040) or produced and frozen in house (*see* **section 3.4.1**.). MEFs may be growth-inactivated by mitomycin C treatment or irradiation after thawing (*see* **section 3.4.2 and 3.4.3**). Alternatively, pre-irradiated MEFs can be purchased (ATCC cat. no. SCRC-1046.1). Most convenient for mESC, if X-ray facilities are available, is to freeze large stocks of irradiated, in-house-derived MEFs and thaw just before use. Mitomycin-C-pretreated MEFs may also be used as frozen stocks for mESCs (but attach less well to the substrate and are not suitable for hESC cut and paste: *see* **section 2.2.1**.).
3. MEF medium (*see* **Table 1**).
4. mESC culture medium = MEF medium + leukemia inhibitory factor (LIF) (10^3 U/ml) + β-mercaptoethanol (0.1 mM) (MEF medium ++) (*see* **Table 1**).
5. EB culture medium: MEF medium + β-mercaptoethanol only (MEF medium −/+).
6. Trypsin, Versene, chick Plasma (TVP) ×10 (*see* **Table 2**).
7. PBSθ: Phosphate-buffered saline, without Ca^{++} and Mg^{++}.
8. Multiwell tissue-culture plastics coated with 0.1% gelatin.
9. Bacteriological dish (10-cm diameter) (Greiner, cat. no. 633181).
10. Tissue-culture dish (3 cm), coated with 0.1% gelatine.
11. Cell counter.
12. Centrifuge and centrifuge tubes (250 × *g*).
13. Trypsin-ethylenediaminetetraacetic acid (EDTA) (×1) for passaging MEFs (GIBCO Invitrogen Corp. 25300-054, 100 ml).

Table 1
Composition of Basic Media

1.	MEF medium		
	×1 Glasgow minimum essential medium (GMEM) (BHK-21)	GIBCO Invitrogen 21710-025	500 ml
	L-Glutamine 200 mM (×100)	GIBCO Invitrogen 25030-024	5 ml
	Sodium pyruvate 100 mM	GIBCO Invitrogen 11360-039	5 ml
	Non-essential amino acids ×100	GIBCO Invitrogen 11140-035	5 ml
	FCS (always batch tested for optimal growth of ESCs)	Greiner	50 ml
	Penicillin/streptomycin (×100)	GIBCO Invitrogen Corp. 15070-063	5 ml
2.	MESC medium/MEF medium ++ (for passaging mESC on MEFs)		
	MEF medium		40 ml
	β-Mercaptoethanol (0.1 M)		40 μl (0.1 mM)
	LIF (10^6 U/ml)	ESGRO-LIF CHEMICON International, Inc., ESG 1107	40 μl (10^3 U/ml)
3.	β-Mercaptoethanol (×1000)		
	Stock solution 0.1 M (filter through 0.22-μm MILLEX filter prewashed with distilled water, make aliquots)		
	β-Mercaptoethanol, 14.2 M	Merck 15433	0.1 ml
	Distilled water		14.1 ml
	Working concentration, 0.1 mM		

ESC, embryonic stem cell; FCS, fetal calf serum; LIF, leukemia inhibitory factor; MEF, mouse embryonic fibroblast feeder cell.

14. Gelatin 0.1%: Dissolve 0.5 g gelatin (Sigma, G1890) in 25 ml distilled H_2O and autoclave. Add the hot gelatin solution to 475 ml distilled H_2O and filter through STERICUP 0.22 μm (Millipore SCGV05RE).
15. DMSO: Sigma, D-2650, 500 ml. Store in aliquots at −20°C.

Table 2
Composition TVP

1. TVP (culture of mESC on BRL-conditioned medium, without MEFs; *see* **note 8**)		
PBS θ		191 ml
EDTA 40 mM (0.6 g/40 ml) (filter through 0.22-μm MILLEX filter unit, prewashed with distilled water)		5 ml
Chicken serum	GIBCO Invitrogen Corp. 16110-082	2 ml
Trypsin 2.5%	GIBCO Invitrogen Corp. 25090-028	2 ml
2. TVP (×10) (culture of mESC on MEFs)		
9 ml TVP + 1 ml trypsin 2.5%		

EDTA, ethylenediaminetetraacetic acid; mESCs, mouse embryonic stem cells; BRL, Buffalo Rat Liver; MEF, mouse embryonic fibroblast; PBSθ, phosphate-buffered saline.

2.1.2. mESC-Derived Cardiomyocyte Dissociation

1. Dissociation medium: Low Ca^{2+} medium supplemented with 30 μM Ca^{2+} and 1 mg/ml collagenase B (Boehringer Mannheim, cat. no. 1088840).
2. Low Ca^{2+} medium: 120 mM NaCl, 5.4 mM KCl, 5 mM $MgSO_4$, 5 mM Na-pyruvate, 20 mM glucose, 20 mM taurine, 10 mM HEPES, pH7.4 (*see* **Table 3**).
3. Tissue-culture dish (3-cm diameter) or 0.1% gelatine-coated coverslips.
4. 15-ml plastic centrifuge tube (Corning, cat. no. r 430791).
5. Microscalpel (Fine Science Tools, cat. no. 10055-12).
6. P1000 pipette and blue pipette tips.
7. Wide mouth pipette, 10 ml (10-ml plastic disposable pipette of which the tip has been briefly heated in a gas flame to smoothen the pipette opening).

2.2. hESC Cardiomyocyte Differentiation

2.2.1. hES2 and hES3 Passaging Using the Cut-and-Paste Method

1. Newly prepared MEF feeder dishes (*see* **section 3.4.**), freshly treated with mitomycin C.
2. Standard hESC medium for hES2 and hES3 culture on MEFs (*see* **Table 4**).
3. Glass needles: These needles are used to cut the hESC colonies into pieces for transfer to new MEF feeder dishes. They are prepared by heating a glass pipette (Harvard Apparatus, 500 pcs borosilicate glass capillaries, GC100T-15, 1.0 mm OD × 0.78 mm ID) over a (yellow) flame and pulling it into a thin thread until it

Table 3
Low-Ca^{++} Medium for Dissociation of Mouse Embryonic Stem Cell (mESC)-Derived Cardiomyocytes

120 mM NaCl
5.4 mM KCl
5 mM $MgSO_4$
5 mM Na-pyruvate
20 mM glucose
20 mM taurine
10 mM HEPES
pH 7.4 (NaOH)

breaks. Both ends can be used. The needle is put in a holder to make holding and handling easier.

4. Dispase: 10 mg/ml in standard hESC medium, freshly prepared and filter-sterilized (Invitrogen, cat. no. 17105 041, 5 g).
5. PBS+ (containing Ca^{++} and Mg^{++}) Invitrogen, cat. no. 14040-091.
6. 35-mm diameter dishes with PBS+ at 37°C.
7. Organ culture dishes: Falcon 35-3037 Centre well organ culture dish.
8. 100-μl pipette.
9. Dissecting microscope.
10. Heated stage (optional).

Table 4
Human Embryonic Stem Cell (hESC) (hES2 and hES3) Standard Medium: Composition of Standard Culture Medium for hES2 and hES3 Cell Lines

DMEM* (high glucose)	Invitrogen, cat. no. 11960-044
L-Glutamine 200 mM (1:100)	Invitrogen, cat. no. 25030-024
Penicillin/streptomycin (1:250)	Invitrogen, cat. no. 15070-022
Non-essential amino acids (1:100)	Invitrogen, cat. no. 11140-035
Insulin, transferrin, selenium (ITS) (1:100)	Invitrogen, cat. no. 41400-045
2-Mercaptoethanol (1.8 μl/ml medium)	Invitrogen, cat. no. 31350-010
20% fetal calf serum	Hyclone/Perbio

*DMEM, Dulbecco's modified Eagle's medium.
Once mixed, filter medium through Stericup-GV filter unit (Millipore: SC GVU05RE).

2.2.2. hES2 and hES3 Cardiomyocyte Differentiation by END2 Co-culture

1. Standard hESC medium (*see* **section 2.2.1.**, **Table 4**) with and without FCS.
2. PBSθ Invitrogen, cat. no. 14190-094.
3. PBS+ (with Ca^{++} and Mg^{++}) pre-warmed to 37°C.
4. Dispase (*see* **section 2.2.1.**).
5. 0.1% gelatin for coating 12-well plates (with or without coverslips, as required).
6. 35-mm diameter dishes (for PBS+ wash).
7. Organ culture dishes with hESC colonies on MEFs.
8. 12-well plates containing END2 cells (*see* **section 3.4.4.**).
9. P1000 pipette with blue tips.

2.2.3. Dissociation of hESC-Derived Cardiomyocytes

1. "Spring Scissors" with fine small blades (Fine Science Tools GmbH, Heidelberg, Germany, order no. 15000-08).
2. Dish (for collecting beating areas).
3. Low Ca^{2+}: Buffer 1 (*see* **Table 5**).
4. Enzyme medium: Buffer 2 (*see* **Table 5**).
5. KB medium: Buffer 3 (it see **Table 5**).
6. Standard hESC medium (*see* **Table 4**).
7. Gelatin-coated (0.1%) coverslips in 12-well plates.
8. Parafilm.
9. Non-pivoting shaker.
10. Collagenase A (Roche, order no. 10 103 586 001, 500 mg).
11. Organ dishes (for buffers and medium).

2.3. hESC-derived and mESC-Derived Cardiomyocyte Characterization

2.3.1. Immunofluorescence

1. Cells cultured on microscope coverslips (15-mm diameter) (*see* **Section 3.2.3**).
2. PBS+.
3. 2% paraformaldehyde solution: Dissolve 2% paraformaldehyde in PBSθ (in the presence of Ca^{++} and Mg^{++}, a precipitate forms) at 70°C in a water bath. Cool immediately on ice.
4. 0.1%, Triton-X 100 in PBS+.
5. Blocking buffer: 4% normal goat serum (NGS, X0907, DakoCytomation) in PBS+. An alternative blocking method uses NET-gel: 50 mM Tris pH 8.0, 150 mM NaCl, 1 mM EDTA, 0.2% gelatine.
6. PBS+/0.01% Tween 20.

Table 5
Dissociation Buffers for hESC Cardiomyocytes

	Buffers		
	Low-Ca^{2+}	Enzyme	KB (*see* **notes**[1] and [2])
NaCl	12 ml (1 M)	12 ml (1 M)	–
$CaCl_2$		3 μl (1 M)	
K_2HPO_4	–	–	3 ml (1 M)
KCl	0.54 ml (1 M)	0.54 ml (1 M)	8.5 ml (1 M)
Na_2ATP	–	–	2 mmol/l
$MgSO_4$	0.50 ml (1 M)	0.50 ml (1 M)	0.50 ml (1 M)
EGTA	–	–	0.1 ml (1 M)
Na Pyruvate	0.50 ml (1 M)	0.50 ml (1 M)	0.50 ml (1 M)
Glucose	2 ml (1 M)	2 ml (1 M)	2 ml (1 M)
Creatine	–	–	5 ml (0.1 M)
Taurine	20 ml (0.1 M)	20 ml (0.1 M)	20 ml (0.1 M)
Collagenase A	–	1 mg/ml	–
HEPES	1 ml (1 M)	1 ml (1 M)	–
pH corr.	NaOH	NaOH	
PH	6.9	6.9	7.2

Notes: Volumes are for making 100 ml of each buffer.

Buffers should be sterilized by filtration and can be stored at –20°C.

When making KB, leave out the glucose, otherwise a precipitate forms at –20°C; add glucose just before use.

When making buffer 3, add K_2HPO_4 as the last step, otherwise a precipitate forms.

Table 6
Antibodies Used for (Human) Cardiomyocyte Characterization

Antibody	Species	Source	Dilution
α-Actinin	Mouse monoclonal	Sigma	1:800
Troponin I (Western blot)	Rabbit polyclonal AB1627	Chemicon International	1:100
Tropomyosin	Mouse	Sigma	1:400
Mlc2a	Mouse	Synaptic Systems	1:50
Mlc2v	Mouse	Synaptic Systems	1:50

7. Primary antibodies, for example, mouse anti α-actinin (clone EA-53, Sigma), final dilution 1:800; mouse anti-Troponin-T (clone JLT-12, Sigma), final dilution 1:100 (*see* **Table 6** for alternatives).
8. Secondary antibodies: Fluoresceine or Cy3 labeled secondary antibodies (Jackson Immuno Research Laboratories), final dilution 1:100/1:250.
9. Small humidified box containing an elevated Parafilm-wrapped glass plate.
10. Mowiol (Calbiochem 475904).

2.3.2. Reverse Transcriptase–Polymerase Chain Reaction

1. cDNA from undifferentiated and differentiated stem cells.
2. dNTP mix (10 mM dATP, dCTP, cGTP, and dGTP) (Amersham Biosciences, 27-2035-01).
3. Taq-polymerase (Amersham Biosciences).
4. Taq-buffer (Amersham Biosciences).
5. PCR cycler, for example, Perkin Elmer 9600.
6. Agarose gel system.
7. TBE agarose gel buffer.
8. Ethidium bromide.
9. UV gel imager.
10. Primers (*see* **Table 7**).

2.4. Feeder Preparation

2.4.1. Preparation of Primary MEFs and MEF-F1 from Midgestation Mouse Embryos

1. Culture medium: Dulbecco's modified Eagle's medium (DMEM) (high glucose) + L-glutamine (1:100), penicillin/streptomycin (1:200) + MEM (100×) non-essential amino acids 10% FCS (Cambrex) is used to derive MEFs from 129 sv (or ICR) mice used for the hESC cultures. MEF medium (*see* **Table 1**) is used for MEF-F1 feeders (from F1 embryos of CBA and C57 Bl/6j crosses) for mESC cultures. In our hands, MEF-F1 feeders are easier to derive but do not support long-term self-renewal of hESCs.
2. Mouse embryos collected at embryonic day (E)13.5 (morning of the plug is E0.5).
3. PBSθ.
4. 3-cm diameter bacteriological dishes.
5. Scalpels.
6. Trypsin/EDTA.
7. Syringe.
8. 18-G and 21-G needle.
9. Tube containing 5 ml MEF medium/embryo.
10. Clean tube.
11. 175-cm^2 tissue-culture flasks (1 embryo/flask).

Table 7
PCR Primer Pairs Used for Characterization of ES Derived Cardiomyocytes

Name	Forward/reverse (5′ → 3′)	Annealing (°C)	Product (bp)	Number of cycles
Mouse primers				
ß-Tubulin	TCACTGTGCCTGAACTTACC	55	319	27
	GGAACATAGCCGTAAACTGC			
Cardiac actin	TGTTACGTCGCCTTGGATTTTGAG	54	494	29
	AAGAGAGAGACATATCAGAAGC			
MLC2v	GCCAAGAAGCGGATAGAAG	54	499	34
	CTGTGGTTCAGGGCTCAGTC			
KCNH2 B	ATGGCGATTCCAGCCGGGAA	55	361	35
Splice variant	ATGTCCACGATGAGGTC			
Human primers				
ß-Tubulin	TGGCTTTGCCCCTCTCACCA	61	369	27
	CGGCGGAACATGGCAGTGAA			
ANF	GAACCAGAGGGGAGAGACAGAG	61	406	30
	CCCTCAGCTTGCTTTTTAGGAG			
α-Actinin	GGCGTGCAGTACAACTACGTG	56	580	31
	AGTCAATGAGGTCAGGCCGGT			
KCND3	CACCCCAGAAGAGGAGCACAT	55	322	35
	AGTAGCTGGCAGGTTAGAATT			

12. Cryovials (two per embryo).
13. Freezing medium: MEF medium + 20% DMSO (1:1; final concentration DMSO 10%).

2.4.2. Irradiation and Stock Freezing MEF-F1 Feeders for mESC Culture

1. One vial of primary MEF-F1.
2. 175-cm^2 tissue-culture flask.
3. MEF medium (*see* **Table 1**).
4. PBSθ.
5. Trypsin/EDTA.
6. One hundred cryovials for freezing irradiated cells.
7. Centrifuge $250 \times g$.
8. DMSO.

2.4.3. Mitomycin C Treatment of MEFs for hESC Culture

2.4.3.1. Preparation of Mitomycin C Stocks

1. 2 mg mitomycin C (Sigma, cat. no. M0503 $5\,mg \times 2\,mg$).
2. 18-G needles.
3. $1 \times 0.22\,\mu m$ syringe filters.
4. 2.5-ml syringe.
5. 1.5-ml cryotube.
6. Sterile PBSθ.

2.4.3.2. Mitomycin Treatment

1. MEF medium (*see* **Section 2.4.1**.).
2. Sterile PBSθ.
3. Mitomycin C stock (*see* **section 2.4.3**., item 1).
4. Trypsin/EDTA.
5. Organ dishes (0.1% gelatin-coated).
6. Standard hESC medium (*see* **Table 4**) (for refreshment after 24 and 48h).

2.4.4. Preparation and Mitomycin Treatment of END2 Co-Culture Cells

1. END2 cells: Can be obtained from Christine Mummery at the Hubrecht Laboratory after completion of a Material Transfer Agreement (www.niob.knaw.nl), email: christin@niob.knaw.nl.
2. END2 culture medium: Invitrogen DMEM/F12 (1:1) + L-glutamine + penicillin/streptomycin +7.5% serum (Cambrex).
3. Mitomycin C stock (*see* **section 2.4.3**., item 1).
4. PBSθ.
5. Trypsin/EDTA.
6. 25-cm^2 tissue-culture flask coated with 0.1% gelatin.

7. 175-cm^2 flask coated with 0.1% gelatin.
8. 12-well plates coated with 0.1% gelatin.
9. Cover slips treated with 0.1% gelatin, in a 12-well plate.

3. Methods

3.1. mESC Cardiomyocyte Differentiation (see Section 2.1.1.)

3.1.1. mESC Cardiomyocyte Differentiation in EBs

3.1.1.1. Day 1. Plating of MEF-F1 Feeders

1. Thaw one ampoule irradiated MEFs (*see* **section 3.4.2.**) and transfer the content to a centrifuge tube with 5 ml MEF medium.
2. Spin 5 min at 250 × *g* to remove DMSO and suspend in MEF medium.
3. Plate MEFs at the same density at which they were frozen. Add 0.25 ml medium/cm^2 culture surface.
4. Incubate overnight in incubator at 37°C, 5% CO_2.

3.1.1.2. Day 2. Plating of mESC on MEF Feeder

5. Thaw mESC and transfer to a centrifuge tube containing 5 ml mESC medium (MEF++) (*see* **Table 1**). Plate at approximately the same density at which they were frozen.
6. mESC on MEFs are not usually counted for passaging, because they contain a mixture of mESC and MEFs. Instead, they are trypsinized and replated at dilutions between 1:5 and 1:8, every 48 h. Always try two densities and select the cultures with most round colonies and fewest differentiated (flat or epithelial) cells.

3.1.1.3. Day 5, 6, or 7: "Hanging Drop" Cultures to Induce ESC Aggregation and Differentiation as EBs

7. For trypsinization, select wells as in 6.
8. Remove medium and wash cells twice with PBSθ.
9. Add 50 μl TVP × 10/cm^2 confluent culture and incubate for 2–5 min at room temperature.
10. Add nine volumes mESC medium without LIF (MEF medium −/+, *see* **Section 2.1.1., item 5 and Table 1**). To remove feeders, pipette cells up and down until a single-cell suspension is obtained. Plate cell suspension on to gelatine-coated tissue-culture wells of twice the surface area they are derived from. Return to incubator for at least 1 h and then examine culture under the microscope. The MEFs should be strongly adherent to the bottom of the well. Gently remove the medium. Add 1 ml EB medium (MEF medium −/+) and resuspend the loosely adherent mESC into a single-cell suspension by pipetting up and down with a blue tip.
11. Count cells in cell suspension.
12. Make a dilution of 40,000 cells/ml in EB medium.

13. Fill the base of a 10-cm diameter bacterial dish with PBS (for humidification).
14. Make 20 µl drops of mESC in suspension (equivalent to 800 cells/20 µl) on the lid of the bacteriological dish, about 50 drops/lid. The more drops the better, because it prevents drying out.
15. Invert lid and place on the PBS-filled base.
16. Incubate at 37°C in 5% CO_2 for 5 days. Single EBs will form in each of the hanging drops.

3.1.1.4. Day 10 Transfer of EBs to Tissue-Culture Plates

17. Transfer EBs to gelatin-coated tissue-culture dishes or plates with EB medium using a blue tip (P1000 pipette). Approximately 1–2 EBs/cm^2 in 0.5 ml (10 EBs/3-cm diameter dish).

3.1.1.5. Day 14 and Onwards: Observation of EBs for Beating Areas

18. Daily observation: Areas containing beating cells become evident in EBs at increasing frequency.
19. Refresh the differentiation medium every 2 days until the cardiomyocytes are used.

3.1.2. mESC-Derived Cardiomyocyte Dissociation (see **notes 10 and 11**)

1. Excise clusters of beating mESC (*see* **section 3.1.1.5.**, item 18) under a stereo microscope in a laminar flow cabinet, using a sterile microscalpel, or collect whole beating EBs with a P1000 pipette with a blue tip.
2. Transfer to low-Ca^{2+} medium in a 15-ml centrifuge tube. Allow excised clusters and/or EBs to sink by gravity.
3. After maximally 30 min, remove excess medium and add low-Ca^{2+} medium supplemented with 30 mM Ca^{2+} and 1 mg/ml collagenase B.
4. Incubate at 37°C for 45 min with intermittent, repetitive up-and-down titration, using a P1000 pipette to resuspend cell clusters (~every 5 min).
5. Collect dissociated cells at the bottom of the tube by 5 min centrifugation 800 × *g*.
6. Remove as much dissociation solution as possible.
7. Carefully resuspend cells in EB medium using a wide-mouth pipette.
8. Transfer cells to tissue-culture dishes or 0.1% gelatine-coated coverslips.

3.2. hESC Cardiomyocyte Differentiation (see *notes 10 and 12*) (Section 2.2)

3.2.1. hES2 and hES3 Passaging Using the Cut-and-Paste Method (See Section 2.2.1)

1. Passage HES2 cells once a week. (Colonies have usually enlarged to 0.5–2-mm diameter but will not become confluent.)
2. Sterilize glass needles by autoclaving and just before passaging, heat the needles over a flame, pull, and break to give usable cutting ends.

3. Inspect and select plates with undifferentiated hESC colonies.
4. Use the glass needles to cut and select undifferentiated "pieces" of the colonies (as cutting a pizza; *see* **Fig. 1**) for passaging. Removal of the differentiated areas is essential for long-term propagation of undifferentiated cells. Work with ×4 magnification on a (stereo) dissecting microscope.
5. Remove medium and add 0.5 ml dispase solution. Place the plate in the incubator for approximately 2 min or leave on a heated stage if available.
6. Take two 35-mm diameter dishes with pre-warmed PBS+ from the incubator.
7. Pick up undifferentiated colony pieces from dispase-treated plates using a P100 pipette with yellow tip and transfer them to the first dish of PBS+ and subsequently to the second dish. This results in two washes with PBS+.
8. Place the colony pieces evenly distributed on to a newly prepared MEF feeder dish (*see* **section 3.4.3.**) with 1 ml standard hESC medium (*see* **Table 4**), 9 pieces/plate. Do not put the pieces too close to the side of the dish or one another, because the colonies need sufficient space to grow and access for next passage may become more difficult.
9. Place the organ dish carefully in an incubator at 37°C and 5% CO_2 and refresh daily with standard hESC medium.

3.2.2. hES2 and hES3 Cardiomyocyte Differentiation by END2 Co-Culture (See Section 2.2.2.)

1. Refresh mitomycin-C-treated END2 cells (*see* **section** 3.4.4.) in 12-well plates with hESC medium without FCS at least 1.5 h before plating the hESC cell pieces.

Fig. 1. Stages of passaging hES2 cells by "cut and paste."

2. Prepare dishes for washing undifferentiated hESC colony pieces by filling (a) six 35-mm diameter dishes with PBS+ and (b) two organ dish with 1 ml hESC medium with FCS; place in incubator at 37°C and 5% CO_2.
3. Starting material for two 12-well hESC-END2 co-culture plates is 12 organ culture dishes each with 9–10 colonies of either hES2 or hES3. Detach hESC colonies from MEFs by adding 0.5 ml dispase and placing in the incubator for 7 min.
4. Collect all undifferentiated hESC colonies from the 12 organ dishes using a P1000 pipette with a blue tip and distribute them for washing over three dishes of PBS+ prepared previously.
5. Transfer the colonies to three new PBS+ dishes (to remove MEFs attached to the colonies).
6. Transfer the colonies to two organ dishes prepared previously containing 1 ml standard hESC medium (*see* **Table 4**).
7. Break colonies into pieces by firmly pipetting up and down (2–3 times depending on the size of the colonies) against the bottom of the dish using a P1000 pipette.
8. Transfer small *clumps* of hESC cells to 2 × 12-well plates containing confluent mitomycin-C-treated END2 cells (*see* **section 3.2.3.**, **Fig. 2**).
9. Refresh medium on days 5, 9, and 12.
10. Score beating areas by microscopic examination 12 days after plating.

3.2.3. Dissociation of hESC-Derived Cardiomyocytes (see *note 13*)

1. Isolate beating areas from co-culture plates by cutting them out with scissors (see Section 2.2.3.) and collect excised tissue in standard hESC medium with serum. (see Section 3.2.1.)

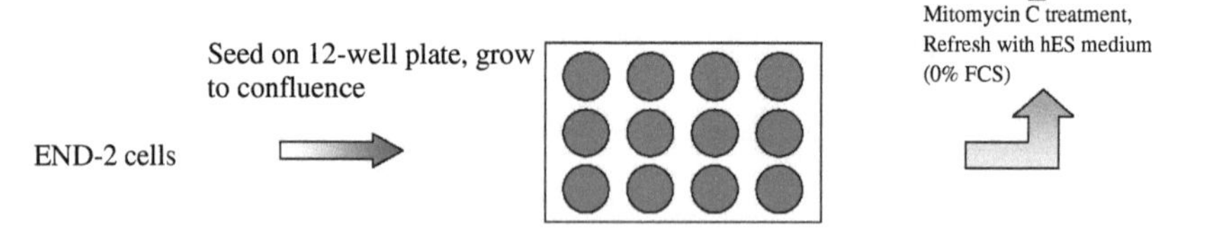

Fig. 2. Induction of cardiomyocyte differentiation by human embryonic stem cell (hESC)-END2 co-culture in serum-free medium. Source: Passier et al. (2005) *Stem Cells* **23**, 772–780.

2. To start dissociation, transfer excised tissue pieces to a dish or well of a 12-well plate with low-calcium (buffer 1; table 5) using a P1000 pipette with blue tip; leave for 30 min at room temperature.
3. Transfer cell clumps from low-calcium buffer into enzyme buffer (buffer 2; table 5) and incubate at 37°C for approximately 45 min (cover dish with Parafilm before transferring to the CO_2 incubator).
4. Transfer cell clumps from enzyme buffer into KB buffer (buffer 3; table 5). Shake gently in KB buffer (buffer 3) at room temperature for 1 h at 100 rpm on non-pivoting shaker.
5. Transfer cell clumps from KB buffer into hESC culture medium with 20% FCS to promote attachment and survival. Break up the cell clumps by pipetting up and down (2–4 times) against the bottom of the dish using a P1000 pipette. The degree of dissociation required depends on the particular experiments that will be done with the cardiomyocytes:

   i. For transplantation into animals or immunofluorescent staining, it is sufficient to obtain a mixture of cell clumps and single cells.
   ii. For electrophysiology, single cells are required.

   Note: cardiomyocytes are very fragile! If pipetting is too rigorous, cells will fail to recover either in culture or in vivo.
6. For electrophysiology and immunofluorescent staining, seed dissociated beating areas in standard hESC medium (*see* **section 2.2.1.**), on gelatine-coated (0.1%) coverslips or culture plates at 37°C, as required.
7. Allow cells on coverslips to recover for at least 2 days and up to a week for electrophysiology experiments.

3.3. mESC-Derived and hESC-Derived Cardiomyocyte Characterization

3.3.1. Immunofluorescence (see Section 2.3.1)

1. Seed cells on 15-mm coverslips, usually in 12-well tissue-culture plates, as described above (*see* **section 3.1.2** or **3.2.1**).
2. Rinse cells once with PBS+.
3. Fix with 2% paraformaldehyde for 25 min.
4. Remove paraformaldehyde.
5. Wash three times with PBS+.
6. Incubate in 0.1% Triton-X 100 for 5 min to permeabilize the cells.
7. Wash three times with PBS+.
8. Block samples in PBS/4% NGS for 60 min at room temperature or using NET-gel (*see* **section 2.3.1.**). The choice depends on the individual antibody combination and the level of background fluorescence that results.

9. Carefully transfer coverslips onto the Parafilm-wrapped glass plate, cells upwards, in the humidified box and add 50 ml diluted antibody (diluted in PBS+/4% NGS) on top of the coverslip.
10. Close the humidified box and incubate at room temperature for at least 60 min.
11. Transfer coverslips to wells of a 12-well plate (remove antibody solution by touching the edge of the coverslip with a Kleenex tissue).
12. Wash coverslips in the 12-well plate three times with PBS+/0.01% Tween for about 20 min each time.
13. Dilute secondary antibody in PBS+/4% NGS.
14. Add 50 ml diluted secondary antibody solution on top of the coverslips and incubate for 60 min at room temperature in a humidified box (in the dark!).
15. Remove antibody solution with a Kleenex tissue as above.
16. Transfer coverslips to wells of a 12-well plate (keep in the dark, e.g., wrap in aluminum foil between washes).
17. Wash three times for 10 min each by adding PBS+/0.01% Tween.
18. Mount the samples. Remove all solution from the coverslip by touching the edge of the coverslip with a Kleenex tissue, dip the coverslips 2–3 times in distilled water and carefully invert the slip top-down into a 15-μl drop of Mowiol solution. Dry the coverslips overnight at room temperature and view the slides under an epifluorescent microscope (coverslips can be stored at 4°C).
19. Indirect staining of cardiac marker proteins can also be carried out on standard paraffin sections or cryosections of EBs.

3.3.2. Reverse Transcriptase–Polymerase Chain Reaction

1. Use cDNA, for example, from undifferentiated cells and cells at different stages of differentiation. Standard controls are similar samples without prior RT reaction to exclude DNA contamination of the RNA sample and water.
2. Make 25 μl reaction mixture by combining: 0.5–1 μl cDNA, 2.5 μl × 10 PCR buffer, 0.5 μl dNTP set, 0.25 μM sense primer, 0.25 μM antisense primer 1 unit Taq-polymerase, H_2O.
3. Amplify using the following conditions: 5 min 95°C, followed by 27–35 cycles of 30 s 95°C, 45 s annealing temperature, 1 min 72°C. Following the last cycle, continue for 5 min at 72°C. Note: this depends on the particular primer pairs used.
4. Add 5 μl × 6 concentrated DNA sample buffer.
5. Analyze PCR products by gel electrophoresis on ethidium-bromide stained 1% agarose gel.

3.4. Feeder Preparation (see ***notes 3–5)***

3.4.1. Derivation of MEFs and MEF-F1 Feeder Cells from Mouse Embryos (see ***Section 2.4.1.)***

1. Euthanize female mice on E13.5 of pregnancy.

2. Isolate embryos and wash once in PBSθ, dissect individual embryos to remove the head and soft tissues (liver, heart, and other viscera), and wash the carcasses twice in PBSθ.
3. Transfer the carcasses to 3-cm diameter bacteriological dishes, with two in each dish.
4. Mince two embryo carcasses together in a very small volume PBSθ into fine pieces with two scalpels.
5. Per embryo add 1 ml Trypsin/EDTA ×1.
6. Leave for 10 min at room temperature while gently swirling the dish.
7. First use a syringe with a pink 18-G needle and subsequently with a green 21-G needle to dissociate the digested tissue into a single-cell suspension. Transfer the cell suspension to a tube containing 5 ml MEF medium/embryo (*see* **Table 1, section 2.1.1. and 2.4.1.**).
8. Allow the large pieces of tissue debris to settle and transfer the supernatant into a clean tube. Spin down at 250 × *g* for 5 min.
9. Resuspend the cell pellet in MEF medium and plate the suspension in 175-cm^2 tissue-culture flasks with approximately 1 embryo/flask.
10. Culture the cells for 24–48 h, trypsinize the flasks, and freeze the cells at passage 1 in two vials per original embryo.
11. Freezing medium is MEF medium with DMSO; final concentration 10% (Sigma, D-2650).

3.4.2. Irradiation and Stock Freezing MEF-F1 Feeders for mESC Culture

1. MEF pre-irradiation and use of frozen stocks support robust self-renewal of mESC but are less suited in our hands for prolonged self-renewal of hESC (hES2 and hES3). For hESC, frozen, irradiated human fibroblasts or freshly mitomycin-c treated MEFs are best.
2. Confluent flasks with MEFs are usually irradiated at passage 4.
3. Thaw one vial of primary MEFs and start culture in a 175-cm^2 tissue-culture flask using 30 ml MEF medium (*see* **Table 1**.).
4. Culture the cells to confluence and passage 1:5 by washing the cells once with PBSθ, trypsinizing the cells with trypsin/EDTA, 3 ml/flask for 5 min at room temperature, neutralizing with 3 ml MEF medium, and splitting the cells over five new flasks.
5. Culture the cells again to confluence and passage 1:5 (25 flasks). The cells will tend to grow somewhat slower and sometimes have difficulty reaching confluence. It should take approximately 5–7 days to reach confluence in this phase.
6. Label 100 cryovials.
7. Prepare 25 ml MEF medium containing 20% DMSO. Cool on ice.
8. Trypsinize the cells (*see* **section 3.4.2.,** item 4). Collect the trypsin-neutralized batches and spin for 5 min at 250 × *g*. Remove the supernatants and resuspend the cells in 25 ml MEF medium.
9. Cool the cell suspension on ice.

10. Place the cell suspension into a gamma irradiation chamber and irradiate with 2500 rad. (Time depends on strength of gamma source, usually calibrated annually e.g., a 30 Gy source will take about 4 min.)
11. Irradiated cells are frozen in ampoules. Each ampoule contains all cells from 45-cm^2 flask surface. In general, cells are not counted.
12. Mix the cell suspension with 25 ml MEF medium containing 20% DMSO by adding it drop wise to the cell suspension. Final cocentration 10% DMSO.
13. Freeze the cells in 0.5 ml/vial = approximately 45 cm^2/vial.
14. After thawing an ampoule for use, cell culture is started again on 45-cm^2 flask surface per ampoule.

3.4.3. Mitomycin C Treatment of MEFs for hESC Culture Preparation of Mitomycin C Stocks (see ***Section 2.4.3.****)*

1. An ampoule containing 2 mg Mitomycin-C is dissolved in 1 ml PBSθ.
2. Using an 18-G needle, add 1 ml PBSθ to the mitomycin C ampoule and make sure that all powder is dissolved.
3. Pre-wet a 0.22-μm syringe filter with 10 ml PBSθ.
4. Remove all mitomycin solution with a 2.5-ml syringe and an 18-G needle.
5. Filter from the syringe into a sterile 1 ml cryovial.
6. This stock can be stored at 4°C for up to 4 weeks.
7. To commence with mitomycin-c start with confluent 75-cm^2 flasks, passage 4–6 MEFs (from strain 129 sv or ICR mice).
8. Add mitomycin C at a concentration of 10 μg/ml to culture medium (add 5 μl mitomycin-C stock per ml of medium).
9. Leave for at least 2.75 h, maximum 3 h in the incubator at 37°C (5% CO_2).
10. Remove medium containing mitomycin-C (*waste is toxic*).
11. Wash 1× with MEF medium (*see* **section 2.4.1**.) and twice with PBSθ (about 10–15 ml/T75 flask) (waste is still toxic).
12. Trypsinize cells and seed organ dishes with 175,000 cells/dish in MEF medium.
13. 24 h after plating MEFs, refresh medium with hESC medium with FCS (*see* **section 2.2.1**.).
14. 48 h after plating is the day of transfer of hESC colonies to the MEFs: refresh medium with 1 ml hESC medium + FCS.

3.4.4. Preparation and Mitomycin C Treatment of END2 Co-Culture Cells (see ***Section 2.4.4.****)*

1. On day 1 (preferably Monday), seed a 25-cm^2 tissue-culture flask coated with 0.1% gelatine with END2 cells in END2 culture medium (*see* **section 2.4.4**). END2 cells should be split 1:8 from a confluent flask.
2. On day 5, seed a 175-cm^2 flask coated with 0.1% gelatine with END2 cells using all the cells from the previous 25-cm^2 flask.
3. On day 8, the 175-cm^2 flask should be 100% confluent and is ready for mitomycin C treatment, as described for MEFs (*see* **section 3.4.3.**).

4. Add 5 μl mitomycin-C/ml medium from 10 μg/ml stock solution to the culture medium concentration (*see* **section 3.4.3.**).
5. Incubate flasks at 37°C for at least 2.75 h, maximally 3 h.
6. Aspirate medium, wash wells once with END2 culture medium, followed by two washes with PBSθ. Caution: the waste containing mitomycin C is highly toxic.
7. Trypsinize and count the cells.
8. Plate END2 cells at a density of 175,000 cells/ml on 15-mm diameter cover slips treated with 0.1% gelatine in 12-well plates in END2 culture medium.

Notes

1. The WiCell lines (H1, H7, H9, H13, and H14) are most frequently described in the literature. They can be enzymatically passaged and form EBs, containing derivatives of all three germ layers, in suspension. Differentiation to cardiomyocytes in EBs is "spontaneous" (10^7 ES cells are usually transferred to 100-mm non-adherent Petri dish and cultured for 4–5 days before plating on gelatin-coated dishes). H1, H7, H9, and H14 lines can be cultured feeder-free ***(18)***. They have been described as being relatively easy to transfect (transfection efficiencies 30–80%), using adenoviral vectors, Fugene or Lipofectamine 2000, and stably transduced by murine stem cell virus (MSCV)-based retroviral vectors.
2. hES2 and hES3 cells ***(19)*** differentiate into mesodermally derived cardiomyocytes in co-culture with END2 cells under conditions we have described. Both hES2 and hES3 cells also form EBs/aggregates in suspension (*see* **note 12**). Both are preferably passaged by *cut-and-paste* method for optimal karyotypic stability but can be enzymatically passaged.
3. Gamma-irradiated feeders can be frozen and stored. For hESC lines passaged by cut-and-paste method, mitomycin-C-treated feeders are used fresh. For hESC, feeder dishes and plates should be used after 48 h but not used for transferring hESC colonies if older than 4 days. The passage number up to which feeder cells can be used differs per lab. In general (low), passage 4 cells are preferable, although some labs are able to use feeder cells up to passage 7.
4. We generally use one feeder density (170,000–175,000/2.45 cm^2), although lower densities have been described. Feeder density has been suggested to influence colony morphology: on high-density feeders, colonies may appear "domed" rather than flat.
5. Preparation of MEF feeders for hESC culture may be variable and depend on breeding efficiency and the number of embryos per mouse (129 strain mice have small litters) which affects subsequent growth of the isolated primary fibroblast cells. Cell growth rate determines feeder quality and the number of confluent flasks that are converted into MEF cryovials up to passage 4–5. The vials are thawed each week as required, grown to confluence prior to mitomycin C treatment and then used for hESC passaging.

6. To avoid the use of feeders, some hESC lines have also been cultured on Matrigel or extracellular matrix layers, for example, laminin or fibronectin, using MEF-conditioned medium (+FGF) ***(13,21,22)***. Cells for conditioning medium can be used repeatedly for 7–10 days and the conditioned medium frozen and stored until use.
7. mESC can be obtained from various sources, for example, ATCC. The mESC D3 line (ATCC no. CRL-1934), E14 and R1 all differentiate efficiently to cardiac myocytes in EBs.
8. Mouse embryonic stem cell lines depend on either MEF or STO fibroblast feeder layers (most often irradiated or mitomycin C treated as here) or addition of LIF to the culture medium (Glasgow MEM/20% FCS), to retain an undifferentiated, pluripotent state. A combination of feeder cells and LIF addition is often used or conditioned medium from Buffalo Rat Liver (BRL) cells which contains high concentrations of LIF ***(23)***. BRL-conditioned medium can also be used without MEFs; additional LIF may be added if spontaneous differentiation is significant. Use of BRL-conditioned medium reduces cost substantially.
9. Undifferentiated mESC express the marker protein Oct 4, which can be easily stained in cell cultures to check for residual undifferentiated cells in the differentiated cell cultures (2,13).
10. The efficiency of differentiation toward cardiomyocytes in mouse and hESCs may be variable. Important factors are the initial number of cells per EB and the FCS batch even if the differentiation is not in serum-free medium. It is advisable to test different batches of FCS for differentiation efficiency and choose the batch, which performs best for all subsequent differentiation assays.
11. mESCs also respond to END2 co-culture. Co-culture of mESC with END-2 cells starts with *single-cell suspensions*: Cells start to aggregate and after about 3 days aggregates attach to END2 cells and grow out and differentiate. Beating areas appear after about 8 days ***(2)***. Co-culture with mESC is usually in serum-containing medium, which improves survival.
12. Some hESC lines respond best in END2 co-cultures, others respond better when grown as aggregates in END2-conditioned medium, as described previously for P19EC cells ***(14)***. hESC medium is conditioned for 4 days by a confluent END2 cell monolayer.
13. hESC: The efficiency of dissociation of the beating clusters into individual cardiomyocytes may depend on the batch of collagenase A used and batch testing may be advisable. We use collagenase from Roche.
14. Many studies have demonstrated that mESC-derived and hESC-derived cardiomyocytes have a fetal or immature phenotype (2,13).

References

1. Boheler. K. R., Czyz, J., Tweedie, D., Yang, H. T., Anisimov, S. V., and Wobus, A. M. (2002) Differentiation of pluripotent embryonic stem cells into cardiomyocytes. *Circ. Res.* **91**, 189–201.

2. Mummery, C., Ward, D., van den Brink, C. E., Bird, S. D., Doevendans, P. A., Opthof, T., Brutel de la Riviere, A., Tertoolen, L., van der Heyden, M., and Pera, M. (2002) Cardiomyocyte differentiation of mouse and human embryonic stem cells. *J. Anat.* **200**, 233–242.
3. Rathjen, J., Lake, J. A., Bettess, M. D., Washington, J. M., Chapman, G., and Rathjen, P. D. (1999) Formation of a primitive ectoderm like cell population, EPL cells, from ES cells in response to biologically derived factors. *J. Cell Sci.* **112**, 601–612.
4. Rathjen, J. and Rathjen, P. D. (2001) Mouse ES cells: experimental exploitation of pluripotent differentiation potential. *Curr. Opin. Genet. Dev.* **11**, 587–594.
5. Hescheler, J., Fleischmann, B. K., Lentini, S., Maltsev, V. A., Rohwedel, J., Wobus, A. M., and Addicks, K. (1997) Embryonic stem cells: a model to study structural and functional properties in cardiomyogenesis. *Cardiovasc. Res.* **36,** 149–162.
6. Maltsev, V. A., Ji, G. J., Wobus, A. M., Fleischmann, B. K., and Hescheler, J. (1999) Establishment of beta-adrenergic modulation of L-type Ca2+ current in the early stages of cardiomyocyte development. *Circ. Res.* **84**, 136–145.
7. Kehat, I., Kenyagin-Karsenti, D., Snir, M., Segev, H., Amit, M., Gepstein, A., Livne, E., Binah, O., Itskovitch-Eldor, J., and Gepstein, L. (2001) Human embryonic stem cells can differentiate into myocytes with structural and functional properties of cardiomyocytes. *J. Clin. Invest.* **108**, 407–414.
8. Thomson, J. A., Itskovitch-Eldor, J., Shapiro, S. S., Waknitz, M. A., Marshall, V. S., and Jones, J. M. (1998) Embryonic stem cell lines derived from human blastocysts. *Science* **282**, 1145–1147.
9. Xu, C., Police. S., Rao. N., and Carpenter, M. K. (2002) Characterization and enrichment of cardiomyocytes derived from human embryonic stem cells. *Circ. Res.* **91**, 501–508.
10. He, J. Q., Ma, Y., Lee, Y., Thomson, J. A., and Kamp, T. J. (2003) Human embryonic stem cells develop into multiple types of cardiac myocytes: action potential characterization. *Circ. Res.* **93**, 32–39.
11. Zeng, X., Miura, T., Luo, Y., Bhattacharya, B., Condie, B., Chen, J., Ginis, I., Lyons, I., Mejido, J., Puri, R. K., Rao, M. S., and Freed, W. J. (2004) Properties of pluripotent human embryonic stem cells BG01 and BG02. *Stem Cells* **22**, 292–312.
12. Denning, C., Allegrucci, C., Priddle, H., Barbadillo-Munoz, M. D., Anderson, D., Self, T., Smith, N. M., Parkin, C. T., and Young, L. E. (2006) Common culture conditions for maintenance and cardiomyocyte differentiation of the human embryonic stem cell lines, BG01 and HUES-7. *Int. J. Dev. Biol.* **50**, 27–37.
13. Mummery, C., Ward-van Oostwaard, D., Doevendans, P., Spijker, R., van den Brink, S., Hassink, R., van der Heyden, M., Opthof, T., Pera, M., de la Riviere, A. B., Passier, R., and Tertoolen, L. (2003) Differentiation of human embryonic stem cells to cardiomyocytes: role of coculture with visceral endoderm-like cells. *Circulation* **107**, 2733–2740.

14. van den Eijden-vanRaaij, A. J., van Achterberg, T. A., van der Kruijssen, C. M., Piersma, A. H., Huylebroeck, D., de Laat, S. W., and Mummery, C. L. (1991) Differentiation of aggregated murine P19 embryonal carcinoma cells is induced by a novel visceral-endoderm specific FGF-like factor and inhibited by activin A. *Mech. Dev.* **33**, 157–165.
15. Pera, M. F., Andreade, J., Houssami, S., Reubinoff, B., Trounson, A., Stanley, E. G., Ward-van Oostwaard, D., and Mummery, C. (2004) Regulation of human embryonic stem cell differentiation by BMP-2 and its antagonist noggin. *J. Cell Sci.* **117**, 1269–1280.
16. Sachinidis, A., Gissel, C., Nierhoff, D., Hippler-Altenburg, R., Sauer, H., Wartenberg, M., and Hesheler, J. (2003) Identification of platelet-derived growth factor BB as cardiogenesis-inducing factor in mouse embryonic stem cells under serum free conditions *Cell Physiol. Biochem.* **13**, 423–429.
17. Passier, R., Ward-van Oostwaard, D., Snapper, J., Kloots, J., Hassink, R., Kuijk, E., Roelen, B., Brutel de la Riviere, A., and Mummery, C. (2005) Increased cardiomyocyte differentiation from human embryonic stem cells in serum-free cultures. *Stem Cells* **23**, 772–780.
18. Lebkowski, J. S., Gold, J., Xu, C., Funk, W., Chiu, C. P., Carpenter, M. K. (2001) Human embryonic stem cells: culture, differentiation, and genetic modification for regenerative medicine applications. *Cancer J.* **7**(Suppl. 2), S83–S93.
19. Reubinoff, B. E., Pera, M. F., Fong, C. Y., Trounson, A., and Bongso, A. (2002) Embryonic stem cell lines from human blastocysts: somatic differentiation in vitro. *Nat. Biotechnol.* **18**, 399–404.
20. Cowan, C. A., Klimanskaya, I., McMahon, J., Atienza, J., Witmyer, J., Zucker, J. P., Wang, S., Morton, C. C., McMahon, A..P, Powers, D., and Melton, D. A. (2004). Derivation of embryonic stem-cell lines from human blastocysts. *N. Engl. J. Med.* **350**, 1353–1356.
21. Rosler, E. S., Fisk, G. J., Ares, X., Irving, J., Miura, T., Rao, M. S., and Carpenter, M. K. (2004) Long-term culture of human embryonic stem cells in feeder-free conditions. *Dev. Dyn.* **229**, 259–274.
22. Xu, C., Inokuma, M. S., Denham, J., Golds, K., Kundu, P., Gold, J. D., and Carpenter, M. K. (2001) Feeder-free growth of undifferentiated human embryonic stem cells. *Nat. Biotechnol.* **19**, 971–974.
23. Mummery, C. L., Slager, H., Kruijer, W., Feijen, A., Freund, E., Koornneef, I., and van den Eijnden-van Raaij, A. J. (1990) Expression of transforming growth factor ß2 during the differentiation of murine embryonal carcinoma and embryonic stem cells. *Dev. Biol.* **137**, 161–170.

15

Myocardial Restoration and Tissue Engineering of Heart Structures

Theo Kofidis, Knut Müller-Stahl, and Axel Haverich

Summary

The restoration or de novo engineering of heart structures poses a challenge because of the unique structure and physical properties of the heart. The heart is a heterogeneous, complex helical structure with asymmetric and anisotropic features. Hence, it is variably built and consists of spiraling muscle bands (first described by Torrent-Guasp), valves, coronary vessels, and a conduction system. Therefore, successful construction of muscular or valvular grafts needs to meet specific prerequisites, such as mechanical stability, optimal porosity, and contractile function. A series of studies have reported on various mixtures of scaffolds and (stem) cells and subsequent production of spontaneously contractile cardiac grafts in vitro. In vivo studies have provided evidence of engraftment of such bioartificial myocardial grafts and of improved heart function of the host. Two fundamental types of bioartificial/engineered valves have been reported: decellularized xenogeneic valves and recellularized valves with the recipient's own cells. Large-scale clinical results are awaited. Finally, bioartificial vessels are being produced, either through de novo construction from collagens and cells or from previously harvested recipient's own fibroblasts and endothelial cells. The main goal envisioned here is long-term patency following implantation in vivo. This review surveys upon recent developments and indicates caveats in the field of tissue engineering of cardiac, valvular, and vascular structures.

Key Words: Tissue engineering; Bioartificial heart muscle; Bioartificial valves; Coronary artery disease; Scaffolds; Stem cells; Bioartificial vessels.

1. The Uniqueness of the Heart as a Target for Restorative Procedures Using Bioartificial Tissues: Possibilities and Limitations

As opposed to static structures of the body, that is, bones and cartilages, the heart constitutes a dynamic organ with a complex mechanical function. For

From: *Methods in Molecular Medicine, 2nd ed.: Tissue Engineering*
Edited by: H. Hauser and M. Fussenegger © Humana Press Inc., Totowa, NJ

the fulfillment of this function, the heart utilizes a unique geometry—that of a helix—and interaction of its composing elements *(1,2)* (*see* **Fig. 1**). If the heart were composed of symmetrically and circularly arranged muscular elements, its strong ejection fraction and therefore its pumping function would be diminished. The key mechanism underlying the heart's durable and strong performance

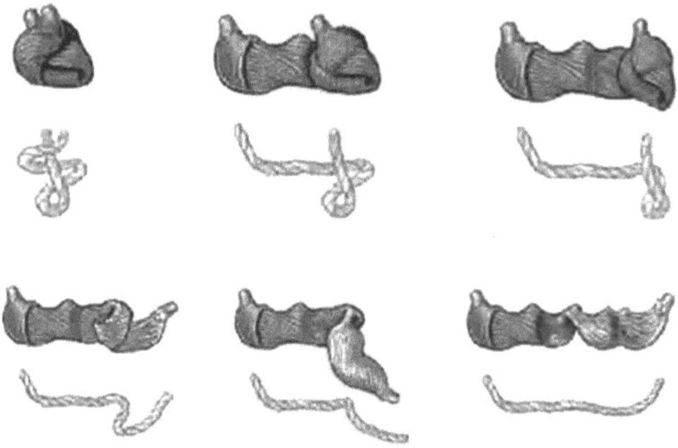

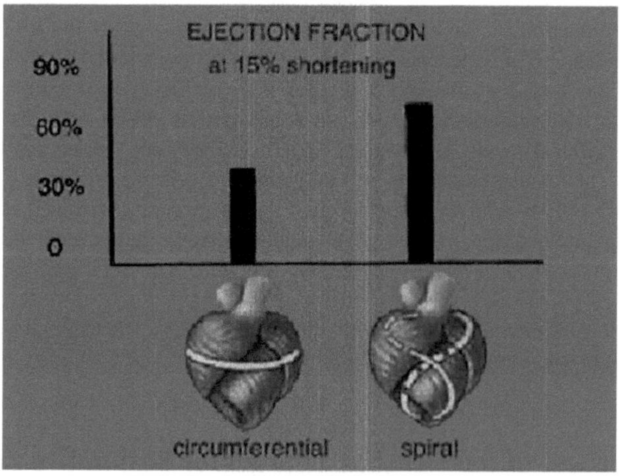

Fig. 1. Fig. 1(a). Unfolding of the muscular bands of the heart resembles unfolding of a rope (as described by F. Torrent-Guasp). (b) The spiral arrangement of the heart's muscle bands generates strong vortex forces in systole (Buckberg, G. (2002) The helix and the heart. *J. Thorac. Cardiovasc. Surg.* **124**, 863–883). This mechanism underlines the difficulty in manufacturing heart tissue in vitro with the appropriate fidelity.

is the vortex function of the myocardial muscle spiral during systole. Thus, the heart is a sophisticated, helical, asymmetrical structure involving muscular, valvular, and vascular elements.

Furthermore, the myocardium is a highly anisotropic structure, that is, the orientation of muscle fibers may be varying from site to site, as do the vascular density and the contractile pattern. Finally, the heart is angiotrophic: densely vascularized and with high metabolic demands. The optimal heart muscle replacement graft has not been constructed yet. A solid and robust tissue graft may be unsuitable for cell seeding and incapable of generating contractile force. Furthermore, cardiac myocytes lack proliferation potential and do not regenerate *(3)*. Therefore, they cannot serve as a direct and one-to-one source of myocardial tissue grafts. Autologous mesenchymal cells or bone marrow stem cells did not exhibit sufficient differentiation potential toward cardiomyocytes. Finally, most of the cells implanted into the area of myocardial lesion, both direct and within a matrix, die within a few days in vivo because of the hypoxic and hostile environment.

Consequently, following caveats and requirements arise for myocardial restoration using bioartificial, cell-enriched tissues:

1. Is the production of natural and efficient heart muscle possible with the currently available tools? If so, what type of cell and what kind of bioartificial matrix is better suited for the production of bioartificial heart tissue?
2. The engineered tissue graft should exhibit strong viability and stability for a sustained myocardial replacement.
3. The inoculated cells should preferably bear differentiation potential to cardiomyocytes or other contractile elements. They should be able to engraft into the microenvironmental niche and integrate in a way that allows for healing and development of functional properties. Paracrine or angiogenic engraftment processes have not been sufficiently elucidated yet.
4. The engineered tissue should be conservable in a viable and functional state to be implanted into the diseased heart muscle at a later time point and in the frame of an elective and well-controlled procedure.
5. The myocardial tissue construct should encompass vessel and neuronal networks to allow for perfusion and signal conduction.
6. The bioartificial matrix, which forms the framework for cellular seeding, has to exhibit adequate porosity and biodegradability to facilitate cell engraftment and infiltration, without losing integrity in a given time span following implantation in vivo.

These so-called hard problems in the manufacture of a myocardial tissue equivalent set the coordinates for future efforts to improve what has already been achieved in the field. In the following, we survey the most recent accomplishments in generating bioartificial heart tissue and restoring injured heart muscle.

2. Tissue Engineering-Based Myocardial Restoration

Heart muscle cells particularly neonatal ones are capable of contracting under specific tissue-culture conditions. This was recognized early and stimulated scientists to populate three-dimensional scaffolds with cardiomyocytes decades ago. In the late 1950s, Moscona et al. ***(4,5)*** generated spheroid aggregates from chick embryonic cells to find out that they assumed cardiomyocyte phenotypes more frequently and contracted better than cells plated two-dimensionally. Moreover, cardiomyocytes were able to assemble in layers and contract rhythmically. The contractile force was a fraction of that of native myocardial tissue. Therefore, further attempts focused on seeding cells in collagen matrices or porous networks.

In the further process, compounds of polymers and cells were generated and supplemented by growth factors. The goals were to refine the construct's physical and engraftment properties ***(6,7)***. In the meantime, various materials were in use: fibrin glue, polylactide/polyglycotide compounds, peptide nanofibers, gelatine hydrogels, hyaluronic acid (HA) gels, and many more ***(8)***. Substantial limitations of all these scaffolds are the lack of vascularization, the stimulation of inflammatory processes, and toxic degradation in the host.

Also, a variety of bioreactor systems and tissue-culture flasks have been introduced ***(9,10)***. Their structure and function vary greatly. Static cultures are being performed in flasks with culture medium circulation ***(11)***. In other systems, the cell–matrix compound matures under permanent rotation or pulsatile perfusion ***(12,13)***. Mechanical bioreactors exist, which aim at applying stretch to the bioartificial myocardial construct, to precondition it for its future in vivo purpose ***(14)***.

The most frequently used cells seeded in matrices to produce bioartificial heart tissue were neonatal rat cardiomyocytes, embryonic cardiomyocytes, embryonic stem cells and their derivatives, various subpopulations of bone marrow stem cells, myoblasts, fibroblasts, and endothelial progenitor cells. Following epimyocardial transplantation, the heart function was reported to improve in animal models of acute myocardial injury. Eschenhagen et al. ***(15)***constructed in vitro spontaneously beating tissue, which displayed superb integrity in vitro. Kofidis et al. ***(16)*** have further treated the tissue, achieving a homogeneous distribution of cells throughout the thickness, circumventing the issue of impaired viability and perfusion in the core of the construct.

One of the limitations of solid myocardial constructs is the deficient healing at the borders of implantation because of the limited regenerative potential of heart muscle. An innovative approach developed by the authors that aims at circumventing this problem involves liquid, injectable myocardial cell–matrix compounds, which consolidate once implanted at the site of myocardial injury. In this approach, the left ventricular geometry and architecture is not distorted ***(17)***. The first natively vascularized myocardial replacement graft was introduced by Haverich et al. based on autologous, decellularized bowel. The autologous graft retains its vascular tree and can be anastomosed to the patient's aorta and right atrium to facilitate right atrial and ventricular replacement. It has already been implanted in two patients.

More analytical approaches emerge that focus on improving the specific parts and function which compose an efficient bioartificial myocardial tissue construct and on facilitating a better microenvironment for the seeded cells. These approaches are summarized in the sections 2.1–2.6.

2.1. Manipulation of Cell Adhesion Mechanisms

The enhancement of cardiomyocyte and endothelial cell adhesion within three-dimensional scaffolds proved beneficial to cell survival. Cannizzaro et al. ***(18)*** added biotin–streptavidin bonds on copolymer structures and achieved a broader and more homogeneous distribution of endothelial cells in the final product. Lee et al. ***(19)*** manipulated the nano-organization of the scaffold to abate intercellular inhibition and reached a better proliferation rate within the construct. Finally, proteins such as osteopontin or integrins were coated on the surface of reticular matrices or were overexpressed by the inoculated cells. This way, a denser tissue mass was achieved ***(20–23)***.

2.2. Manipulation of Biodegradation

In the optimal case, a biopolymer or engineered construct degrades as slow as necessary to allow for engraftment first. For heart muscle replacement, it has to retain its stability until infiltrated or replaced by host tissue and cells because of the danger of disintegration, rupture of the myocardial wall, and fatal bleeding. Controlled degradation rates were achieved by distinct variations of the copolymer composition used: for copolymers such as in the case of polylactide/polyglycotide or HA hydrogels, degradation decelerated from 1 to 38 days ***(24,25)***. Lee et al. ***(26,27)*** accelerated degradation by controlling oxidative processes while stability was retained.

2.3. Manipulation of Porosity

The pore size and connectivity of the matrix are very decisive factors for myocardial tissue engineering. Too small pores impair cell migration and integration *(28)*. On the contrary, too large pores jeopardize angiogenesis and endothelialization. Endothelial cells and cardiomyocytes can only bridge a limited space following inoculation into a scaffold to assume contact with surrounding cells. Furthermore, the pore size has an impact on the mechanical stability of the bioartificial graft. Moreover, small pores may inhibit nutrient and oxygen supply to the interior of the implanted graft and thereby limit cell viability and overall contractility.

2.4. Manipulation of the Protein Content and Protein Delivery

The controlled delivery of proteins, for instance growth factors, from the implanted scaffold can have beneficial trophic, integrative, and antioxidant effects. Depending on the factor or substance, which has been coated or loaded on the scaffold, specific functions can be accomplished. Loading of the matrix with vascular endothelial growth factor (VEGF) induces recruitment of neighboring endothelial cells and promotes angiogenesis, that is, vessel sprouting to the core of the graft *(29,30)*. The enrichment of a bioartificial structure with endothelial cells has a positive impact on electrical conduction and signal transfer between cardiomyocytes that are sharing the same space *(31)*. Marui et al. *(32)* loaded collagen microspheres with a combination of fibroblast growth factor-2 (FGF-2) and hepatocyte growth factor (HGF). Following implantation into muscle tissue, these microspheres promoted tissue perfusion, as shown by LASER-based flow evaluation.

2.5. "Smart Biomaterials" and Nanotechnology in Myocardial Restoration

The development of the so-called smart biomaterials and nanoparticles is still in a nascent state. Future trends in myocardial restoration will eventually involve controlled protein and cytokine release as well as guided in vivo maturation of three-dimensional tissue. Nanoparticles are capable of releasing the appropriate signal at the appropriate time point for a better engraftment and differentiation of implanted tissue *(33,34)*. They may alter remodeling pathways and slow down scar formation in the area of injury and subsequent tissue implantation.

2.6. Concepts of Myocardial Enhancement Using Bioartificial Tissue

Recently, a series of novel concepts of myocardial repair using tissue have been introduced. Bursac et al. and Carrier et al. analyzed their bioartificial constructs—consisting of neonatal rat or chick embryonic cardiomyocytes embedded in polyglycolic acid matrices—electrophysiologically in bioreactor systems *(35,36)*. Leor et al. inoculated rat neonatal cardiomyocytes in an alginate matrix *(7,37)*. Eschenhagen and Zimmermann seeded neonatal rat cardiomyocytes in a liquid matrix cocktail of collagens and Matrigel. They manufactured tissue in various sizes and forms and assessed the pharmacological responsivity of the tissue *(38,39)*. Li et al. seeded the same type of cells in gelatine matrix (Gelfoam®). Our group established three-dimensional cultures of cardiomyocytes in clinically used collagen scaffolds, such as tissue fleece, which is being used for hemostasis *(40–42)*. Haverich et al. developed an autologous vascularized matrix from decellularized bowel (BIOVAM®), thereby maintaining the supplying vessels (*see* **Fig. 2**).

The first combinatory approach of myocardial tissue generation and concurrent seeding of stem cells was carried out by Krupnick et al. *(43)*, who populated fetal liver cells into a collagen mesh. There is a variety of in vivo implantation setups for myocardial restoration. Leor and Cohen *(7)* implanted their alginate construct in infarcted heart muscle of Sprague–Dawley rats and observed the generation of granulation tissue. However, there was only a slight deceleration of heart function deterioration compared with that of untreated, infarcted rats.

Li et al. *(44)* found rhythmical contractions in subcutaneously implanted bioartificial myocardial tissue made from gelfoam and neonatal rat cells. He replaced a portion of the right ventricular outflow tract (RVOT) and found a vigorous endothelial cell infiltration from the surrounding tissue. Zimmermann, Eschenhagen et al. implanted their product (EHT, engineered heart tissue) in syngeneic rats in two different forms: a flat and square one (implanted subcutaneously) and bioartificial rings (implanted epicardially). *(40)* Four weeks later, and under immunosuppression, contractility and angiogenesis could be observed *(45)*. This group was the first to introduce graft implantation in three overlying layers to imitate the distinct structure of the heart muscle, which comprises three muscle sheets in complex interaction. Shimizu et al. *(46)* implanted a tissue patch generated by thermoregulated layers under the skin of syngeneic rats and demonstrated long-term survival and organization of the grafts.

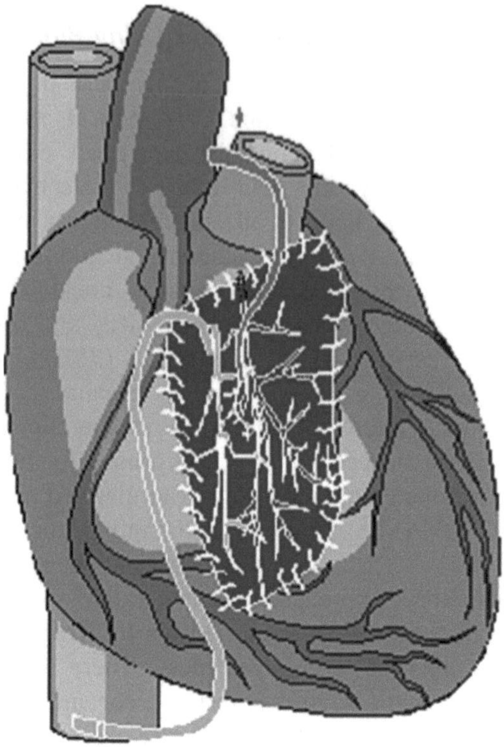

Fig. 2. The autologous matrix developed by Haverich et al. (BIOVAM®) is a decellularized intestinal matrix with retained arterial supply and venous return vessels. It can be anastomosed to the aorta and right atrium or superior vena cava, respectively, and thereby facilitates transmural right atrial or right ventricular replacement.

3. Evaluation of Implanted Grafts and Functional Improvement of the Heart: Cell Tracking, Bioimaging, and Functional Studies

Parallel to the evolution of restorative approaches, there is a notable progress of bioimaging techniques that are crucial for the evaluation of the utilized restorative techniques. First prerequisite of the restorative success is the identification of the graft and the characterization of the engraftment process. This can be achieved by labeling the cells prior to their inoculation in the matrix. The most efficient method involved is the transfection of cells with genes that encode for a fluorescent or bioluminescent protein, which causes the cells to emit light at a specific wavelength and be captured by sensitive microscopes or cameras. Because the surrounding host tissue is not labeled, it remains dark, and the spatial distribution and viability of the implanted graft can be discriminated from the host. The most frequently used method is the detection of green fluorescent protein (GFP) expression in transfected cells ***(47)***. Through concurrent staining with antibodies that bind on specific cell structures, the grafted cells can be closely characterized in their phenotype and function. Sensible confocal microscopes provide high-resolution analysis of the colocalized stains (emitted colors) ***(48,49)***. These methods are widely used both in direct stem cell transfer and in tracking and identification of cells within the vehicle matrix they had been implanted with.

Modern methods of cell labeling encompass ferromagnetic loading of stem cells that can be detected by magnetic resonance imaging (MRI). The downside of using ferromagnetic particles is that they persist beyond the in vivo demise of the cells. Hence, the presence of a signal on MRI does not necessarily represent a viable graft. Advanced positron-emission tomography (PET) or single photon emission computer tomography (CT-SPECT)-fusion methodology conquers the scientific field and facilitates spatial and functional conclusions. In vivo fluorescence and bioluminescence detect viable grafts in vivo and may also identify cell migration in the living animal following tissue engraftment. They enable scientists to track the long-term fate of cells and grafts in experimental animals without the need to kill them and harvest the tissue for histology. Latter methods still need to be amplified because of the limited tissue penetration associated with their use, in order to allow for large animal and human studies ***(50–55)***.

Functional studies focus on describing the effect of transplanted bioartificial grafts on myocardial contractility and ejection fraction in comparison with control groups, which had just been myocardially injured but not treated. Echocardiography is a widely used method regardless of the size of the animal because of the availability of ultrasound probes of various frequencies. More invasive methods require catheterization of the graft recipient but provide profound hemodynamic information. The MRI is gaining significance among bioimaging methods because of its morphological and functional output. The introduction of powerful magnets allows for imaging of small animals as it does for human studies (7.5 T) ***(15,17,56)***.

4. Tissue Engineering of Heart Valves

State-of-the-art heart valve replacement exhibits disadvantages such as mechanical failure, structural deterioration, and the need for permanent anticoagulation. Although animal-derived heart valves do not require life-long anticoagulant treatment, they are prone to calcification and failure 10–15 years following implantation ***(57)***. Allogenic homografts are reported to be more efficient in their long-term performance but are also not free of morphological changes, such as dilatation, regurgitation, or structural failure *(58)*.

Therefore, scientists and surgeons have shifted their interest in generating a valve, which can be engineered and costumed to the patient's needs in vitro and which can be implanted into the host and grow with him. It should require few or no anticoagulation and exhibit enough mechanical stability to withstand the hemodynamic load at the outflow tract. There are two main approaches to tissue engineering heart valves: regeneration and repopulation. Regeneration is based on implantation of a biological or otherwise resorbable matrix with advantageous biodegradation characteristics, such that they allow for gradual remodeling and cell engraftment from the host, before final dissolution of the native structure. Repopulation involves implantation of a decellularized animal valve. This valve has been previously cleaned from the native animal cells, so that only the connective tissue matrix is left for implantation. In the process of in vivo persistence, cells of the host should populate the matrix and align along the free surface of the valve to form an endothelial layer. Both methods bear advantages and disadvantages, both at the manufacture and at the in vivo behavior level. Other most recent approaches involve repopulation of polymeric scaffolds with endothelial cells, endothelial progenitors, or mesenchymal stem cells. The need for manufacturing tissue-engineered valves is defined by the number of patients requiring heart valve replacement, the current reports of prosthetic failure of fabricates destined for open heart surgery and also for deployment, and the sizable pediatric market. Latter cannot be covered by the usual mechanical or biological prostheses because of size constraints *(59)*.

First clinical attempts were made using glutaraldehyde-treated biological matrices, which, however, remained cell free in vivo, most probably due to the chemical treatment. Grimm et al. ***(60,61)*** deactivated the aldehyde component of such constructs and managed to cultivate endothelial cells on the valvular surface. The first attempts to decellularize xenografts for valvular tissue engineering were made by Brendel and Duhammel of the University of Arizona and patented in 1984. The same approach was used in Europe by Steinhoff et al. *(62)* from Rostock, Germany, Stock and colleagues *(63)* from Jena, Germany, and Haverich et al. from Hannover, Germany. Multiple reports later emerged from the Netherlands, United Kingdom, Italy, and recently Asia *(64–67)*. The basic idea of the xenograft method is that once cells are removed from the valve, immunogenicity is eliminated. Furthermore, the valve lies in a position where deposition of cytokine-producing inflammatory cells is unlikely. It is then hoped that the decellularized xenograft structure will be repopulated with circulating endothelial cells and progenitors from bloodstream. The first large-scale industrial production of such valve substitutes was performed by CryoLife in Europe, and first implantations were carried out in children. Many of the children experienced severe regurgitation, and some of them died. Consequently, CryoLife withdrew the product from the market. Histological evaluation displayed inflammation, fibrosis, encapsulation, and perforation. Stock et al. ***(69)*** had previously described such phenomena in the sheep implantation model. Dohmen and Konertz ***(68)*** went on implanting an analogous product (Auto Tissue GmbH) in 50 patients, some of them did not escape adverse effects and death. Another approach at the authors' institution involves decellularization of human homograft valves and reseeding with the recipient's venous endothelial cells in custom-made bioreactors until the cells form stable layers on the valvular surface. Implantations in humans have not been performed yet. Also, Rieder et al. ***(70)*** have demonstrated that the lowest level of immune system stimulation was with decellularized human tissues. The decellularized porcine leaflets triggered a much stronger macrophage response than the extracts of human native pulmonary cusps that had not been decellularized. Thus, they favor the human allograft, not xenografts, as a basis for decellularization technologies designed to obtain functioning valve extracellular matrix (ECM) scaffolds for tissue engineering of heart valves.

An alternative approach, polylactic acid is resorbable synthetic scaffolds. The dominant material is polyglycolic (PLA/PGA), initially used for skin substitutes, which is FDA approved. Vacanti et al. ***(71)*** of Harvard Children's Hospital utilized a more compliant polyhydroxyalkanoate scaffold. Another material that allows for cell embedding and cell–matrix interaction is HA. HA is a polysaccharide that upon modification can be photopolymerized into hydrogels. Masters et al. ***(72)*** have demonstrated that HA hydrogels are suitable scaffolds for the culture and proliferation of valvular interstitial cells (VICs), the most prevalent cell type in native heart valves. The physical properties of HA gels are easily modified through alteration in material crosslink density or by copolymerizing with other reactive macromolecules.

Sutherland et al. created a valve prosthesis using autologous mesenchymal stem cells harvested from bone marrow seeded into a biodegradable scaffold manufactured from a resorbable synthetic biomaterial. The construct was implanted in the pulmonary valve position of sheep and monitored for 8 months. Three valve-typical layers equivalent to the zona ventricularis, the zona spongiosa, and the zona fibrosa developed. The valves exhibited a continuous endothelial layer and normal coaptation. The adequate adaptation of the cells is encouraging that they will eventually retain the capacity to metabolize calcium and avoid the long-term problems of mineralization ***(73)***.

Hybrid heart valve scaffolds were also fabricated, as a combination of the two main principles mentioned above. Porcine aortic heart valve matrices were decellularized and enhanced with bioresorbable polymers: (i) dip coating of lyophilized decellularized matrices and (ii) impregnation of wet decellularized matrices. Tensile and suture retention strengths were tested. A pulse duplicator system was used for functional testing of the valves under physiological systemic load conditions and compared with unseeded synthetic scaffolds and xenografts. Finally, Ramamurthi and Vesely *(74)* have coated non-biodegradable scaffolds with elastin sheets to produce elastin–hyaluronan composites resembling the glycosaminoglycan-rich and elastin-rich layers of the native aortic valve cusp. Neonatal rat aortic smooth muscle cells were cultured atop hylan gel-produced ECM. This group has demonstrated that hylan gels are useful as substrates to induce elastin synthesis in culture to obtain structures that resemble the elastin matrix of the native aortic valve. Finally, Mol et al. *(75)* have indicated the significance of mechanical strain for the maturation and processing of tissue-engineered heart valves. The valve scaffolds used were prepared from a PGA mesh, and human saphenous vein cells were coated on them using fibrin as a cell carrier. Static strains due to tissue contraction occured while dynamic pressure gradients were applied on the valve. Strained valves contained more tissue, which displayed higher strength and integrity and a more homogeneous cellular organization.

5. Tissue Engineering of Vascular Grafts

L'Heureaux, MacAllister et al. *(76)* described the first entirely autologous vascular interposition graft. Clinical studies are ongoing, and the major limitation is the thrombotic occlusion of the graft. Such constructs could be used as bypass grafts or dialysis shunts. Alternatively, synthetic grafts made of polytetrafluorethylene (PTFE) and other organic rigid matrices could be endothelialized to provide a smooth lumina surface following implantation. It remains a challenge to meet all required criteria for the perfect tissue-engineered artery *(77)*, such as burst strength and compliance mismatch at the same time. As Isenberg et al. surveyed in their work, not a single approach has yet produced key features of the media, such as circumferential alignment of Smooth muscle cells (SMCs), collagen fibers, and elastin lamellae. On the contrary, it would be a lengthy process of weeks to isolate and expand the Endothelial cells (ECs) of a patient to the numbers required for seeding a construct of useful length, with circulating EC progenitor cells. The use of stem cells or stem cell-derived cells along with high-output bioreactor systems, which will allow for vessel preconditioning, will eventually yield improvements.

6. Perspective

We anticipate a generation of structured cardiovascular tissues with the highest possible fidelity through the emerging science of nanobiology and nanotechnology. The culture and expansion of high number of pluripotent stem cells with adequate phenotypic features will provide the substrate for cell seeding in preformed scaffolds. Although a transmural myocardial graft may still require vigorous perfusion to achieve survival and maximal contractility, valve and vascular substitutes will greatly benefit from improved cell seeding and adhesion methods, as well as from long-term large animal preclinical studies.

References

1. Buckberg, G. D. (2002) Basic science review: the helix and the heart. *J. Thorac. Cardiovasc. Surg.* **124**, 863–883.
2. Buckberg, G. D., Coghlan, H. C., and Torrent-Guasp. F. (2001) The structure and function of the helical heart and its buttress wrapping. V. Anatomic and physiologic considerations, in the healthy and failing heart. *Semin. Thorac. Cardiovasc. Surg.* **13**, 358–385 (review).
3. Pasumarthi, K. B. and Field, L. J. (2002) Cardiomyocyte cell cycle regulation. *Circ. Res.* **90**, 1044–1054.
4. Moscona, A. (1959) Tissues from dissociated cells. *Sci. Am.* **200**, 132–134.
5. Moscona, A. and Moscona, H. (1952) The dissociation and aggregation of cells from organ rudiments of the early chick embryo. *J. Anat.* **86**, 287–301.
6. Eschenhagen, T. and Zimmermann, W. H. (2005) Engineering myocardial tissue. *Circ. Res.* **97**(12), 1220–1231.
7. Leor, J. and Cohen, S. (2004) Myocardial tissue engineering: creating a muscle patch for a wounded heart. *Ann. N. Y. Acad. Sci.* **1015**, 312–319 (review).
8. Leor, J., Amsalem, Y., and Cohen, S. (2005) Cells, scaffolds, and molecules for myocardial tissue engineering. *Pharmacol. Ther.* **105**(2), 151–163 (review).
9. Barron, V., Lyons, E., Stenson-Cox, C., McHugh, P. E., and Pandit, A. (2003) Bioreactors for cardiovascular cell and tissue growth: a review. *Ann. Biomed. Eng.* **31**(9), 1017–1030 (review).
10. Shachar, M. and Cohen, S. (2003) Cardiac tissue engineering, ex-vivo: design principles in biomaterials and bioreactors. *Heart Fail. Rev.* **8**, 271–276 (review).
11. Sikavitsas, V. I., Bancroft, G. N., and Mikos, A. G. (2002) Formation of three-dimensional cell/polymer constructs for bone tissue engineering in a spinner flask and a rotating wall vessel bioreactor. *J. Biomed. Mater. Res.* **62**, 136–148.
12. Bursac, N., Papadaki, M., White, J. A., Eisenberg, S. R., Vunjak-Novakovic, G., and Freed, L. E. (2003) Cultivation in rotating bioreactors promotes maintenance of cardiac myocyte electrophysiology and molecular properties. *Tissue Eng.* **9**, 1243–1253.

13. Kofidis, T., Lenz, A., Boublik, J., Akhyari. P., Wachsmann. B., Mueller-Stahl. K., Hofmann, M., and Haverich, A. (2003) Pulsatile perfusion and cardiomyocyte viability in a solid three-dimensional matrix. *Biomaterials* **24**, 5009–5014.
14. Akhyari, P., Fedak, P. W., Weisel, R. D., Lee, T. Y., Verma, S., Mickle, D. A., and Li, R. K. (2002) Mechanical stretch regimen enhances the formation of bioengineered autologous cardiac muscle grafts. *Circulation* **106** (12 Suppl. 1), I137–I142.
15. Zimmermann, W. H., Schneiderbanger, K., Schubert, P., Didie, M., Munzel, F., Heubach, J. F., Kostin, S., Neuhuber, W. L., and Eschenhagen, T. (2002) Tissue engineering of a differentiated cardiac muscle construct. *Circ. Res.* **90**, 223–230.
16. Kofidis, T., Akhyari, P., Boublik, J., Theodorou, P., Martin, U., Ruhparwar, A., Fischer, S., Eschenhagen, T., Kubis, H. P., Kraft, T., Leyh, R., and Haverich, A. (2002) In vitro engineering of heart muscle: artificial myocardial tissue. *J. Thorac. Cardiovasc. Surg.* **124**, 63–69.
17. Kofidis, T., Lebl, D. R., Martinez, E. C., Hoyt, G., Tanaka, M., and Robbins R. C. (2005) Novel injectable bioartificial tissue facilitates targeted, less invasive, large-scale tissue restoration on the beating heart after myocardial injury. *Circulation* **112**(9 Suppl.), I173–I177.
18. Cannizzaro, S. M., Padera, R. F., Langer, R., Rogers, R. A., Black, F. E., Davies, M. C., Tendler, S. J., and Shakesheff, K. M. (1998) A novel biotinylated degradable polymer for cell interactive applications. *Biotechnol. Bioeng.* **58**, 529–535.
19. Lee, C. R., Grad, S., Gorna, K., Gogolewski, S., Goessl, A., and Alini, M. (2005) Fibrin-polyurethane composites for articular cartilage tissue engineering: a preliminary analysis. *Tissue Eng.* **11**, 1562–1573.
20. Jaquiery, C., Schaeren, S., Farhadi, J., Mainil-Varlet, P., Kunz, C., Zeilhofer, H. F., Heberer, M., and Martin, I. (2005) In vitro osteogenic differentiation and in vivo bone-forming capacity of human isogenic jaw periosteal cells and bone marrow stromal cells. *Ann. Surg.* **242**, 859–867, discussion 867–868.
21. Lim, J. Y., Taylor, A. F., Li, Z., Vogler E. A., and Donahue H. J. (2005) Integrin expression and osteopontin regulation in human fetal osteoblastic cells mediated by substratum surface characteristics. *Tissue Eng.* **11**, 19–29.
22. Mauney, J. R., Jaquiery, C., Volloch, V., Heberer, M., Martin, I., and Kaplan D. L. (2005) In vitro and in vivo evaluation of differentially demineralized cancellous bonescaffolds combined with human bone marrow stromal cells for tissue engineering. *Biomaterials* **26**, 3173–3185.
23. Rosso, F., Giordano, A., Barbarisi, M., and Barbarisi, A. (2004) From cell-ECM interactions to tissue engineering. *J. Cell Physiol.* **199**, 174–180 (review).
24. Eschenhagen, T. and Zimmermann, W. H. (2005) Engineering myocardial tissue. *Circ. Res.* **97**, 1220–1231.
25. Rabkin, E. and Schoen, F. J. (2002) Cardiovascular tissue engineering. *Cardiovasc. Pathol.* **11**, 305–317.

26. Bouhadir, K. H., Lee, K. Y., Alsberg, E., Damm, K. L., Anderson, K. W., and Mooney, D. J. (2001) Degradation of partially oxidized alginate and its potential application for tissue engineering. *Biotechnol. Prog.* **17**, 945–950.
27. Khang, G., Lee, S. J., Lee, J. H., Kim, Y. S., and Lee, H. B. (1999) Interaction of fibroblast cells on poly(lactide-co-glycolide) surface with wettability chemogradient. *Biomed. Mater. Eng.* **9**, 179–187.
28. Karageorgiou, V. and Kaplan, D. (2005) Porosity of 3D biomaterial scaffolds and osteogenesis. *Biomaterials* **26**, 5474–5491 (review).
29. Kelm, J. M., Diaz Sanchez-Bustamante, C., Ehler, E., Hoerstrup, S. P., Djonov, V., Ittner, L., and Fussenegger, M. (2005) VEGF profiling and angiogenesis in human microtissues. *J. Biotechnol.* **118**, 213–229.
30. Yao, C., Prevel, P., Koch, S., Schenck, P., Noah, E. M., Pallua, N., and Steffens, G. (2004) Modification of collagen matrices for enhancing angiogenesis. *Cells Tissues Organs* **178**, 189–196.
31. Narmoneva, D. A., Vukmirovic, R., Davis, M. E., Kamm, R. D., and Lee, R. T. (2004) Endothelial cells promote cardiac myocyte survival and spatial reorganization: implications for cardiac regeneration. *Circulation* **110**, 962–968.
32. Marui, A., Kanematsu, A., Yamahara, K., Doi, K., Kushibiki, T., Yamamoto, M., Itoh, H., Ikeda, T., Tabata, Y., and Komeda, M. (2005) Simultaneous application of basic fibroblast growth factor and hepatocyte growth factor to enhance the blood vessels formation. *J. Vasc. Surg.* **41**, 82–90.
33. Ito, A., Ino, K., Hayashida, M., Kobayashi, T., Matsunuma, H., Kagami, H., Ueda, M., and Honda, H. (2005) Novel methodology for fabrication of tissue-engineered tubular constructs using magnetite nanoparticles and magnetic force. *Tissue Eng.* **11**, 1553–1561.
34. Shimizu, K., Ito, A., and Honda, H. (2005) Enhanced cell-seeding into 3D porous scaffolds by use of magnetite nanoparticles. *J. Biomed. Mater. Res. B Appl. Biomater.* October 21 (Epub ahead of print).
35. Papadaki, M., Bursac, N., Langer, R., Merok, J., Vunjak-Novakovic, G., and Freed, L. E. (2001) Tissue engineering of functional cardiac muscle: molecular, structural, and electrophysiological studies. *Am. J. Physiol. Heart Circ. Physiol.* **280**, H168–H178.
36. Bursac, N., Papadaki, M., Cohen, R. J., Schoen, F. J., Eisenberg, S. R., Carrier, R., Vunjak-Novakovic, G., and Freed, L. E. (1999) Cardiac muscle tissue engineering: toward an in vitro model for electrophysiological studies. *Am. J. Physiol.* **277** (2 Pt. 2), 433–444.
37. Cohen, S. and Leor, J. (2004) Rebuilding broken hearts. Biologists and engineers working together in the fledgling field of tissue engineering are within reach of one of their greatest goals: constructing a living human heart patch. *Sci. Am.* **291**, 44–51.
38. Zimmermann, W. H. and Eschenhagen, T. (2003) Cardiac tissue engineering for replacement therapy. *Heart Fail. Rev.* **8**, 259–269 (review).

39. Zimmermann, W. H., Schneiderbanger, K., Schubert, P., Didie, M., Munzel, F., Heubach, J. F., Kostin, S., Neuhuber, W. L., and Eschenhagen T. (2002) Tissue engineering of a differentiated cardiac muscle construct. *Circ. Res.* **90**, 223–230.
40. Ozawa, T., Mickle, D. A., Weisel, R. D., Matsubayashi, K., Fujii, T., Fedak, P. W., Koyama, N., Ikada, Y., and Li, R. K. (2004) Tissue-engineered grafts matured in the right ventricular outflow tract. *Cell Transplant.* **13**, 169–177.
41. Ozawa, T., Mickle, D. A., Weisel, R. D., Koyama, N., Ozawa, S. and Li, R. K. (2002) Optimal biomaterial for creation of autologous cardiac grafts. *Circulation* **106**(12 Suppl. 1), I176–I182.
42. Kofidis, T., Akhyari, P., Boublik, J., Theodorou, P., Martin, U., Ruhparwar, A., Fischer, S., Eschenhagen, T., Kubis, H. P., Kraft, T., Leyh, R., and Haverich, A. (2002) In vitro engineering of heart muscle: artificial myocardial tissue. *J. Thorac. Cardiovasc. Surg.* **124**, 63–69.
43. Krupnick, A. S., Balsara, K. R., Kreisel, D., Riha, M., Gelman, A. E., Estives, M. S., Amin, K. M., Rosengard, B. R., and Flake A. W. (2004) Fetal liver as a source of autologous progenitor cells for perinatal tissue engineering. *Tissue Eng.* **10**, 723–735.
44. Li, R. K., Jia, Z. Q., Weisel, R. D., Mickle, D. A., Choi, A., and Yau T. M. (1999) Survival and function of bioengineered cardiac grafts. *Circulation* **100**(19 Suppl.), II63–II69.
45. Zimmermann, W. H., Melnychenko, I., and Eschenhagen T. (2004) Engineered heart tissue for regeneration of diseased hearts. *Biomaterials* **25**, 1639–1647 (review).
46. Shimizu, T., Yamato, M., Isoi, Y., Akutsu, T., Setomaru, T., Abe, K., Kikuchi, A., Umezu, M., and Okano T. (2002) Fabrication of pulsatile cardiac tissue grafts using a novel 3-dimensional cell sheet manipulation technique and temperature-responsive cell culture surfaces. *Circ. Res.* **90**, e40.
47. Chalfie, M., Tu, Y., Euskirchen, G., Ward, W., and Prasher D. (1994) Green fluorescent protein as a marker for gene expression. *Science* **263**, 802–805.
48. Orlic, D., Hill, J. M., and Arai A. E. (2002) Stem cells for myocardial regeneration. *Circ. Res.* **91**, 1092–1102 (review).
49. Melton, L. (2005) Imaging: the big picture. *Nature* **437**, 775–779.
50. Saeed, M., Saloner, D., Weber, O., Martin, A., Henk, C., and Higgins, C. (2005) MRI in guiding and assessing intramyocardial therapy. *Eur. Radiol.* **15**, 851–863.
51. De, A. and Gambhir, S. S. (2005) Noninvasive imaging of protein-protein interactions from live cells and living subjects using bioluminescence resonance energy transfer. *FASEB J.* **19**, 2017–2019.
52. Rodriguez-Porcel, M., Gheysens, O., Chen, I. Y., Wu, J. C., and Gambhir, S. S. (2005) Image-guided cardiac cell delivery using high-resolution small-animal ultrasound. *Mol. Ther.* **12**, 1142–1147.
53. Michalet, X., Pinaud, F. F., Bentolila, L. A., Tsay, J. M., Doose, S., Li, J. J., Sundaresan, G., Wu, A. M., Gambhir, S. S., and Weiss, S. (2005) Quantum dots for live cells, in vivo imaging, and diagnostics. *Science* **307**, 538–544 (review).

54. Wu, J. C., Chen, I. Y., Wang, Y., Tseng, J. R., Chhabra, A., Salek, M., Min, J. J., Fishbein, M. C., Crystal, R., and Gambhir, S. S. (2004) Molecular imaging of the kinetics of vascular endothelial growth factor gene expression in ischemic myocardium. *Circulation* **110**, 685–691.
55. Su, H., Forbes, A., Gambhir, S. S., and Braun, J. (2004) Quantitation of cell number by a positron emission tomography reporter gene strategy. *Mol. Imaging Biol.* **6**, 139–148.
56. Balsam, L. B., Wagers, A. J., Christensen, J. L., Kofidis, T., Weissman, I. L., and Robbins, R. C. (2004) Haematopoietic stem cells adopt mature hematopoietic fates in ischemic myocardium. *Nature* **428**, 668–673.
57. Stein, P. D., Riddle, J. M., Kemp, S. R., Lee, M. W., Lewis, J. W., and Magilligan, D. J., Jr. (1985) Effect of warfarin on calcification of spontaneously degenerated porcine bioprosthetic valves. *J. Thorac. Cardiovasc. Surg.* **90**, 119–125.
58. Staab, M. E., Nishimura, R. A., Dearani, J. A., and Orszulak T. A. (1998) Aortic valve homografts in adults: a clinical perspective. *Mayo Clin. Proc.* **73**, 231–238 (review).
59. Neuenschwander, S. and Hoerstrup, S. P. (2004) Heart valve tissue engineering. *Transpl. Immunol.* **12**, 359–365 (review).
60. Grimm, M., Grabenwoger, M., Eybl, E., Moritz, A., Bock, P., Muller, M. M., and Wolner, E. (1992) Improved biocompatibility of bioprosthetic heart valves by L-glutamic acid treatment. *J. Card. Surg.* **7**, 58–64.
61. Leukauf, C., Szeles, C., Salaymeh, L., Grimm, M., Grabenwoger, M., Losert, U., Moritz, A., and Wolner, E. (1993) In vitro and in vivo endothelialization of glutaraldehyde treated bovine pericardium. *J. Heart Valve Dis.* **2**, 230–235.
62. Steinhoff, G., Stock, U., Karim, N., Mertsching, H., Timke, A., Meliss, R. R., Pethig, K., Haverich, A., and Bader A. (2000) Tissue engineering of pulmonary heart valves on allogenic acellular matrix conduits: in vivo restoration of valve tissue. *Circulation* **102**(19 Suppl. 3), III50–III55.
63. Stock, U. A., Nagashima, M., Khalil, P. N., Nollert, G. D., Herden, T., Sperling, J. S., Moran, A., Lien, J., Martin, D. P., Schoen, F. J., Vacanti, J. P., and Mayer, J. E., Jr. (2000) Tissue-engineered valved conduits in the pulmonary circulation. *J. Thorac. Cardiovasc. Surg.* **119** (4 Pt 1), 732–740.
64. Schenke-Layland, K., Vasilevski, O., Opitz, F., Konig, K., Riemann, I., Halbhuber, K. J., Wahlers, T., and Stock, U. A. (2003) Impact of decellularization of xenogeneic tissue on extracellular matrix integrity for tissue engineering of heart valves. *J. Struct. Biol.* **143**, 201–208.
65. Schenke-Layland, K., Opitz, F., Gross, M., Doring, C., Halbhuber, K. J., Schirrmeister, F., Wahlers, T., and Stock, U. A. (2003) Complete dynamic repopulation of decellularized heart valves by application of defined physical signals – an in vitro study. *Cardiovasc. Res.* **60**, 497–509.

66. Korossis, S. A., Booth, C., Wilcox, H. E., Watterson, K. G., Kearney, J. N., Fisher J., and Ingham, E. (2002) Tissue engineering of cardiac valve prostheses II: biomechanical characterization of decellularized porcine aortic heart valves. *J. Heart Valve Dis.* **11**, 463–471.
67. Zhao, D. E., Li, R. B., Liu, W. Y., Wang, G., Yu, S. Q., Zhang, C. W., Chen, W. S., and Zhou, G. X. (2003) Tissue-engineered heart valve on acellular aortic valve scaffold: in-vivo study. *Asian Cardiovasc. Thorac. Ann.* **11**, 53–56.
68. Konertz, W., Dohmen, P. M., Liu, J., Beholz, S., Dushe, S., Posner, S., Lembcke, A., and Erdbrugger, W. (2005) Hemodynamic characteristics of the matrix P decellularized xenograft for pulmonary valve replacement during the Ross operation. *J. Heart Valve Dis.* **14**, 78–81.
69. Stock, U. A., Skamoto, T., Hatsuoka, S., Martin, D. P., Nagashima, M., Moran, A., Moses, M. A, Khalil, P. N., Schoen, F. J., Vacanti, J. P., and Mayer, J. E. (2000) Patch augmentation of the pulmonary artery using autologous cells and biodegradable polymers. *J. Thorac. Cardiovasc. Surg.* **120**, 1158–1168.
70. Rieder, E., Seebacher, G., Kasimir, M. T., Eichmair, E., Winter, B., Dekan, B., Wolner, E., Simon, P., and Weigel, G. (2005) Tissue engineering of heart valves: decellularized porcine and human valvescaffolds differ importantly in residual potential to attract monocytic cells. *Circulation* **111**, 2792–2797.
71. Sodian, R., Sperling, J. S., Martin, D. P., Egozy, A., Stock, U., Mayer, J. E., Jr., and Vacanti, J. P. (2000) Fabrication of a trileaflet heart valve scaffold from a polyhydroxyalkanoate biopolyester for use in tissue engineering. *Tissue Eng.* **6**, 183–188.
72. Masters, K. S., Shah, D. N., Walker, G., Leinwand, L. A., and Anseth K. S. (2004) Designing scaffolds for valvular interstitial cells: cell adhesion and function on naturally derived materials. *J. Biomed. Mater. Res. A.* **71**, 172–180.
73. Sutherland, F. W., Perry, T. E., Yu, Y., Sherwood, M. C., Rabkin, E., Masuda, Y., Garcia, G. A., McLellan, D. L., Engelmayr, G. C., Jr., Sacks, M. S., Schoen, F. J., and Mayer, J. E., Jr. (2005) From stem cells to viable autologous semilunar heart valve. *Circulation* **111**, 2783–2791.
74. Ramamurthi, A. and Vesely, I. (2005) Evaluation of the matrix-synthesis potential of crosslinked hyaluronan gels for tissue engineering of aortic heart valves. *Biomaterials* **26**, 999–1010.
75. Mol, A., Driessen, N. J., Rutten, M. C., Hoerstrup, S. P., Bouten, C. V., and Baaijens, F. P. (2005) Tissue engineering of human heart valve leaflets: a novel bioreactor for a strain-based conditioning approach. *Ann. Biomed. Eng.* **33**, 1778–1788.
76. L'Heureux, N., Dusserre, N., Konig, G., victor, B., Keire, P., Wight, T.N., Chronos, N.A., kyles, A.E., Gregory, C.R., Hoyt, G., Robbins, R.C., and McAllistu T.N. (2006) Human tissue-engineered blood vessels for adult arterial revascularization. *Nat.Med.* (**12**), 361–365.
77. Isenberg, B. C., Williams, C., and Tranquillo, R. T. (2006) Small-diameter artificial arteries engineered in vitro. *Circ. Res.* **98**, 25–35.

16

Practical Aspects of Cardiac Tissue Engineering With Electrical Stimulation

Christopher Cannizzaro, Nina Tandon, Elisa Figallo, Hyoungshin Park, Sharon Gerecht, Milica Radisic, Nicola Elvassore, and Gordana Vunjak-Novakovic

Summary

Heart disease is a leading cause of death in western society. Despite the success of heart transplantation, a chronic shortage of donor organs, along with the associated immunological complications of this approach, demands that alternative treatments be found. One such option is to repair, rather than replace, the heart with engineered cardiac tissue. Multiple studies have shown that to attain functional tissue, assembly signaling cues must be recapitulated in vitro. In their native environment, cardiomyocytes are directed to beat in synchrony by propagation of pacing current through the tissue. Recently, we have shown that electrical stimulation directs neonatal cardiomyocytes to assemble into native-like tissue in vitro. This chapter provides detailed methods we have employed in taking this "biomimetic" approach. After an initial discussion on how electric field stimulation can influence cell behavior, we examine the practical aspects of cardiac tissue engineering with electrical stimulation, such as electrode selection and cell seeding protocols, and conclude with what we feel are the remaining challenges to be overcome.

Key Words: Cardiac tissue engineering; Cardiomyocytes; Bioreactors; Biomaterial scaffolds; Electrical stimulation; Electrodes; Electrochemical impedance spectroscopy.

1. Introduction

The overall goal of cardiac tissue engineering is to direct the cells to (re)establish the structure and function of the native tissue being repaired over clinically relevant thicknesses, approximately 1 cm for adult human heart *(1)*. The utility of engineered cardiac grafts depends on cell survival, integration, functionality, and electrical coupling *(2)*. To engineer functional

From: *Methods in Molecular Medicine, 2nd ed.: Tissue Engineering*
Edited by: H. Hauser and M. Fussenegger © Humana Press Inc., Totowa, NJ

myocardium, we used a "biomimetic" approach that involves the cultivation of cardiac cell populations on scaffolds (designed to provide a structural and logistic template for tissue formation), using bioreactors (designed to provide environmental control and biophysical signaling). This approach involves three related aspects: (a) establishment and maintenance of physiologic density of viable cells, (b) convective–diffusive oxygen supply to the cells using channeled scaffolds (to mimic the capillary flow) and oxygen carriers in culture medium (to mimic the role of hemoglobin), and (c) electrical field stimulation (to induce excitation–contraction cell coupling). In this chapter, we focus on the electrical stimulation aspect of cardiac tissue engineering using established methods in our laboratory (*see* **note 1**).

Native heart tissue has low-resistance pathways for electrical signal propagation due to the presence of gap junctions and high cell density. Individual cells are packed tightly together and held in place by tight junctions, such that the myocardium acts as a synctium *(3–5)*. To induce synchronous contractions of cultured cardiac constructs, we apply electrical signals designed to mimic those orchestrating the synchronous contractions of cells in native heart. Over only 8 days in vitro, electrical field stimulation can induce cell alignment and coupling, increase the amplitude of synchronous construct contractions, and result in a remarkable level of ultrastructural organization. Electrical stimulation can promote cell differentiation and coupling, as evidenced by the presence of striations and gap junctions, and it results in concurrent development of conductive and contractile properties of cardiac constructs (*see* **Fig. 1**) *(6)*.

2. Materials

2.1. Bioreactor

1. 60-mm glass Petri dish (VWR, West Chester, PA).
2. 6-well polystyrene plate (NalgeNunc International, Rochester, NY).
3. 3.0-mm biopsy punch (MedExSupply, Monsey, NY).
4. Poly-dimethylsiloxane (PDMS; Sylgard 184, Dow Corning, MI).
5. Stainless steel push pins (size 000, Fine Science Tools, North Vancouver, Canada).
6. Electrodes, all in the form of a 1/8-inch rod:
   i. Carbon (12 inch × 0.120 inch; Ladd Research, Williston, NY).
   ii. Stainless steel, type 303 (McMaster-Carr, Atlanta, GA).
   iii. Titanium, grade 2 (McMaster-Carr).
   iv. Titanium nitride (TiN)-coated titanium, grade 2 (Eclat Industries, Levittown, NY).
7. Platinum wire (0.005 inch in diameter; Ladd Research).

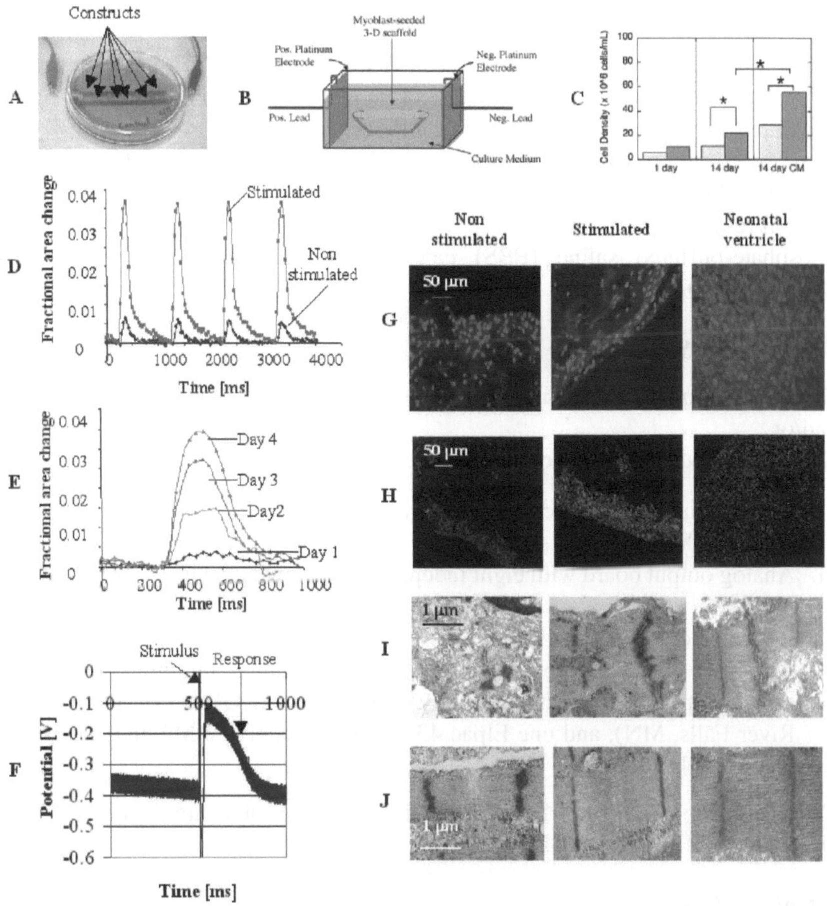

Fig. 1. Electrical stimulation promoted myocyte differentiation and function. (a) Experimental setup for supra-threshold stimulation of cardiac myocytes cultured on an elastic, porous scaffold. A Petri dish is fitted with two stimulating electrodes that generate electrical field gradients with a supra-threshold amplitude; results are shown in d–j. (b) Experimental setup for stimulation of endothelial cells on a polymer scaffold. A chamber is fitted with two platinum electrodes generating electrical field stimuli with a sub-threshold amplitude; results are shown in c. (c) Enhanced cell proliferation. ∗ denotes statistically significant difference (P 0.05; Jukey's post hoc test with one-way ANOVA, n 5–10) samples per group and time point). (d) Contraction amplitude was seven times higher in stimulated than in non-stimulated constructs. (e) Contraction amplitude progressively increased with time of stimulation. (f) Contractile activity was associated with the propagation of action potential at cell membranes. Expression of (alpha) and (beta) isoforms of myosin heavy chain, MHC, (g) and the gap-junctional

2.2. Electrical Characterization of Bioreactor

1. Electrochemical impedance spectroscopy (EIS): Solartron 1287 and Solartron 1250 FRA (Solartron Analytical, Oak Ridge, TN) interfaced with computer using ZPlot software (Scribner Associates, Southern Pines, NC).
2. Digital oscilloscope with storage capability: TDS3000B (Techtronix, Richardson, TX).
3. Alligator clips and test leads (Mouser, Mansfield, TX).
4. 10 ohm resistors (Mouser).
5. Phosphate-buffered saline (PBS) calcium and magnesium free (Gibco/BRL, Bethesda, MD).

2.3. Electrical Stimulation

1. Dual-output square pulse stimulator (Grass S88X; Astro-Med, West Warwick, RI) and/or
2. Custom computer-controlled stimulator consisting of the following:

   i. Data acquisition and control software: Labview 7.1 (National Instruments, Austin, TX).
   ii. Analog output board with eight independent channels: NI 6713 (National Instruments).
   iii. I/O connector block and cable: SCB 68 and R6868 (National Instruments).
   iv. Amplifier board and power supply: for each channel, one high-current operational amplifier connected in unity-gain mode (TI OPA-551; Digikey, Thief River Falls, MN), and one Elpac 43 W power supply (WM220-1; Digikey) for all channels.

3. Flat ribbon cable (16 wires, 26 AWG) and IDC ribbon cable interconnects (Mouser).

2.4. Cell Culture

1. Hanks' balanced salt solution (HBSS) and PBS with 1 M HEPES buffer (Gibco/BRL).
2. Trypsin (U.S. Biochemicals, Cleveland, OH).
3. Collagenase type II (Worthington, Freehold, NJ).
4. Matrigel® (Becton Dickinson, Bedford, MA).
5. Ultrafoam hemostat collagen sponge (Davol, Cranston, RI).

Fig. 1. protein connexin 43, Cx43, (h), were markedly higher in stimulated group. Stimulated constructs had well-developed myofibrils with parallel sarcomeres and intercalated discs placed symmetrically between the Z lines. (i) Intercalated discs. (j) Sarcomeres.

6. Cardiac growth medium: Dulbecco's modified Eagle's medium (DMEM), 10% fetal bovine serum (FBS), 1% HEPES, and 1% penicillin (10,000 U/ml)/streptomycin (10, 000 μg/ml) (all from Gibco/BRL).

2.5. Histological Analysis

1. 10% buffered formalin (Sigma Diagnostic, St. Louis, MO).
2. Horse serum (Vector Laboratories, Burlingame, CA).
3. Tween 20 (Sigma, St. Louis, MO).
4. Avidin–biotin complex agent (Sigma).
5. 3,39'-Diaminobenzidine (Sigma).
6. Mouse anti-cardiac troponin I (TnI; Biodesign, Saco, ME), 1:150 dilution in PBS containing 0.5% Tween 20 and 1.5% horse serum.
7. Rabbit anti-connexin 43 (Cx43; Chemicon International, Temecula, CA), 1:150 dilution in PBS containing 0.5% Tween 20 and 1.5% horse serum.
8. Horse anti-mouse IgG, Standard Elite ABC kit (Vector Laboratories), 1:200 dilution in PBS containing 0.5% Tween 20 and 1.5% horse serum.
9. Fluorescein-conjugated goat anti-rabbit IgG (Sigma), 1:200 dilution in PBS containing 0.5% Tween 20 and 1.5% horse serum.

2.6. Ultrastructure Analysis

1. Karnovsky's reagent: 0.1 M sodium cacodylate with 2% paraformaldehyde and 2.5% glutaraldehyde, pH 7.4.
2. Epoxy embedding kit (Epon 812; SPI Supplies, West Chester, PA).
3. 1% osmium tetroxide in veronal-acetate buffer.
4. Graded ethanol in propylene oxide.
5. Lead citrate.
6. Uranyl acetate.

3. Methods

3.1. Bioreactor Assembly

To assess cell alignment after stimulation, it is desirable to maintain a constant alignment of scaffolds with respect to the direction of the electric field gradient while not restricting the contractions of the tissue construct or the ability to observe the constructs with a microscope. We accomplish this task using stainless steel pins, held in place by a thin layer of PDMS at the bottom of a Petri dish (*see* **Fig. 2**). Below is protocol for bioreactor assembly, whereas scaffold positioning is covered in **section 3.4**.

1. Prepare 10 g 10:1 mixture of PDMS and initiator. Pour 7 g mixture into a 60-mm glass Petri dish and de-gas under vacuum for 1 h. Cure in oven at 65°C for 2 h.

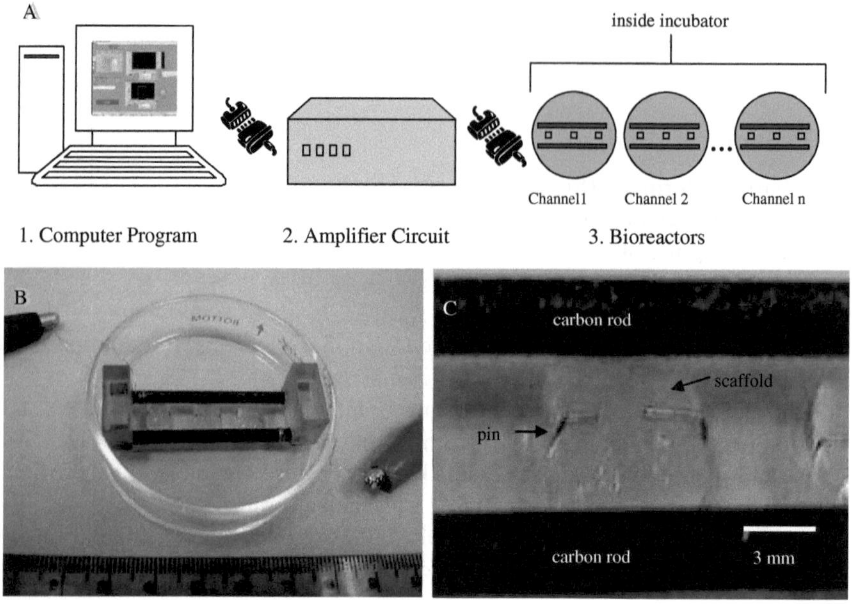

Fig. 2. Experimental setup for supra-threshold stimulation of cardiac myocytes. (a) Electrical stimulation voltages are set using a computer program, output through eight-channel AO card, amplified, and interfaced with bioreactors. (b) A 60-mm Petri dish with carbon electrodes spaced 1 cm apart. (c) Close-up view of scaffold positioned between electrodes and held in place with two stainless steel pins.

The final thickness of PDMS is 0.3 cm, thick enough to hold pins in place but thin enough to observe scaffolds with an inverted microscope.

2. Prepare an additional 60 g 10:1 mixture of PDMS and initiator, add 10 g to each well of a 6-well plate, and cure. The final thickness of PDMS in each well will be 1.0 cm.
3. Remove PDMS discs (3.5 cm in diameter) from well and cut in half. Use a 3.0 mm biopsy punch to make two holes along linear surface, 1.0,mm from bottom edge and spaced 1.0 cm apart.
4. Cut electrode rods to length of 5.0 cm.
5. Wrap wire multiple times around the ends of each rod. For carbon rods, drill a small hole at the end of the rod and insert wire before wrapping. Leave an additional 10 cm wire attached to each electrode.
6. Insert two rods into punched holes of one PDMS half-disc. On opposing end, insert rods into a second half-disc. Length of electrode exposed to electrolyte is 4.0 cm.
7. Place electrode/PDMS block into the center of the Petri dish with PDMS base. Electrode supports may need to be trimmed to fit.

3.2. Electrical Characterization of Bioreactor

Stimulation efficiency is determined by the ability to attain a desired physiological response with minimal damage to the surrounding tissue. For each application, electrical stimulation conditions should be optimized by not only choosing appropriate electrode geometry but also studying electrode material properties and charge-transfer characteristics at the electrode–electrolyte interface. Electrodes must be biocompatible to avoid toxic or immune responses in the adjacent tissue or medium, and they should efficiently transfer charge from the electrode material where it is carried by *free electrons* to the medium or tissue where it is carried by *ions*. Charge transfer can occur through three mechanisms: (a) non-faradaic charging/discharging of the electrochemical double layer, (b) reversible faradaic reactions, and (c) non-reversible faradaic reactions. The first two mechanisms are desirable, whereas the last should be avoided because it is associated with electrode degradation and harmful byproducts. The relative presence of each mechanism can be assessed using EIS, from which an equivalent circuit of the stimulation system can be constructed.

3.2.1. EIS Measurements of Bioreactor

In our laboratory, EIS measurements are taken with an electrochemical interface (Solartron 1287) and a frequency response analyzer (FRA; Solartron 1250) controlled by a computer with ZPlot software. Equivalent circuits and associated parameters are determined using ZView 2.5b. Please see *(7–9)* for additional information on EIS measurement of common electrode materials.

1. Take EIS measurements of electrodes in a Petri dish with 20 ml PBS (*see* **note 2**). Acquire EIS spectra over a frequency range from 1×10^6 to 1×10^{-2} Hz with a perturbation amplitude of 10 mV.
2. Record for each frequency the real (resistive) and imaginary (capacitive) components of the impedance response, Z' and Z'', respectively.
3. Evaluate data in ZView to generate Nyquist and Bode plots for each condition (*see* **note 3**).
4. Create an "equivalent circuit" of the system using resistors and capacitors in series and in parallel (*see* **note 4**). Calculate the values for constant-phase element (CPE), polarization resistance (R_p), and η using instant fit functions in ZView software. An example of EIS analysis of common electrode materials (carbon, stainless steel, titanium, and TiN) is shown in **Fig. 3**, and extended studies on carbon are shown in **Fig. 4**.

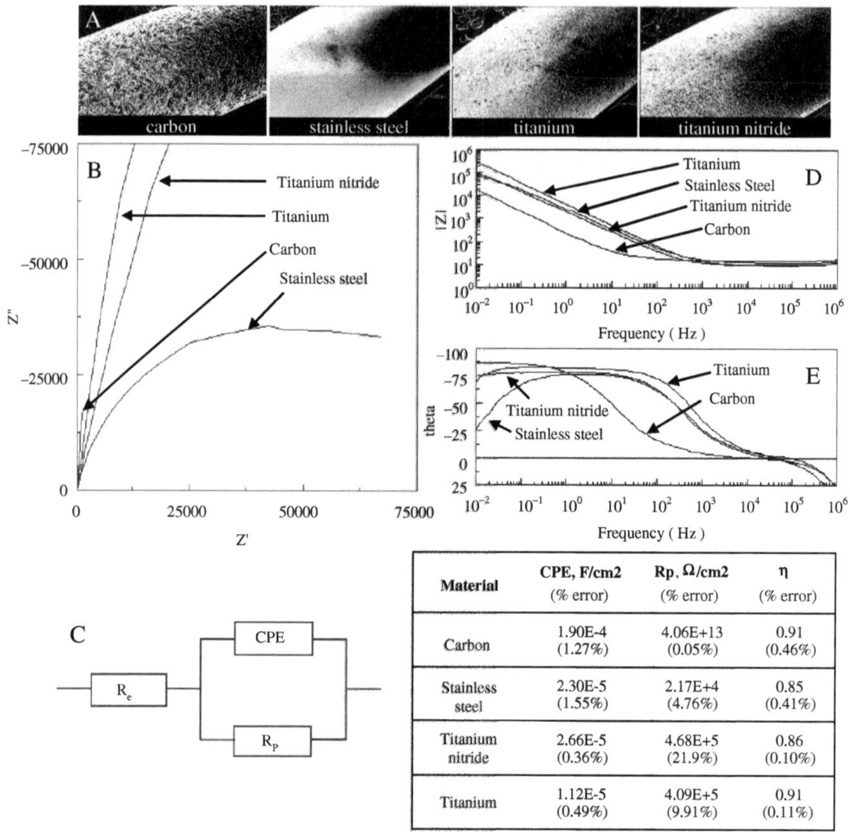

Material	CPE, F/cm2 (% error)	Rp, Ω/cm2 (% error)	η (% error)
Carbon	1.90E-4 (1.27%)	4.06E+13 (0.05%)	0.91 (0.46%)
Stainless steel	2.30E-5 (1.55%)	2.17E+4 (4.76%)	0.85 (0.41%)
Titanium nitride	2.66E-5 (0.36%)	4.68E+5 (21.9%)	0.86 (0.10%)
Titanium	1.12E-5 (0.49%)	4.09E+5 (9.91%)	0.91 (0.11%)

Fig. 3. Electrical characterization of stimulation electrodes with electrochemical impedance spectroscopy (EIS). ×30 scanning electron microscopy (SEM) images of electrodes, 1/8 inch in diameter (a). White bar corresponds to 500 μm. The semicircular shape of stainless steel in the Nyquist plot (b) suggests the presence of reactions, whereas titanium, titanium nitride (TiN), and carbon have progressively more linear profiles associated with high polarization resistance (R_p). When the electrode–electrolyte interface is modeled with an equivalent circuit (c), where R_p is the R_p, CPE a constant-phase element, and R_e electrolyte resistance, the relatively low value of R_p for stainless steel confirms that it is more susceptible to faradaic reactions, and hence corrosion. At the other extreme, the carbon electrodes are best suited for electrical stimulation: their very high R_p value minimizes faradaic reactions and the relatively high CPE value indicates that the electrode transfers more charge to electrolyte, and hence tissue construct. Furthermore, as indicated in the Bode plots (d and e), carbon has the lowest impedance modulus |Z| across all frequencies. Experimental setup was as shown in **Fig. 2** with 4 cm electrodes and 10 mV perturbation.

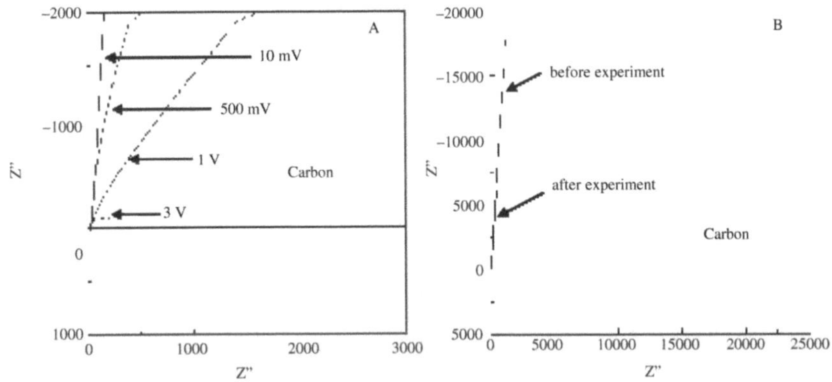

Fig. 4. (a) Nyquist plot of carbon electrodes for increasing perturbation voltage. As potential moves further from equilibrium, the plots become curved and faradaic reactions become of increasing importance. (b) Nyquist plot of carbon electrodes measured before an experiment and after 5 days of electrical stimulation (5 V, 2 ms, 1 Hz). Although at low frequencies the aged electrode had decreased impedance, its resistance to reactions remained unchanged, that is, R_p values were equal. Experimental setup was as shown in **Fig. 2** with 4 cm electrodes.

3.2.2. Current Measurement in the Bioreactor

The amount of injected charge is calculated simply by measuring the potential drop across a resistor placed in series in the stimulation loop.

1. Place a 10-ohm resistor in series with the bioreactor in the stimulation loop. The value of this resistance is high enough to allow measurement of a voltage across the electrode, even with small current.
2. Apply the stimulus (e.g., 5 V, 2 ms square wave). Check with oscilloscope that intended waveform is faithfully applied to bioreactor. If not, current limit of stimulator may be exceeded and corrective action must be taken (e.g., electrode area or medium conductivity reduced or current rating of stimulator increased).
3. Record over time the voltage across the resistor with the oscilloscope, and for each time point, calculate the value of current by dividing the voltage across the resistor by the resistance value (Ohm's law).
4. Calculate the total amount of injected charge by integrating the current profile for the duration of the stimulus. Current profiles and calculated injected charge for electrodes of different materials (carbon, stainless steel, titanium, and TiN) are shown in **Fig. 5**.

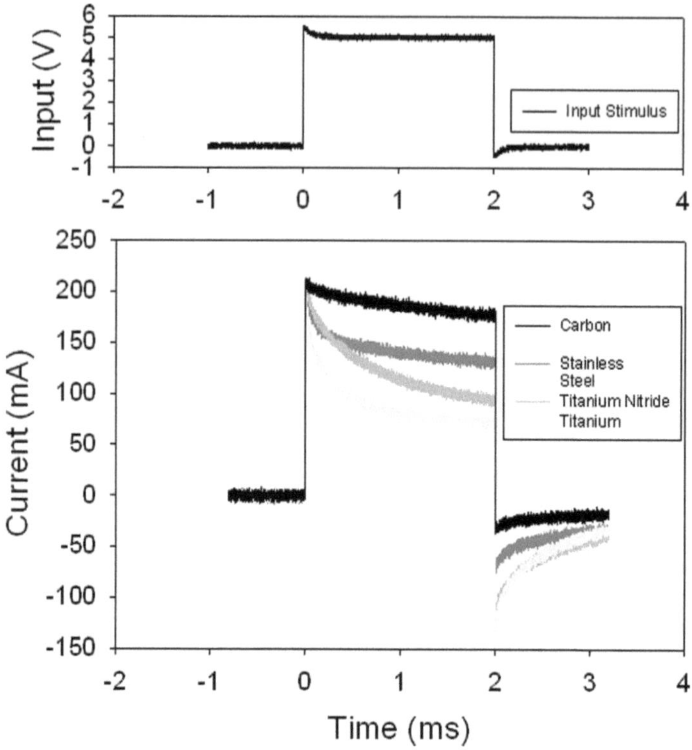

Fig. 5. Current measured over 5 V, 2 ms square wave pulse. The total amount of injected charge in coulombs is equal to area beneath each curve or 2.3×10^{-4} C for carbon, 1.79×10^{-4} C for stainless steel, 1.63×10^{-4} C for titanium nitride (TiN), and 1.20×10^{-4} C for titanium. Experimental setup was as shown in **Fig. 2** with 4 cm electrodes.

3.2.3. Choice of the Electrode Material

The design considerations required for electrical stimulation of engineered cardiac tissue are many and interdependent. Factors to consider include the duration and shape of the stimulus waveform, the size of the tissue construct in question, the electrical property being exploited (e.g., generation of reactive oxygen species versus eliciting action potential), the duration of the experiment, and the mechanical properties required of the electrode in the bioreactor setup. For our work, we started with a desired stimulus waveform, examined the electrical characteristics of that stimulus, and based subsequent design decisions with this in mind.

1. Calculate the power spectral density of the stimulus waveform to determine the frequency band that concentrates most of the energy in the pulse. For 5 V, 2 ms square pulses delivered at 1 Hz, it is the frequency band below 1 kHz (*see* **note 5** and **Fig. 6**).
2. Analyze the Bode plot to determine the behavior of the electrode material within this frequency band. For example, the carbon electrodes we use have a corner frequency of approximately 10 Hz, and so the electrode behaves both as a capacitor below this frequency and as a resistor above this frequency. Therefore, we must consider both the electrode's CPE and R_p values (*see* **Fig. 3**).
3. Compare CPE, R_p, and injected charge values calculated from EIS for different electrodes. Choose electrode material with high CPE to increase charge injection and high R_p to reduce harmful reactions.

3.3. Cell Line and Culture

The following protocols are for neonatal rat cardiomyocytes, which are readily available at most institutions. However, the general methods of electrical stimulation described here are applicable to all electrically excitable cells.

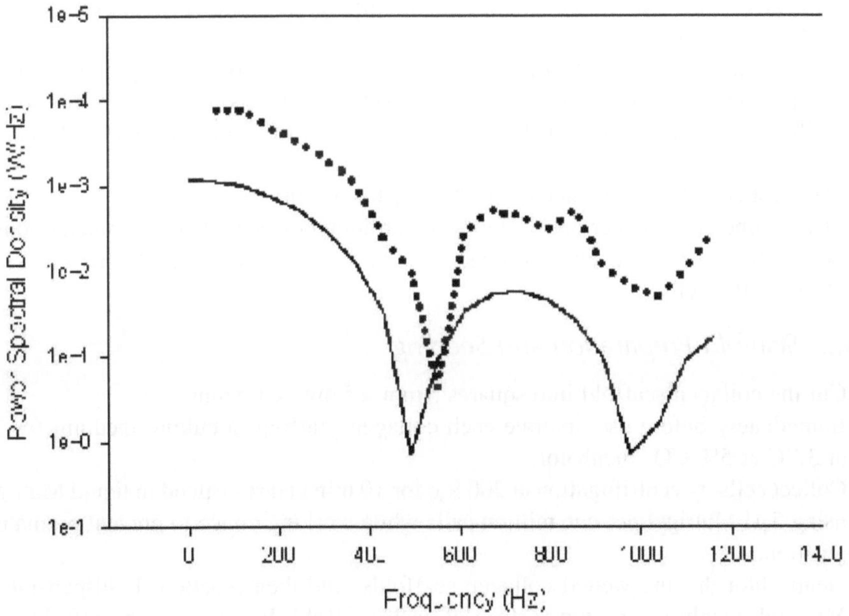

Fig. 6. Power spectral density for 5 V, 2 ms square wave pulse. Solid line is theoretical spectrum (*see* **note 4**), and dotted line is measured spectrum.

3.3.1. Rat Cardiomyocyte Isolation

Heart cells are obtained from 2-day-old neonatal Sprague–Dawley rats, using a protocol approved by our Committee on Animal Care [see *(6)* for additional details].

1. Excise a whole heart from the rat and place it in cold HBSS buffer.
2. Remove the left and right atria and blood vessels by using small scissors and forceps.
3. Quarter remaining heart ventricles and wash in HBSS buffer.
4. Digest quartered hearts in 25 ml 0.06% (w/v) trypsin in HBSS buffer by shaking overnight at 4°C.
5. Stop trypsin digestion by adding 10 ml cardiac growth medium followed by incubation at 37°C for 4 min with shaking at 150 rpm.
6. After discarding supernatant, add 10 ml 0.1% (w/v) type II collagenase in HBSS to the tissue and incubate at 37°C for 4 min with shaking at 150 rpm.
7. Collect the cell suspension from the digestion and wash the pellet with 10 ml HBSS buffer and collect the supernatant.
8. Repeat steps 6 and 7 until no more tissue remains.
9. Centrifuge at $150 \times g$ for 10 min the pooled cell suspension obtained in step 8 and then wash the cell pellet with 25 ml cell culture medium to remove residual collagenase (*see* **note 6**).
10. Resuspend the cells in the cell-culture medium and pre-plate for 1 h to enrich cell suspension with cardiomyocytes by removing fibroblasts. In this step, fibroblasts will adhere to the culture dish within 1 h, but most of cardiomyocytes will remain in suspension.
11. Harvest cells by centrifugation at $200 \times g$ for 5 min.
12. Determine cell number and viability of cardiomyocytes by hemocytometer counts using trypan blue to exclude dead cells. Twenty neonatal rat hearts will yield $8–12 \times 10^7$ cells.

3.3.2. Scaffold Preparation and Seeding

1. Cut the collagen scaffold into squares 5 mm × 5 mm × 1.5 mm.
2. Immediately before use, hydrate each collagen scaffold in culture medium for 2 h at 37°C at 5% CO_2 incubator.
3. Collect cells by centrifugation at $200 \times g$ for 10 min and resuspend in liquid Matrigel using 5 μl Matrigel per one million cells while working on ice to prevent premature gelation.
4. Gently blot dry pre-wetted collagen scaffolds, and then pipette cell suspension in Matrigel evenly on the top surface of each scaffold. Inoculate each scaffold with freshly isolated heart cells at a density 1.35×10^8 cells/cm^3.
5. Gelation is complete within 15 min in a 37°C incubator, and inoculated scaffolds are then transferred into a Petri dish.

3.4. Electrical Stimulation of Constructs

The objective of electrical stimulation is to deliver enough current to cells to depolarize membrane and elicit an action potential. In addition, there is some evidence that electrical stimulation aids in cardiomyocyte alignment. For scaffolds of cardiomyocytes seeded with Matrigel, we have previously determined that the optimal time to begin electric field stimulation is 3 days after scaffolds are seeded and to assess contractile function after 8 days *(6)*.

1. Place three scaffolds between electrodes in a 60-mm Petri dish as described in **section 3.1**. Fix scaffold position with two stainless steel pins bent over scaffold and into PDMS base layer (*see* **Fig. 2**). These pins should not touch each other or the electrodes to minimally interfere with the electric field.
2. Add 15 ml cardiac medium to the Petri dish and place in incubator (37°C, 5% CO_2).
3. Make electrical connections with an effort to prevent any undesired electrical connections via metal or electrolyte. Place a layer of autoclave paper beneath all bioreactors to prevent unintentional electrical connections via moisture in the bioreactor or via metal parts within the incubator.
4. Stimulate with 5 V, 2 ms square pulses delivered at 1 Hz (*see* **note 7**). Monophasic pulses were chosen for their simplicity and compatibility with carbon electrodes (*see* **note 8**).
5. Start electrical stimulation 3 days after seeding scaffolds with isolated heart cells.
6. Stop electrical stimulation 8 days after seeding scaffolds (5 days of electrical stimulation).

3.5. Scaffold Characterization

3.5.1. Contractile Activity

Contractile function of engineered cardiac constructs is evaluated online (during culture) by measuring contractile activity in response to electrical field stimulation.

1. Place each construct between two carbon electrodes connected to a cardiac stimulator (either Grass S88X or custom system) in a 60-mm Petri dish filled with 15 ml cardiac growth medium.
2. Maintain the temperature of the Petri dish at 37°C using heating tape fixed at the bottom of the Petri dish and connected to a temperature controller (*see* **note 9**).
3. Place the entire setup on an optical microscope and monitor contractile responses to electrical stimuli (rectangular pulses, 2 ms duration) using ×2 magnification.
4. Increase signal amplitude in 0.1 V increments up to 10 V and the stimulation frequency up to 400 beats per minute (bpm).
5. For each combination of voltage and frequency, look for the presence of random, spontaneous contractions (an indicator of immature tissue) versus synchronous

contractions in response to electrical pacing (an indicator of more mature tissue containing electromechanically coupled cells).

6. Measure two parameters to evaluate the contractile behavior in response to electrical stimulation: *excitation threshold*, ET (the minimum voltage of electrical stimulation required to elicit sustained synchronous contractions of tissue constructs at a frequency of 60 bpm), and *maximum capture rate*, MCR (the maximum frequency of sustained synchronous contractions that can be achieved at a stimulation voltage corresponding to 1.5 ET).

3.5.2. Histological Analysis

Fix cell-scaffold constructs in 10% buffered formalin for 24 h, dehydrate, embed in paraffin, bisect in cross section through the center, and section to 5-μm thickness. Sections are stained with hematoxylin and eosin for general evaluation and stained with cardiac-specific antibodies to assess the distribution of cardiac markers. Use a humidified chamber for all incubation steps. Neonatal rat heart and bovine tendons serve as positive and negative controls, respectively.

1. Deparaffinize sections to retrieve antigen by heat treatment for 20 min at 95°C in a decloacking chamber (Biocare Medical, Concord, CA).
2. Subsequently, block sections with 10% horse serum for 30 min at room temperature in a humidified chamber.
3. Incubate the sections for 1 h at 37°C with mouse anti-cardiac TnI and rabbit anti-Cx43 diluted in PBS containing 0.5% Tween 20 and 1.5% horse serum.
4. For TnI, incubate sections at room temperature, first for 30 min with the secondary antibody (horse anti-mouse IgG), then for 30 min with an avidin–biotin complex agent for 30 min, and finally for 15 min with 3,39'-diaminobenzidine (Sigma).
5. For Cx43, use fluorescein-conjugated goat anti-rabbit IgG (1:200).
6. Assess construct architecture and cell distribution from the stained sections using fluorescent microscope (Axioplan, Zeiss, Thornwood, NY).

3.5.3. Ultrastructure Analysis

Transmission electron microscopy allows observation of gap junction and Z-band formation.

1. Fix constructs in Karnovsky's reagent (0.1 M sodium cacodylate with 2% paraformaldehyde and 2.5% glutaraldehyde, pH 7.4), post-fix in 1% osmium tetroxide in veronal-acetate buffer, dehydrate in graded ethanol in propylene oxide, and embed in Epon 812 (Polysciences).
2. Cut thin sections (70 nm) using a Leica Ultra Cut and a diamond knife.
3. Stain sections with lead citrate and uranyl acetate.

4. Examine sections for ultrastructural properties relevant to cardiac tissue (volume fraction and developmental stage of sarcomeres; *Z* and *M* lines; H, I, and A bands; gap junctions; and T tubules) using a Philips EM410 transmission electron microscope operated at 80 kV (JEOL-100CX, JEOL).
5. Morphometric analysis was performed on 20–46 randomly taken transmission electron micrographs at magnification of 100,000. A test grid with uniform squares (0.26 cm × 0.26 cm) was superimposed onto the micrographs, and the area covered by sarcomeres, mitochondria, and nuclei, and the overall area of the cells were determined. The volume fraction of each organelle was determined as previously described ***(10)***. The frequency of intercalated discs and gap junctions (number per μm^2) was determined by counting. A total of 20 micrographs of non-stimulated constructs, 46 micrographs of stimulated constructs, and 42 micrographs of neonatal ventricles were evaluated with respect to each structural parameter by two independent observers.

Notes

1. Looking forward:
   i. Better analytical tools required such as closed-loop feedback systems for capture rate. Off-the-shelf pacemakers may be useful in this regard.
   ii. High-throughput approaches are needed to determine optimal stimulation parameters.
   iii. More advanced bioreactors that integrate electrical stimulation with medium perfusion are needed to engineer tissue of sufficient thickness for clinical applications.
2. Measurements are made in PBS to facilitate comparison with EIS data in the literature and from our own laboratory. The conductivity of PBS (15 mS/cm) is similar to that of culture medium with 10% FBS (~14 mS/cm).
3. In a Nyquist plot, the imaginary component of the impedance Z'' is plotted against the real part Z' at each frequency. A Bode plot gives the logarithm of impedance, $|Z|$, and the phase angle versus the logarithm of frequency.
4. The electrode–electrolyte system can be described by a Randles cell that consists of electrolyte resistance R_e, CPE, and charge transfer or polarization resistance R_p. The CPE is a mathematical description of the double-layer capacitance and accounts for non-ideal capacitive behavior of the electrochemical double layer with the non-dimensional η term *(8)*.
5. The power spectral density is a term for the distribution of signal power in the frequency domain. In general, the analytical solution for the power spectral density of a signal is calculated by squaring the magnitude of the continuous Fourier transform of the signal. More practically, power spectral density may be calculated using software including MATLAB® or SigmaPlot® or in real time with an oscilloscope. For **Fig. 6**, SigmaPlot® software was utilized.

6. It is important that collagenase removal by washing is complete to prevent degradation of collagen scaffold during cell seeding.
7. Electric field stimulation may be delivered from a commercially available stimulator or through custom-designed hardware controlled by a computer. Although computer-controlled stimulation allows additional flexibility, it requires some expertise in circuit design and software programming.
8. Waveform selection (e.g., among monophasic, biphasic, and other wave shapes) depends on the electrode material characteristics, as biphasic pulses can aid in reversing reversible faradaic reactions but at the same time may be undesirable because biphasic pulses may inhibit action potential.
9. In a very nice microelectrode array (MEA) study on cardiomyocyte activity, Giovangrandi et al. ***(11)*** clearly showed how cardiomyocyte beat rate closely follows imposed temperature profile (beat rate increases as temperature increases and vice versa). In an earlier study by the same group ***(12)***, medium pH and osmolarity were also considered.

Acknowledgments

The authors thank NIH for financial support of this work (P41 EB002520-01, R01 HL076485-01) and Eclat Industries for supplying TiN-coated electrodes.

References

1. Zammaretti, P. and Jaconi, M. (2004) Cardiac tissue engineering: regeneration of the wounded heart. *Curr. Opin. Biotechnol.* **15,** 430–434.
2. Zimmermann, W. H., Didie, M., Wasmeier, G. H., Nixdorff, U., Hess, A., Melnychenko, I., Boy, O., Neuhuber, W. L., Weyand, M., and Eschenhagen, T. (2002) Cardiac grafting of engineered heart tissue in syngenic rats. *Circulation* **106,** I151–I157.
3. Merrill, D. R., Bikson, M., and Jefferys, J. G. R. (2005) Electrical stimulation of excitable tissue: design of efficacious and safe protocols. *J. Neurosci. Methods* **141,** 171–198.
4. Basser, P. J. and Roth, B. J. (2000) New currents in electrical stimulation of excitable tissues. *Annu. Rev. Biomed. Eng.* **2,** 377–397.
5. Malmivuo, J. and Plonsey, R. (1995) *Bioelectromagnetism – Principles and Applications of Bioelectric and Biomagnetic Fields*, Oxford University Press, New York.
6. Radisic, M., Park, H., Shing, H., Consi, T., Schoen, F. J., Langer, R., Freed, L. E., and Vunjak-Novakovic, G. (2004) Functional assembly of engineered myocardium by electrical stimulation of cardiac myocytes cultured on scaffolds. *Proc. Natl. Acad. Sci. U. S. A.* **101,** 18129–18134.
7. Norlin, A., Pan, J., and Leygraf, C. (2004) Investigation of Pt, Ti, TiN, and nanoporous carbon electrodes for implantable cardioverter-defibrillator applications. *Electrochim. Acta* **49,** 4011–4020.

8. Norlin, A., Pan, J., and Leygraf, C. (2005) Investigation of electrochemical behavior of stimulation/sensing materials for pacemaker electrode applications I. Pt, Ti, and TiN coated electrodes. *J. Electrochem. Soc.* **152,** J7–J15.
9. Norlin, A., Pan, J., and Leygraf, C. (2005) Electrochemical behavior of stimulation/sensing materials for pacemaker electrode applications III. Nanoporous and smooth carbon electrodes. *J. Electrochem. Soc.* **152,** J110–J116.
10. Nuccitelli, R. (1992) Endogenous ionic currents and DC electric fields in multicellular animal tissues. *Bioelectromagnetics* **Suppl. 1,** 147–157.
11. Giovangrandi, L., Gilchrist, K. H., Whittington, R. H., and Kovacs, G. T. A. (2006) Low-cost microelectrode array with integrated heater for extracellular recording of cardiomyocyte cultures using commercial flexible printed circuit technology. *Sens. Actuat. B Chem.* **113,** 545–554.
12. Gilchrist, K. H., Giovangrandi, L., Whittington, R. H., and Kovacs, G. T. A. (2005) Sensitivity of cell-based biosensors to environmental variables. *Biosens. Bioelectron.* **20,** 1397–1406.

17

Biological Scaffolds for Heart Valve Tissue Engineering

Artur Lichtenberg, Serghei Cebotari, Igor Tudorache, Andres Hilfiker, and Axel Haverich

Summary

Tissue engineering might bear promising solutions for overcoming the limitations of biological and mechanical heart valve substitutes. The concept of heart valve tissue engineering bases on decellularized biological matrices, as removal of cellular components might reduce immunological reactions, which are thought to be responsible for accelerated valvular graft deterioration and their subsequent repopulation with autologous cells, which leads to tissue integrity and continuous remodeling. Here, we report a method for efficient removal of cells from ovine heart valve tissue (including cusp, wall, and myocardial cuff) by detergents for the generation of biological valvular scaffolds.

Key Words: Cardiac surgery; Pulmonary valve; Heart valve substitutes; Tissue engineering; Extracellular matrix.

1. Introduction

Despite ongoing efforts to improve longevity of biological heart valves, the long-term results of xenogenic grafts remain unsatisfactory *(1)*. Furthermore, cryopreserved allogenic valvular graft conduits did not fulfill the expectations in regard to long-term stability. Although cellular integrity and function as well as careful preservation of matrix components seem to be important determinants for long-term function of allograft heart valves, it has been shown that endothelial cells express major histocompatibility complex (MHC) class I and II molecules, representing a potential immunogenic surface and stimulate a donor-specific immune response that can cause the degeneration of the implanted valve *(2)*. Owing to high immunological competence in children,

From: *Methods in Molecular Medicine, 2nd ed.: Tissue Engineering*
Edited by: H. Hauser and M. Fussenegger © Humana Press Inc., Totowa, NJ

early allograft valve failure often occurs in the pediatric age population *(3)*. The concept of tissue engineering is characterized by two fundamental steps. First, decellularization of matrices because of the removal of cellular components might reduce immunological reactions, which are thought to be responsible for accelerated graft deterioration *(4)*. Second, in vitro and in vivo repopulation of biological scaffolds with autologous cells for maintenance of continuous remodeling and growth of the implanted valvular prosthesis *(5)*. Thus, the turnover into autogenous tissue might result in long-term durability and the ability to grow, and therefore, tissue engineering might bear promising solutions for overcoming the limitations of biological heart valve substitutes *(5–7)*. Here, we describe our method of efficient cell removal from the valvular conduit tissue (valvular cusp, wall, and myocardial cuff) that allows optimal preservation of extracellular matrix components. This method includes the usage of a mixture of two detergents: sodium deoxycholate and sodium dodecyl sulfate (SDS) *(8,9)*. The efficiency of cell removal from the tissue is controlled by standard hematoxylin & eosin (H&E) staining and DNA assay. Optimal preservation of scaffold components is investigated by revealing specific extracellular matrix proteins, in particular collagen and elastic fibers (H&E stain, Elastica van Gieson stain, and immunohistological stain for collagen-I). Preservation of the basement membrane (BM) can be visualized by scanning electron microscopy (SEM) and staining for specific proteins (immunohistological stain for collagen-IV and immunofluorescence stain for perlecan). The sheep is a widely accepted standard animal model for heart valve durability research *(5)*. Thus, we describe an efficient method for decellularization in ovine heart valves.

2. Materials

2.1. Valve Preparation

1. Earle's medium 199 (PAA Laboratories, storage, 2–8°C).
2. Penicillin–streptomycin (storage, −20°C).
3. Phosphate-buffered solution (PBS) (storage, 2–8°C).
4. Iodine solution, "Braunol" (B. Braun).

2.2. Decellularization of Valve Tissue

1. Deoxycholic acid (Sigma). Store at room temperature. Harmful if swallowed. Irritating to eyes and skin. Do not inhale dust.
2. SDS ultra pure (Carl Roth). Harmful if swallowed. Irritating to eyes and skin. Do not inhale dust.

2.3. Stains

2.3.1. Hematoxylin & Eosin

1. Acetone (J.T. Baker).
2. Hematoxylin (Merck).
3. Eosin (Merck).
4. Ethanol.
5. Xylol.
6. Eukitt (Sigma).

2.3.2. Elastica van Gieson

1. Acidified potassium permanganate (Riedel-de Haen).
2. Oxalic acid 1% (J.T. Baker).
3. Weigerts Resorcin Fuchsin (Merck).
4. Van Gieson (Merck).
5. Eukitt (Sigma).

2.3.3. Immunohistology and Immunofluorescence

1. Collagen-I: Mouse monoclonal immunoglobulin G_1 (IgG_1) antibody (Chemicon) that binds an epitope near the c-terminal of type I collagen, ovine reactive, is used as primary antibody for immunohistology. Store at 2–8°C.
2. Collagen-IV: Mouse monoclonal anti-human IgG_1 antibody (clone CIV 22; Dako) that reacts with ovine collagen IV is used as primary antibody for immunohistology. This antibody is a useful tool for the identification of BMs. For continuous use, store at 2–8°C for up to 1 month. For extended storage, freeze in working aliquots at −20°C.
3. Perlecan: Polyclonal rat IgG_2 antibody anti-Perlecan (clone A7L6; Research Diagnostics) that is used as primary antibody for immunofluorescence. Store at 2–8°C. Perlecan is a major heparan-sulfate proteoglycan (HSPG) within all known BMs.
4. Biotinylated horse anti-mouse IgG (H+L; Vector Laboratories) is used as secondary antibody for immunohistology. Refrigerate for long-term storage, aliquots may be stored frozen.
5. Cy™3-conjugated AffiniPure donkey anti-rat IgG (H+L; Jackson ImmunoResearch) is used as secondary antibody for immunofluorescence. Undiluted, this antibody is stable for several weeks at 2–8°C; however, working dilutions should be used not longer than 1 day.
6. Wash buffer: To make 1000 ml buffer, dilute 876 mg NaCl (58.4 g/mol; Merck) in 588 mg Tris–HCl (157.6 g/mol; Merck) and in 100 ml distilled H_2O each, mix and add distilled H_2O up to 1000 ml.

7. Antibody dilution buffer: 1% bovine serum albumin (BSA) in wash buffer.
8. Acetone.
9. Avidin–biotin complex (ABC) (Immunoperoxidase ABC kit, Vector Laboratories).
10. Diaminobenzidine (DAB, Dako).
11. Harris–Hematoxylin (1:3) (Merck).
12. Ethanol.
13. Xylol.
14. Mounting medium (Dako).

2.4. Scanning Electron Microscopy

1. Glutharaldehyde (2.5%; Polysciences). Stored at 4°C in sealed ampullae is stable for 1 year. Harmful if inhaled, absorbed through skin, or swallowed.
2. About 0.1 M sodium cacodylate buffer at 4°C (Merck).
3. Critical Point Dryer (Balzers).
4. SEM Coating System (Polaron Instruments).

2.5. DNA Assay

1. DNeasy Tissue Kit (Qiagen).
2. Ethanol.

3. Methods

3.1. Valve Tissue

Ovine pulmonary valve conduits can be harvested in local slaughter houses. The grafts, including pulmonary valve cusp and pulmonary wall, are dissected from the heart leaving only a thin subvalvular myocardial margin. For transportation, the grafts are maintained at 4°C in Earle's medium 199 enriched with 100 IU/ml penicillin–streptomycin. For disinfection prior to decellularization, valves are treated with povidone–iodine solution for 10 min and ulterior washed in PBS for 2 h at room temperature.

3.2. Decellularization Process

1. Make detergent solution by dissolving sodium deoxycholate (0.5% w/v) and SDS (0.5% w/v) in sterile distilled water (*see* **notes 1 and 2**).
2. Incubate valves under constant agitation in detergent solution for 24 h at room temperature (*see* **note 3**). About 200 ml detergent solution is needed for one valve conduit. The detergent solution should be changed twice.

3. To remove residual detergents and cell debris, decellularized valves are washed under agitation in PBS (200 ml/valve) supplemented with 100 IU/ml penicillin–streptomycin for 3–4 days by 6–8 wash cycles of 12 h each (*see* **note 4**).

3.3. H&E Staining

The quality of decellularization is tested by an H&E overview stain that allows the detection of residual cellular remnants and shows the matrix architectonic structure (nuclei, blue/black and collagen and cytoplasm, red/pink). Samples to investigate are made from small pieces of subvalvular wall that was paraffin embedded and cut with a microtome into sections of 5-μm thickness. The stain is performed as follows:

1. Fix sections of decellularized valve tissues in acetone at −20°C for 8–10 min.
2. Rehydrate in tap water.
3. Incubate in hematoxylin for 8 min.
4. Wash in tap water for 10 min.
5. Incubate in eosin for 5 min.
6. Rinse in tap water.
7. Dehydrate stepwise: 80% ethanol (2 min); 90% ethanol (2 min); and 100% ethanol (2 min).
8. Differentiate in xylol.
9. Dry overnight, place a drop of Eukitt onto the slide before adding the overslip and visualize in a bright field microscope (*see* **Fig. 1**).

3.4. Elastica van Gieson Staining

Visualize elastic tissue with Elastica van Gieson stain (collagens appear in red and elastic fibers in black color).

1. Rehydrate in tap water.
2. Incubate with acidified potassium permanganate for 2 min.
3. Rinse with tap water.
4. Incubate in 1% oxalic acid for 1 min.
5. Rinse in tap water.
6. Wash in 70% alcohol (optional).
7. Stain in Weigerts Resorcin-Fuchsin at room temperature for 30–45 min.
8. Wash in tap water for 10 min.
9. Samples are stained in van Gieson for 5 min.
10. Wash in water.
11. Differentiate in ethanol (80% for 2 min, 90% for 2 min, and 100% for 2 min).

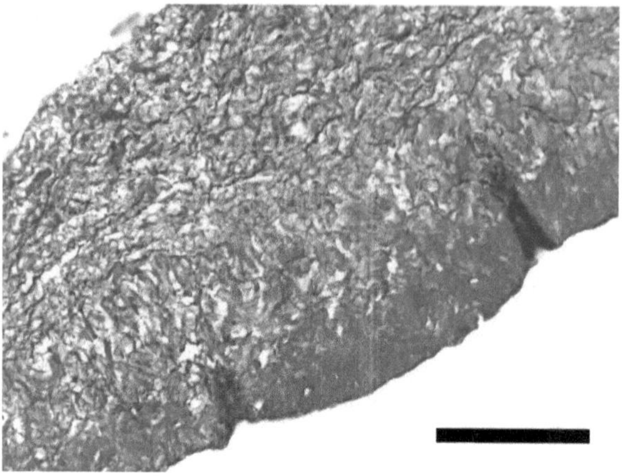

Fig. 1. Representative hematoxylin & eosin (H&E) staining shows a decellularized ovine valve leaflet. Please note that no nuclei (would appear in blue/black) can be detected, indicative for complete decellularization. Scale bar, 50 μm.

12. Dehydrate the samples with xylol.
13. Air dry, place a drop of Eukitt onto the slide before adding the coverslip and visualize in a bright field microscope.

3.5. Immunohistochemical Staining Using ABC

For immunohistochemical stains make kryosections of 5-μm thickness from small subvalvular specimens.

1. Fix kryosections on gelatine-coated slides for 8 min in cold acetone at −20°C.
2. Block in serum, of the same species as secondary antibody is derived from to avoid unspecific binding, for 15 min.
3. Incubate with primary antibody against either collagen-I antibody in dilution buffer (1:50) or collagen-IV antibody in dilution 1:500 for 1 h at room temperature or overnight. Do not rinse sections between blocking step and primary antibody incubation (*see* **note 5**).
4. Rinse in wash buffer.
5. Incubate with biotinylated secondary horse anti-mouse antibody in dilution buffer (1:50) for 30 min at room temperature (*see* **note 6**).
6. Rinse in wash buffer.
7. Incubate in ABC solution for 30 min at room temperature.
8. Stain in DAB as a chromogen for 7 min.
9. Counterstain in Harris–Hematoxylin (1:3) for 10 min.
10. Dehydrate in 95% ethanol for 1 min, then twice in 100% ethanol for 3 min.
11. The results are visualized using a standard bright microscope (*see* **Fig. 2**).

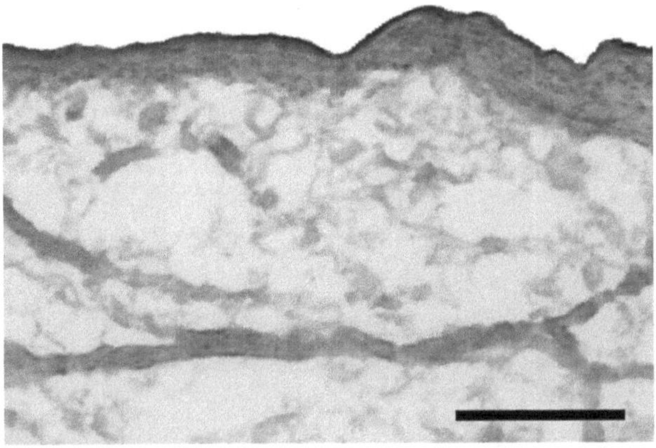

Fig. 2. Immunohistochemistry of a pulmonary valve leaflet shows the presence of collagen-IV after decellularization. Scale bar, 50 μm.

3.6. Immunofluorescence Staining for Perlecan

1. Fix kryosections on gelatine-coated slides for 8 min in cold acetone at −20°C.
2. Block in serum, of the same species as secondary antibody is derived from to avoid unspecific binding, for 15 min.
3. Incubate with primary antibody against perlecan antibody in dilution buffer (1:10,000) for 1 h at room temperature or overnight. Do not rinse sections between blocking step and primary antibody incubation (*see* **note 6**).
4. Wash in PBS.
5. Incubate with fluorescent secondary Cy™3-conjugated AffiniPure donkey anti-rat IgG antibody in dilution buffer (1:500) for 30 min at room temperature (*see* **note 6**).
6. Cover samples with mounting medium, apply coverslip, and analyze by fluorescence microscopy. The perlecan stained by cyanine fluorescent dye 3 (Cy3, red emission) is maximally excited near 550 nm with peak fluorescence near 570 nm.

3.7. Sample Preparation for SEM

1. Fix tissue with 2.5% glutaraldehyde in 0.1 M phosphate buffer (pH 7.2–7.4) for 24–48 h.
2. Incubate in cocadilate for 24–48 h.
3. Dehydrate tissue with acetone stepwise in 40, 50, 70, and 90% twice for 5 min each and in 100% six times.
4. Dehydrate samples using Critical Point Dryer (Balzers).
5. Cover samples with SEM Coating System (Polaron Instruments).
6. Visualize surface of the tissue by SEM (*see* **Fig. 3**).

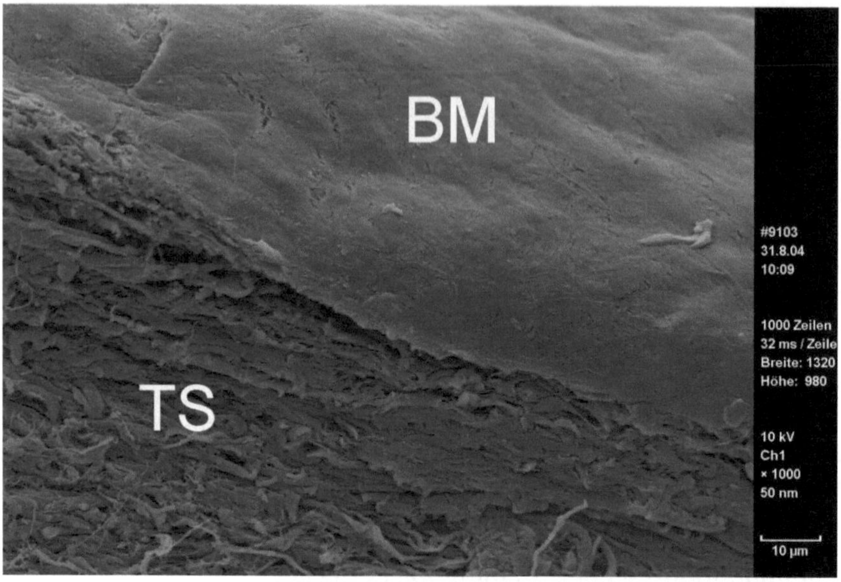

Fig. 3. Scanning electron microscopy (SEM) of a pulmonary valve leaflet shows the presence of the basement membrane (BM) on the luminal surface without cells and transversal structures (TS) at the fracture side. Scale bars, 10 μm.

3.8. DNA Assay

1. Cut decelluarized valve tissue into small pieces.
2. Perform DNA extraction with the DNeasy Tissue kit according to the instruction of the manufacturer.
3. For calculation of the DNA concentration, measure an aliquot at 260 nm for absorption in a photometer.

Notes

1. Detergent solution should be used sterile. Although the detergents interact by solubilizing cellular and nuclear membranes, and may have a bactericide effect, sometimes we observed a suprainfection of detergent solution during the decellularization process. Detergent solution can be easily sterilized by microporous filtration. Sterilization of detergent solution in an autoclave may lead to modification of chemical composition and should be avoided.
2. Detergent solution should be prepared always fresh. Long-term storage before processing may also influence the quality of decellularization.
3. We strongly recommend performing decellularization at room temperature (22–24°C). Decellularization at lower temperatures (e.g., non-heated rooms or cold

room) leads to crystallization of detergent solution and may influence the quality of decellularization. Washing cycles performed at low temperature preclude elimination of crystallized detergents from the tissue.

4. Testing washing solutions for toxic effects on human endothelial cells, we gain experimental evidence that after six wash cycles, the residual concentration of detergents does not provide any toxic effect anymore. Thus, we recommend at least six wash cycles to obtain a non-toxic tissue.
5. Repeated freezing and thawing of antibodies are not recommended.
6. Specificity of the labeling (negative control) is confirmed by omission of the primary antibody and treating with the appropriate isotype IgG. Frozen sections of ovine pulmonary wall tissue serve as a positive control. Negative and positive controls should be run simultaneously with experimental specimens.

References

1. Schoen, F. J. and Levy, R. J. (2005) Calcification of tissue heart valve substitutes: progress toward understanding and prevention. *Ann. Thorac. Surg.* **79**, 1072–1080.
2. den Hamer, I., Hepkema, B., Prop, J., Elzenga, N., and Ebels, T. (1997) HLA antibodies specific for cryopreserved heart valve "homografts" in children. *J. Thorac. Cardiovasc. Surg.* **113**, 417–419.
3. Yankah, A. C., Alexi-Meskhishvili, V., Weng, Y., Berger, F., Lange, P., and Hetzer, R. (1995) Performance of aortic and pulmonary homografts in the right ventricular outflow tract in children. *J. Heart Valve Dis.* **4**, 392–395.
4. Elkins, R. C., Dawson, P. E., Goldstein, S., Walsh, S. P., and Black, K. S. (2001) Decellularized human valve allografts. *Ann. Thorac. Surg.* **71**, S428–S432.
5. Lichtenberg, A., Tudorache, I., Cebotari, S., Suprunov, M., Tudorache, G., Goerler, H., Park, J.-K., Hilfiker-Kleiner, D., Ringes-Lichtenberg, S., Karck, M., Brandes, G., Hilfiker, A., and Haverich, A. (2006) Preclinical testing of tissue engineered heart valves re-endothelialized under simulated physiological conditions. *Circulation.* **114**, 1559–65.
6. Cebotari, S., Mertsching, H., Kallenbach, K., Kostin, S., Repin, O., Batrinac, A., Kleczka, C., Ciubotaru, A., and Haverich, A. (2002) Construction of autologous human heart valves based on an acellular allograft matrix. *Circulation* **106**, I63–I68.
7. Lichtenberg, A., Breymann, T., Cebotari, S., and Haverich, A. (2006) Cell seeded tissue engineered cardiac valves based on allograft and xenograft scaffolds. *Prog. Pediatr. Cardiol.* **21**, 211–217.
8. Lichtenberg, A., Cebotari, S., Tudorache, I., Sturz, G., Winterhalter, W., Hilfiker, A., and Haverich, A. (2006) Flow dependent re-endothelialization of tissue engineered heart valves. *J. Heart Valve Dis.* **15**, 287–294.
9. Lichtenberg, A., Judorache, I., Cebotari, S., Ringes-Lichtenberg, S., Sturz, G., Hoeffler, K., Hurscheler, C., Brandes, G., Hilfiker, A., Haverich, A. (2006) In vitro re-endothelialization of detergent decellularized heart valves under simulated physiological dynamic conditions. *Biomaterials* **27**(23), 4221–9.

18

In Vitro Heart Valve Tissue Engineering

Dörthe Schmidt, Anita Mol, Jens M. Kelm, and Simon P. Hoerstrup

Summary

Heart valve replacement represents the most common surgical therapy for end-stage valvular heart diseases. A major drawback all contemporary heart valve replacements have in common is the lack of growth, repair, and remodeling capabilities. To overcome these limitations, the emerging field of tissue engineering is focusing on the in vitro generation of functional, living heart valve replacements. The basic approach uses starter matrices of either decellularized xenogeneic or biopolymeric materials configured in the shape of the heart valve and subsequent cell seeding. Moreover, in vitro strategies using mechanical loading in bioreactor systems have been developed to improve tissue maturation. This chapter gives a short overview of the current concepts and provides detailed methods for in vitro heart valve tissue engineering.

Key Words: Heart valves; Bioreactor; In vitro tissue maturation; Endothelial cells; Myofibroblasts; Stromal marrow cells; Endothelial progenitor cells; Scaffolds.

1. Introduction

Currently available heart valve prostheses for the treatment of advanced heart valve diseases represent non-living foreign material and therefore are inherently different from the tissue they replace. Thus, they are associated with substantial morbidity and mortality with regard to increased risks of thromboembolism, increased rates of infections, immunological reactions, and related prosthesis malfunctions. Heart valve tissue engineering represents a promising strategy to overcome today's lack of living autologous replacements with the capacity of growth, regeneration, and self-repair.

A series of studies have been undertaken to determine whether tissue engineering principles could be used to develop living valve substitutes with a thromboresistant surface and a viable interstitium with repair, remodeling,

From: *Methods in Molecular Medicine, 2nd ed.: Tissue Engineering*
Edited by: H. Hauser and M. Fussenegger © Humana Press Inc., Totowa, NJ

and growth capabilities. Several groups demonstrated the feasibility of creating living cardiovascular structures by cell seeding on synthetic polymers, collagen, or xenogeneic scaffolds *(1–4)*. As to heart valve tissue engineering, significant milestones comprise the successful replacement of a single pulmonary valve leaflet by a tissue-engineered autologous leaflet *(5)*, followed by in vitro-generated complete autologous trileaflet heart valves based on ovine *(3,6)* and human stromal marrow cells *(7, see* **Fig. 1***)*.

Two principle strategies have been developed to fabricate living autologous heart valve replacements. The first strategy requires an ex vivo phase generating the tissue-engineered heart valve by in vitro cell technologies *(8)*. The second strategy circumvents the in vitro tissue-culture phase by direct implantation of natural tissue-derived heart valve matrices for potential cell ingrowth and remodeling in vivo *(9)*. Matrices used for the latter approach include decellularized tissues derived from pericardium or valves, cell-free porcine small intestine submucosa *(9)*, or synthetic biodegradable polymeric scaffolds, such as collagen or fibrin gels. Although first clinical trials have been initiated *(10,11)* and decellularized scaffolds implanted in humans have demonstrated ingrowth of host cells, a strong inflammatory response and fatal valve failures were reported *(12)*.

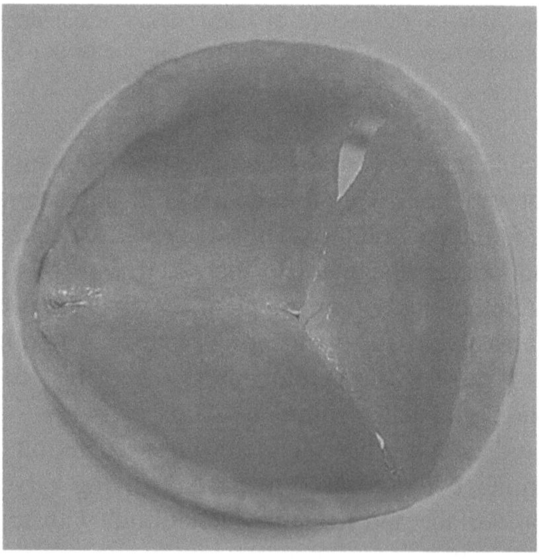

Fig. 1. Heart valve tissue engineered from human marrow stromal cells *(7)*.

The process of in vitro engineering of heart valves (*see* **Fig. 2**) requires the harvest of autologous cells. After isolation, cells are expanded in vitro. After a sufficient number of cells for the tissue engineering approach are obtained, cells are seeded onto biodegradable heart valve scaffolds. To guide tissue maturation and formation, the seeded constructs are implanted into a bioreactor system, applying mechanical stimulation to the growing tissues. Mimicking a physiological environment with regard to flow and pressure in vitro has been shown to result in improved tissue maturation and implantable autologous heart valve (***3,13,14***).

2. Materials

2.1. Cell Culture

1. Medium for myofibroblasts: Dulbecco's modified Eagle's medium (DMEM; Invitrogen, Carlsbad, CA) supplemented with 10% fetal bovine serum (FBS; Invitrogen) and gentamycin (50 µg/ml; Pan Biotech GmbH, Aidenbach, Germany).
2. Medium for endothelial cells: Endothelial basal medium (EBM™-2; Cambrex, Walkersville, MD) containing vascular endothelial growth factor (VEGF), human fibroblast growth factor (hFGF), human recombinant long-insulin-like growth factor-1 (R-3-IGF-1), human epidermal growth factor (hEGF), gentamycin and amphotericin (GA-1000), hydrocortisone, heparin, ascorbic acid (all form Cambrex), and 20% FBS (Invitrogen).

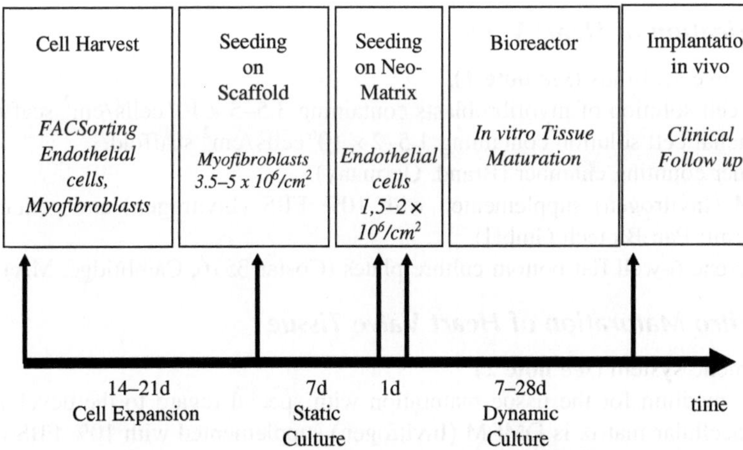

Fig. 2. Schematic of the in vitro protocol for generating heart valves.

3. Medium for bone marrow cell isolation: DMEM (Invitrogen) supplemented with 10% FBS, streptomycin (Invitrogen) and 1000 U heparin (Roche Pharma AG, Reinach, Switzerland).
4. Density gradient for isolation of umbilical cord blood-derived cells or bone marrow-derived cells: Ficoll–Histopaque-1077 (Sigma Chemical Company, St. Louis, MO).
5. Washing solution for tissue biopsies: Washing solution based on phosphate buffer solution (PBS; Kantonsapotheke, Zurich, Switzerland) containing Genatmycin 200 μg/ml (Pan Biotech GmbH) and Amphotericin B 2.5 μg/ml (Pan Biotech GmbH).
6. Collagenase solution for isolation of endothelial cells from vessels: Endothelial basal medium (EBM™-2; Cambrex) containing collagenase (final concentration 2 mg/ml; Sigma Chemical Company).

2.2. Characterization of Cells by Flow Cytometry

1. Fluorescence Activated Cell Sorting (FACS) buffer: PBS (Kantonsapotheke) containing 2% bovine serum albumin.
2. Primary antibodies against CD31 (clone JC/70A), von Willebrand factor (vWF; affinity-purified rabbit antibodies), vimentin (clone 3B4), desmin (clone D33; all from DakoCytomation, Glostrup, Denmark), α-smooth muscle actin (α-SMA; clone Sigma Chemical Company), CD34 (AC136; Miltenyi, Bergisch Gladbach, Germany), and secondary antibody (Fluorescein isothio-cyanate-conjugated (FITC); BD Bioscience, San Jose, CA).
3. 4% formalin (Kantonsapotheke).
4. 0.1% Triton-X 100 solution (Sigma Chemical Company).
5. 5 ml polystyrene round-bottom tubes (Becton Dickinson Labware, Franklin Lakes, NJ).
6. FacsScan (Becton Dickinson, Sunnyvale, CA).

2.3. Fabrication of Heart Valves

1. Heart valve scaffolds (*see* **note 1**).
2. Single-cell solution of myofibroblasts containing 3.5–5 × 10^6 cells/cm^2 scaffold.
3. Endothelial cell solution containing 1.5–2 × 10^6 cells/cm^2 scaffold.
4. Neubauer counting chamber (Brand, Germany).
5. DMEM (Invitrogen) supplemented with 10% FBS (Invitrogen) and gentamycin (50 μg/ml; Pan Biotech GmbH).
6. Polystyrene 6-well flat-bottom culture plates (Costar 3516, Cambridge, MA).

2.4. In Vitro Maturation of Heart Valve Tissue

1. Biomimetic system (*see* **note 2**).
2. Culture medium for the tissue maturation with special regard to the development of extracellular matrix is DMEM (Invitrogen) supplemented with 10% FBS (Invitrogen), gentamycin (50 μg/ml; Pan Biotech GmbH) and ascorbic acid (0.25 mg/ml medium; Sigma Chemical Company).

3. Methods

3.1. Cell Culture

Several alternative human cell sources have been investigated for their use in heart valve tissue engineering. Among the most promising are vascular-derived cells, bone marrow-derived cells, blood-derived cells, and umbilical cord-derived cells (*see* **Fig. 3**), particularly for pediatric applications. In most cardiovascular tissue engineering approaches, cells are harvested from vascular structures, such as mammary, radial artery, and saphenous vein ***(15)***.

The obtained mixed-cell population consisting of myofibroblasts and endothelial cells is sorted by fluorescence-activated cell sorting, and the pure-cell populations are cultivated and used for the fabrication of the replacements ***(16)***. Alternatively, to detach the endothelium from the luminal layer, the vessel biopsies are digested with collagenase, and the endothelial cells are harvested. Afterward, the tissue is minced into small pieces, and outgrowing myofibroblasts are cultured.

3.1.1. Isolation of Human Myofibroblasts

1. The vessel tissue is minced into several pieces (~3 mm × 3 mm).
2. Five to seven tissue pieces are placed into a cell-culture dish (3.5 cm in diameter) and cultured in DMEM medium containing 10% FBS and gentamycin.
3. Outgrowing cells are expanded using standard cell-culture protocols.

Fig. 3. Human endothelial progenitor cells isolated from human umbilical cord blood form cobblestone monolayers after 21 days demonstrate typical characteristics of endothelial cells ***(24)***.

3.1.2. Human Endothelial Cell Isolation

3.1.2.1. Isolation of Human Endothelial Cells from Human Vessels

1. Directly after vessel harvest, the biopsies are rinsed with the washing solution.
2. Afterward, one end of the vessel is closed using clamps, and the collagenase solution is filled into the lumen of the vessel. Afterward, the other opening is closed as well.
3. After 20 min of incubation at 37°C in a humidified incubator (5% CO_2), clamps are removed for liberating the cell suspension containing the endothelial cells.
4. The cell suspension is collected and transferred into a centrifugation tube for centrifugation at 30 × *g* for 8 min at room temperature.
5. After the supernatant is removed carefully, the cell pellet is resuspended, and the isolated cells are expanded in EBM-2 medium containing the supplements, growth factors, and 20% FBS.

3.1.2.2. Isolation of Human Umbilical Cord Blood-Derived Endothelial Progenitor Cells

1. Obtained fresh human umbilical cord blood (~20 ml) has to be layered onto 15 ml density gradient that is placed in a 50-ml centrifugation tube.
2. After centrifugation for 30 min at room temperature, endothelial progenitor cells (EPCs) are isolated from mononuclear cells using antibodies against CD34 and FACS sorting.
3. Isolated cells are cultured in endothelial medium containing the above-mentioned growth factors and supplements but 2% FBS using 6-well plates.
4. After 4 days, attached cells are trypsinized and reseeded, and after 7 days, the medium is changed to endothelial medium containing the same growth factors but 20% FBS (*see* **note 3**).

3.1.3. Isolation of Human Bone Marrow-Derived Cells

1. Human marrow stromal cells are isolated from 10–15 ml bone marrow aspirate (e.g., from iliaca crest) and resuspended in 20 ml bone marrow cell isolation medium.
2. Following the cell suspension is centrifuged over a density gradient at 1500 rpm (100g) for 10 min.
3. The interface fraction is collected and cultured into medium for myofibroblasts (*see* **note 4**).
4. Medium has to be replaced after 24 and 72 h and every further 6 days.

3.2. Characterization of Cells by Flow Cytometry

Isolated cells are characterized using flow cytometry. Therefore, cells are detached from the culture flasks using trypsin.

1. About 500,000 detached cells are needed per sample. All samples are stored at 4°C following trypsinization and placed in a 5 ml polystyrene round-bottom tube.

2. For detection of intracellular antigens, cells are fixed and permeabilized. Therefore, 2 ml 4% formalin is slowly added and cells are left at room temperature for 30 min (*see* **note 5**).
3. After fixation, cells are centrifuged and resuspended in 100 μl 0.1% Triton-X 100.
4. After incubation for 3–5 min at room temperature, cells are washed with 1 ml ddH_2O and centrifuged at 250 × *g* for 6 min. Afterward, the supernatant is removed.
5. Cells are resuspended in the primary antibody solution (primary antibody is diluted 1:100 in 100 μl FACS buffer).
6. After incubating for 30 min, 1 ml FACS buffer is added, and the sample is centrifuged at 250 × *g* for 6 min.
7. As the primary antibody is not directly labeled, steps 6 and 7 have to be repeated using the FITC-labeled secondary antibody (dilution, 1:100) before FACS analysis can be performed on the FacsScan.

3.3. Fabrication of Heart Valves

For the fabrication of heart valves, sterilized scaffolds (*see* **Fig. 4**) are seeded with cells. The seeding procedure is performed sequentially.

1. Human myofibroblasts are trypsinized from the culture flasks, wherein they have been expanded and suspended to create a single-cell solution.
2. The cell number of the solution is determined by direct counting using the Neubauer counting chamber.
3. The cell suspension is centrifuged at 1000 rpm (100g) at 20°C for 5 min, the supernatant is removed, and fresh culture medium is added. The cell concentration has to

Fig. 4. Stented heart valve based on a non-woven PGA coated with 1% poly-4-hydroxybutyrate (P4HB) *(17)*.

be adjusted to the surface area (3.5–5 × 10^6 cells resuspended in 50 μl medium/cm^2 scaffold).
4. Seeding is performed in polystyrene 6-well flat-bottom culture plates (Costar 3516).
5. Therefore, scaffolds are placed in the center of the wells, and the cells are evenly distributed into the scaffolds.
6. Myofibroblasts are seeded into the constructs that are statically cultured followed by endothelial cell seeding after 4 days.
7. After 2 h culturing in a humidified incubator (37°C, 5% CO_2) that allows attachment of the cells to the scaffolds, 5 ml medium/well is added for further culture.
8. After 4 days, endothelial cells are trypsinized and seeded into the constructs using the same seeding protocol, but 1.5–2 × 10^6 cells are resuspended in 50 μl medium/cm^2 scaffold.

3.4. Maturation of Heart Valve Tissue

Tissue maturation of the heart valves can be performed in vitro using biomimetic systems (*see* **Fig. 5**). After a static culture phase, seeded constructs are implanted into the bioreactor and conditioned for 14 days. The bioreactor protocol depends on the requirements that the generated tissue has to match in vivo. For instance, maturation of heart valves for application in a high-pressure system requires other components of the extracellular matrix compared with those for low-pressure systems. Thus, the conditioning protocol has to be adapted to the needs of tissue components. The diastolic pulse duplicator *(17)*, for example, was developed to apply increasing strain onto the growing tissue during culture time (*see* **Fig. 5**), aiming at valve substitutes for systemic pressure application.

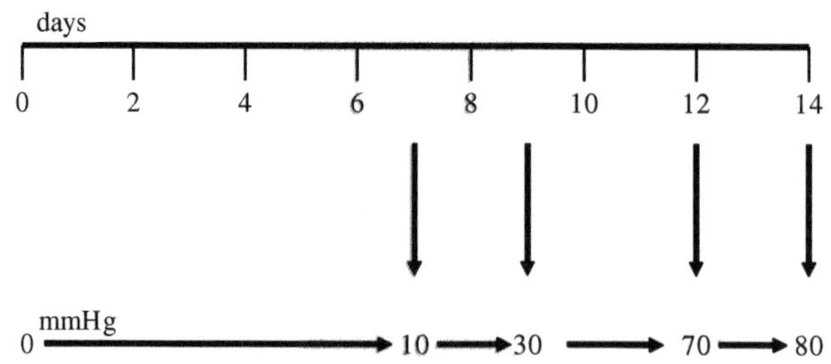

Fig. 5. Conditioning protocol for tissue maturation in the diastolic pulse duplicator based on Mol et al. *(17)*. Dynamic transvalvular pressure increases gradually from 0 to 80 mmHg within 14 days of culturing.

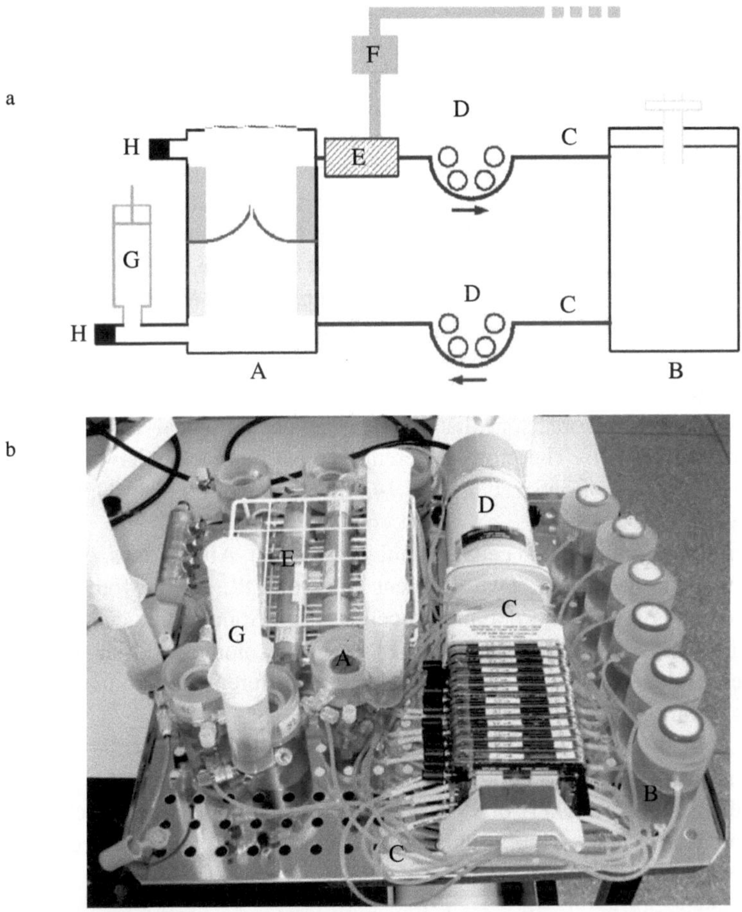

Fig. 6. Example of a bioreactor *(17)*. (a) The schematic drawing and its function of one diastolic pulse duplicator and (b) a photograph of six systems used simultaneously. The system consists of a bioreactor (A), in which the valve is implanted, and a medium container (B). These are connected through two parallel tubing series (C) running through a roller pump (D). For mechanical stimulation, a tube is placed in a polycarbonate cylinder (E) in which compressed air is released through a magnetic valve (F) resulting in compressing and decompressing the tubing and therewith in a pressure difference over the heart valve leaflets. A syringe (G) serves as a compliance chamber. The pressure difference across the leaflets is monitored using pressure sensors (H).

Notes

1. Heart valve scaffolds can be produced based on several matrices including synthetic and biological. Several synthetic biodegradable polymers, such as polyglactin *(1)*, polyglycolic acid (PGA) *(5)*, polylactid acid (PLA) *(18)*, polyhydroxyalkanoates (PHA) *(19)*, and a copolymer of PGA and PLA (PGLA) *(20)*, have been investigated. Another approach for fabricating complete, trileaflet heart valves is the use of PGA coated with poly-4-hydroxybutyrate (P4HB) (***3***,***7***). This biologically derived and rapidly absorbable copolymer is strong and pliable. Owing to its thermoplasticity, it can be molded into almost any shape, and complete degradation has been observed within 8 weeks after implantation *(3)*. Other studies in heart valve tissue engineering were based on biological scaffolds such as xenogenic or allogeneic decellularized fixed heart valves *(21)*.
2. Tissue formation of the seeded heart valves takes place in a biomimetic in vitro culture system (bioreactor) that mimics the physiological environment of their native counterparts. Therefore, different types of bioreactors have been developed, such as pulsatile flow reactors (***13***,***14***,***22***) or diastolic pulse duplicators *(17)* as shown in Fig. 6.
3. After isolation, the plated cells are initially rounded. After 4 days, cells are attached and they form clusters. Two different types of EPC can be found: spindle like-shaped cells (80%) and polymorph cells (20%). Spindle-like cells die out during the first days, whereas polymorph EPCs form colonies and show good growth under in vitro conditions. After 3 weeks, they differentiate into mature endothelial cells forming cobblestone monolayers (***23***,***24***).
4. The non-adherent cells float off, whereas marrow stromal cells adhere, spread, and grow.
5. Antibodies against CD31 or vWF are used for endothelial cell detection and CD34 for progenitor cells. Myofibroblasts are characterized by the intracellular antigens vimentin, α-SMA, and desmin.

Acknowledgements

The authors acknowledge are grateful for the finantial support of NIH (P41 EB002520-01, R01 HL076485-01 to GV-N), Regione Veneto (Azione Biotech II to NE), University of Padova (fellowship to EF), and Fullbright Association (fellowship to NE). We also thank the Eclat Industries for supplying TiN coated electrodes.

References

1. Shinoka, T., Breuer, C. K., Tanel, R. E., Zund, G., Miura, T., Ma, P. X., Langer, R., Vacanti, J. P., and Mayer, J. E. (1995) Tissue engineering heart valves: valve leaflet replacement study in a lamb model. *Ann. Thorac. Surg.* **60**, 513–516.

2. Bader, A., Schilling, T., Teebken, O. E., Brandes, G., Herden, T., Steinhoff, G., and Haverich, A. (1998) Tissue engineering of heart valves – human endothelial cell seeding of detergent acellularized porcine valves. *Eur. J. Cardiothorac. Surg.* **14**, 279–284.
3. Hoerstrup, S. P., Sodian, R., Daebritz, S., Wang, J., Bacha, E. A., Martin, D. P., Moran, A. M., Guleserian, K. J., Sperling, J. S., Kaushal, S., Vacanti, J. P., Schoen, F. J., and Mayer, J. E., Jr. (2000) Functional living trileaflet heart valves grown in vitro. *Circulation* **102**, III44–III49.
4. Elkins, R. C., Goldstein, S., Hewitt, C., Walsh, S. P., Dawson, P. E., Ollerenshaw, J. D., Black, K. S., Clarke, D. R, and O'brien, M. F. (2001) Recellularization of heart valve grafts by a process of adaptive remodeling. *Semin. Thorac. Cardiovasc. Surg.* **13**, 87–92.
5. Shinoka, T., Ma, P. X., Shum-Tim, D., Breuer, C. K., Cusick, R. A., Zund, G., Langer, R., Vacanti J. P., and Meye, J. E., Jr. (1996) Tissue engineered heart valves. Autologous valve leaflet replacement study in a lamb model. *Circulation* **94**, I164–I168.
6. Sutherland, F. W., Perry, T. E., Yu, Y., Sherwood M. C., Rabkin, E., Masuda, Y., Garcia, G. A., McLellan, D. L., Engelmayr, G. C., Jr., Sacks, M. S., Schoen, F. J., and Mayer, J. E., Jr. (2005) From stem cells to viable autologous semilunar heart valves. *Circulation* **111**, 2783–2791.
7. Hoerstrup, S. P., Kadner, A., Melnitchouk, S., Trojan, A., Eid, K., Tracy, J., Sodian, R., Visjager, J. F., Kolb, S. A., Grunenfelder, J., Zund, G., and Turina, M. I. (2002) Tissue engineering of functional trileaflet heart valves from human marrow stromal cells. *Circulation* **106**, I143–I150.
8. Langer, R. and Vacanti, J. P. (1993) Tissue engineering. *Science* **260**, 920–926.
9. Matheny, R. G., Hutchison, M. L., Dryden, P. E., Hilles, M. D., and Shaar, C. J. (2000) Porcine small intestine submucosa as a pulmonary valve leaflet substitute. *J. Heart Valve Dis.* **9**, 769–775.
10. Shinoka, T., Matsumura, G., Hibino, N., Naito, Y., Watanabe, M., Konuma, T., Sakamoto, T., Nagatsu, M., and Kurosawa, H. (2005) Midterm clinical results of tissue-engineered vascular autografts seeded with autologous bone marrow cells. *J Thorac. Cardiovasc. Surg.* **129**, 1330–1338.
11. Shinoka, T., Imai, Y., and Ikada, Y. (2001) Transplantation of a tissue-engineered pulmonary artery. *N. Engl. J. Med.* **344**, 532–533.
12. Simon, P., Kasimir, M. T., Seebacher, G., Weigel, G., Ullrich, R., Salzer-Muhar, U., Rieder, E., and Wolner, E. (2003) Early failure of the tissue engineered porcine heart valve SYNERGRAFT in pediatric patients. *Eur. J. Cardiothorac. Surg.* **23**, 1002–1006.
13. Hoerstrup, S. P., Sodian, R., Sperling, J. S., Vacanti, J. P., and Meyer, J. E., Jr. (2000) New pulsatile bioreactor for in vitro formation of tissue engineered heart valves. *Tissue Eng.* **6**, 75–79.

14. Niklason, L. E., Gao, J., Abbott, W. M., Hirschi, K. K., Houser, S., Marini, R., and Langer, R. (1999) Functional arteries grown in vitro. *Science* **284**, 489–493.
15. Schnell, A. M., Hoerstrup, S. P., Zund, G., Kolb, S., Sodian, R., Visjager, J. F., Grunenfelder, J., Suter, A., and Turina, M. (2001) Optimal cell source for cardiovascular tissue engineering: venous vs. aortic human myofibroblasts. *Thorac. Cardiovasc. Surg.* **49**, 221–225.
16. Hoerstrup, S. P., Zund, G., Schoeberlein, A., Ye, O., Vogt, P. R., and Turina, M. (1998) Fluorescence activated cell sorting: a reliable method in tissue engineering of a bioprosthetic heart valve. *Ann. Thorac. Surg.* **66**, 1653–1657.
17. Mol, A., Driessen, N. J., Rutten, M. C., Horstrup, S. P., Bouten, C. V., and Baaijens, F. P. (2005) Tissue engineering of human heart valve leaflets: a novel bioreactor for a strain-based conditioning approach. *Ann. Biomed. Eng.* **33**, 1778–1788.
18. Shinoka, T., Shum-Tim, D., Ma, P. X., Tanel, R. E., Isogai, N. Langer, R., Vacanti, J. E., and Mayer, J. E., Jr. (1998) Creation of viable pulmonary artery autografts through tissue engineering. *J. Thorac. Cardiovasc. Surg.* **115**, 536–546.
19. Sodian, R., Hoerstrup, S. P., Sperling, J. S., Daebritz, S., Martin, D. P., Moran, A. M., Kim, B. S., Schoen, F. J., Vacanti, J. P., and Mayer, J. E., Jr. (2000) Early in vivo experience with tissue-engineered trileaflet heart valves. *Circulation* **102**, III22–III29.
20. Zund, G., Breuer, C. K., Shinoka, T., Ma, P. A., Langer, R., Mayer, J. E., and Vacanti, J. P. (1997) The in vitro construction of a tissue engineered bioprosthetic heart valve. *Eur. J. Cardiothorac. Surg.* **11**, 493–497.
21. Schenke-Layland, K., Opitz, F., Gross, M., Doring, C., Halbhuber, K. J., Scirmeister, F., Wahkers, T., and Stock, U. A. (2003) Complete repopulation of decellularized heart valves by application of defined physical signals – an in vitro study. *Cardiovasc. Res.* **60**, 497–509.
22. Sodian, R., Lempke, T., Loebe, M. Hoerstrup, S. P., Potapov, E. V., Hausmann, H., Meyer, R., and Hetzer, R. (2001) New pulsatile bioreactor for fabrication of tissue-engineered patches. *J. Biomed. Mater. Res.* **58**, 401–405.
23. Schmidt, D., Mol, A., Neuenschwander, S., Breymann, C., Gössi, M., Zund, G., Turina, M., and Hoerstrup, S. P. (2005) Living patches engineered from human umbilical cord derived fibroblasts and endothelial progenitor cells. *Eur. J. Cardiothorac. Surg.* **27**, 795–800.
24. Schmidt, D., Breymann, C., Weber, A., Guenter, C. I., Neuenschwander, S., Zund, G., Turina, M., and Hoerstrup, S. P. (2004) Umbilical cord blood derived endothelial progenitor cells for tissue engineering of vascular grafts. *Ann. Thorac. Surg.* **78**, 2094–2098.

Index

MIX
Papier aus verantwortungsvollen Quellen
Paper from responsible sources
FSC® C105338

If you have any concerns about our products,
you can contact us on
ProductSafety@springernature.com

In case Publisher is established outside the EU,
the EU authorized representative is:
Springer Nature Customer Service Center GmbH
Europaplatz 3, 69115 Heidelberg, Germany

Printed by Libri Plureos GmbH
in Hamburg, Germany